GRAPE GROWING

GRAPE GROWING

ROBERT J. WEAVER
Department of Viticulture and Enology
University of California, Davis

A WILEY-INTERSCIENCE PUBLICATION

JOHN WILEY & SONS, New York • London • Sydney • Toronto

Library of Congress Cataloging in Publication Data:

Weaver, Robert John, 1917–
Grape growing.

"A Wiley-Interscience publication."
Bibliography: p.
1. Viticulture. 2. Viticulture—California.
3. Wine and wine making. I. Title.

SB389.W4 634′.8′09794 76-22753
ISBN 0-471-92324-9

Printed in the United States of America

10 9 8 7 6 5 4 3 2 1

To the Grape Growers and Winemakers of California

PREFACE

This book is intended primarily as a textbook in grape growing, for use in a freshman or lower-division college course. It should also be very useful for the home grape grower, as an attempt was made to present the material in a clear, concise manner. Much of the book is concerned with explanations of how different operations are carried out. The subject matter is presented in 20 chapters arranged under five main headings: nature of the vine, strategy for grape production, vine management, crop hazards, and grape growing and winemaking at home. A bibliography, glossary, and conversion tables have been added to make the book more useful.

Viticulture (grape growing) and enology (winemaking) have increased greatly in importance and popularity over the past decade. One of the best examples of two different disciplines, namely plant culture and fermentation science, working in conjunction to make a superior product, wine, is the combination of viticulture and enology. The studies of grape growing and winemaking go hand in hand. With the increasing sophistication in techniques for modern viticulture and enology, it becomes more and more important for an enologist or viticulturist to have at least a general knowledge of both fields of endeavor. Thus, although the major thrust of this book concerns grape growing, a chapter on wine has been included.

Although most of the material in this book refers to grape growing in California, unless otherwise stated, I believe that it will serve as a practical guide for grape growers around the world.

Registration of chemicals is in a state of flux because of the great emphasis on improvement of the environment and prevention of pollution. Discussion of uses for chemicals does not imply that the Environmental Protection Agency has approved all usages discussed. The grower should carefully follow the recommendations on the label, and if any question arises the farm advisor should be consulted. I cannot assume responsibility for success or failure of any spray treatments. The use of trade names is for convenience of the reader and does not imply endorsement of any single product.

I appreciate the help rendered by several of my colleagues. Some of the chapters in draft form were reviewed by the following scientists: C. J. Alley, University of California, Davis; H. W. Berg, University of California, Davis; L. P. Christensen, Farm Advisor, Fresno County; H. B. Currier, University of California, Davis; E. M. Gifford, Jr., University of California, Davis; A. C. Goheen, U.S. Department of Agriculture, Davis; F. L. Jensen, Extension Viticulturist, University of California, San Joaquin Valley Agricultural Research Extension Center, Parlier; A. N. Kasimatis, Extension Viticulturist, University of California, Davis; A. H. Lange, Extension Weed Control Specialist, University of California, San Joaquin Valley Agricultural Research Extension Center, Parlier; F. G. Mitchell, Extension Pomologist Marketing, University of California, Davis; V. Petrucci, California State University, Fresno; H. B. Schultz, University of California, Davis; E. M. Stafford, University of California, Davis; and W. E. Wildman, Extension Soils Specialist, University of California, Davis.

ROBERT J. WEAVER

Davis, California
March 1976

CONTENTS

Part 1

NATURE OF THE VINE

Chapter 1
HISTORY AND CLASSIFICATION OF GRAPES

HISTORY OF *VITIS VINIFERA*

The Old World species, *Vitis vinifera,* is the grape of antiquity often mentioned in the Bible. Most table, wine, and raisin grapes are produced from this variety. *V. vinifera* originated in the regions between and south of the Caspian and Black seas in Asia Minor (Winkler et al., 1974), has been carried from region to region by civilized man in all temperate climates, and has been grown more recently in subtropical climates. Several thousand varieties of grapes have been derived from this species. *Vinifera* is also a parent of many hybrid grapes in eastern America, as breeders in this region desired to introduce some of the qualities of *vinifera* into their grapes.

V. vinifera was brought by the Spaniards to Mexico and areas now occupied by California and Arizona. The mission vines that were introduced flourished and some grew to a huge size. English settlers brought the Old World grape with them and made plantings along the Atlantic seaboard in the colonies of Massachusetts, New York, Pennsylvania, Virgina, North Carolina, South Carolina, and Georgia. In spite of repeated attempts, these vineyards were a failure because of the presence of insect phylloxera, and fungus diseases such as black rot, downy mildew, and powdery mildew, as well as the low winter temperatures and the hot humid summers of the eastern states. For these reasons *vinifera* cannot be grown successfully in regions east of the Rocky Mountains, except in limited sites and under special management. In the southern states, scorching of the berries and the leaves makes *vinifera* grape growing impossible. Pierce's disease is also a major limitation.

THE GENERA *VITIS* AND *MUSCADINIA*

The genus *Vitis* has two subgenera: *Euvitis,* the true grapes, and *Muscadinia.* Various species of *Vitis* and *Muscadinia* in different regions of the world are shown in Table 1-1 (Bailey, 1934; H. P. Olmo, private communication). *Muscadinia* can be easily identified by tight bark that does not shed, simple tendrils (that do not fork), nodes without a diaphragm (continuous pith at nodes), and small clusterlets with berries that detach as they mature. *Vinifera* has forked tendrils, bark that sheds, a diaphragm at the nodes, and elongated clusters with berries that adhere to the pedicels at maturity.

Vinifera also has intermittent tendrils, thin, smooth shiny leaves with three, five, or seven lobes, although leaves of young shoots may be downy or hairy. Size of berries varies; and they may be round or oval with edible skins that adhere to the flesh

In the American species, the skins slip from the pulp (slipskin) and are usually not eaten; the berries are always round or nearly round. Many of the American varieties have a characteristic foxy odor and taste.

Except for *V. vinifera,* most species of Euvitis originated in the northern hemisphere and are especially common in North America (Table 1-1). The two American species of *Muscadinia* are *M. rotundifolia* (southern U.S.) and *M. munsoniana* (central and southern Florida).

SPECIES USED FOR GRAPE PRODUCTION

V. vinifera produces over 90% of the world's grapes, which are either pure *vinifera* or *vinifera* hybridized with one or more American species. About 85% of the grapes in the United States, grown mainly in California, are derived from pure *vinifera* varieties.

The most important varieties of pure species in North America used for fruiting, with two varieties of each in parentheses, are the following: *V. labrusca* (Concord,* Niagara), *V. aestivalis* (Norton, Delaware), *V. vulpina* (Elvira, Clinton), *M. rotundifolia* (Scuppernong, Eden), and *V. rupestris* (Rupestris St. George—used mainly as phylloxera-resistant rootstocks).

The Concord variety makes up around 80% of the total production of the American varieties. The New York State wine industry is largely based on the following varieties: Isabella and Catawba (both chance seedlings), Concord, Delaware, Dutchess (contains some labrusca blood), Elvira, and Niagara (Einset and Robinson, 1973).

* Concord is not considered to be pure labrusca since the berries, for example, are too large.

Table 1-1 Some species of *Vitis* and *Muscadinia* found in various regions of the world (prepared by H. P. Olmo)

I Middle Asian and Mediterranean

V. vinifera, Linnaeus. Wine grape, "European" grape.

II North American

V. aestivalis, Michaux. "Summer Grape." New York to Georgia, west to Missouri on sandy or rocky uplands.

V. argentifolia, Munson. "Silverleaf Grape." Dry lands in Massachusetts, Ontario, Canada, Wisconsin, Illinois, Minnesota to South Carolina, and Tennessee.

V. arizonica, Englemann. "Canyon Grape." West Texas, New Mexico, Arizona, and Mexico.

V. berlandieri, Planchon. "Spanish Grape." Limestone hills, and along streams in southwestern Arkansas, central Texas, and northern Mexico.

V. baileyana, Munson. "Possum Grape." Virginia, Kentucky, and south to Georgia and Alabama.

V. californica, Bentham. "Pacific Grape," "California Wild Grape." California and southern Oregon.

V. candicans, Englemann. "Mustang Grape." Texas, Louisiana, Arkansas, Oklahoma, and Mexico. Mostly on limestone soils.

V. champini, Planchon. "Calcaire Grape." Limy soils in central-southern Texas.

V. cinerea, Engelmann. "Grayback Grape." Indiana to Missouri, and south to Gulf of Mexico.

V. cordifolia, Lamarck. "Winter Grape." Pennsylvania to Florida, west to Kansas, Oklahoma, and central Texas.

V. doaniana, Munson. "Panhandle Grape." Southwestern Oklahoma, northern Texas, eastern New Mexico. Mostly on sandy soils.

V. gigas, Fennell. "Florida Blue Grape." Florida.

V. girdiana, Munson. "Valley Grape." Southern California.

V. helleri, Small. "Round-leaf Grape." Southern Texas.

V. Illex, Bailey, "Manatee Grape." Manatee County, Florida.

V. labrusca, Linnaeus. "Fox grape." Northeastern United States.

V. lincecumii, Buckley. "Post-oak Grape." High post-oak lands of eastern Texas, Louisiana, Arkansas, and Missouri.

V. longii, Prince (solonis and novo-mexicana). "Bush Grape." Oklahoma, northern Texas, eastern New Mexico, southwestern Kansas, southeastern Colorado.

V. monticola, Buckley. "Sweet Mountain Grape." Limestone hills in central-southwestern Texas.

Table 1-1 (Continued)

II North American (Continued)

V. novae-angliae, Fernald. "Pilgrim Grape." Maine, New Hampshire, Massachusetts, Connecticut, Rhode Island.

V. palmata (rubra), Vahl. "Cat Grape." Indiana to Missouri, and south to Louisiana and Texas.

V. riparia, Michaux (V. vulpina, Linn.). "Frost Grape." Canada to Tennessee and northern Texas; Montana and Utah to the Atlantic Ocean.

V. rufotomentosa, Small. "Redshank Grape." Florida to Louisiana.

V. rupestris, Scheele. "Sand Grape." Missouri, Illinois, Kentucky, Tennessee, Arkansas, Oklahoma, and east Texas. Mostly on gravelly stream beds.

V. shuttleworthii, House. "Calloosa Grape." Southern Florida.

V. smalliana, Bailey, "Figleaf Grape." Florida.

V. simpsoni, Munson. "Currant Grape." Southern Georgia, Florida.

V. sola, Bailey. "Curtiss Grape." Florida.

V. treleasei, Munson. "Gulch Grape." Texas, New Mexico, Arizona.

III Caribbean

V. indica (tiliafolia, caribeaea). Northern Amazon, Caribbean basin.

IV Asiatic Species

V. coignetiae, Pulliat. Japan, South Korea.

V. flexuosa, Thunberg. Himalaya, Java.

V. pentagona, Diels and Gilg. Central-south China, Laos, North Vietnam.

V. amurensis, Ruprecht. Sakhalin Island, Amur River.

V. embergeri, Golet. Central-east China.

V. betulifolia, Diels and Gilg. Central-east China.

V. reticulata, Pampanini. Central-east China.

V. armata, Diels and Gilg. Central-east China.

V. davidii, Romanet du Caillaud. Central China.

V. lanata, Roxburgh. China, northern India.

V. pedicellata, Lawson. Himalaya, northern India.

V Genus Muscadinia (the Muscadine grapes)

M. rotundifolia, Michaux. "Muscadine Grape." Delaware to Missouri, and south to Texas.

M. munsoniana, Simpson. "Little Muscadine Grape." Central and southern Florida.

M. popenoei, Fennel. "Mexican Muscadine Grape." New Mexico.

HYBRIDS

Many grape varieties have been derived by crossbreeding between native species, between a species and a hybrid previously produced, or by crossing two hybrids (Winkler et al., 1974). Crossbreeding between native species, hybridizing with different representatives of *vinifera,* and interbreeding with grapes thus derived have produced many American grape varieties.

Rootstocks

Because of their inherent resistance to certain destructive insect pests, especially grape phylloxera (*Dactylosphaera vitifoliae*) and nematodes, American species have often proved invaluable where own-rooted vines have failed.

American grapes from the Mississippi Valley region have considerable resistance to phylloxera since this pest is native to the region. The principal rootstock species used for phylloxera-resistant rootstocks are *riparia, berlandieri, rupestris, aestivalis, cordifolia,* and *monticola.* Rootstocks are also essential for the control of nematodes, although in some situations partial control may be obtained with soil fumigants. The main species resistant to nematodes, either in pure form or as hybrid rootstocks, are *rotundifolia, champini, candicans,* and *longii* (Solonis).

USE CLASSIFICATION FOR GRAPES

Grapes are divided into five main classes, depending on their purpose. (Jacob, 1950).

Table Varieties

These grapes are utilized for food and for decorative purposes. They must have an attractive appearance, good eating qualities, good shipping and storage qualities, and be resistant to injury incurred in handling. Large berries of uniform size, with firm pulp, tough skin, a sturdy rachis, and strong adherence of berries to cap stems are desirable, especially in grapes transported by truck, train, sea, or air. In the United States there is a strong preference for seedless grapes. A cluster so compact that it cannot be bent or a loose cluster lacking a full set of berries should be avoided.

Wine Grapes

These varieties can produce satisfactory wine in some localities. The majority of grapes grown throughout the world are utilized for wine making. For dry or table wines, grapes of high acidity and moderate sugar content are desirable, while grapes with high sugar content and moderately low acid are required for sweet or dessert wines. Varietal grapes such as Cabernet Sauvignon, Riesling, and Pinot noir have the outstanding bouquet and flavor essential for production of highest quality premium wines. Thin-skinned berries with very soft pulp may require more care in picking, mechanical harvesting, and transportation than those with a thicker skin and a firmer pulp. For mechanical harvesting grapes should have berries that detach easily from the cap stem.

Raisin Grapes

These may include any dried grapes, although several standards must be met if a suitable dried raisin is to be made. The dried raisins must be soft in texture and should not stick together when stored. Early ripening is important so that berries can be dried in favorable weather. Seedlessness is preferred, and raisins must have a good flavor. The berries should either be large for eating or small for the bakery trade. The vines should be highly productive, and the berries should dry rapidly.

Few varieties can meet all of these criteria. Some of the best and most widely grown grape varieties for raisins are Thompson Seedless, Black Corinth, and Muscat of Alexandria; the latter has seeds that can be removed by machine.

Juice Grapes

In the manufacture of sweet unfermented juice, the clarifying and preserving procedure should not destroy the natural flavor of the grape. In the United States juice is usually pasteurized after it has been clarified, destroying the natural flavor of the *vinifera* varieties in the process and producing a cooked, undesirable taste. Some American varieties do survive processing and pasteurization procedures almost unchanged, and manage to maintain their strong foxy flavor. In the United States grape juice is usually produced from Concord grapes or a blend of Concord and other varieties. When close filtration is used for sterilization, *vinifera* can be used for unfermented juice grapes.

Canning Grapes

Only seedless grapes are suitable for use in canned fruit. The Thompson Seedless variety is most commonly used, alone or in combination with other fruits as fruit salad or fruit cocktail.

Chapter 2
VINE STRUCTURE

The grapevine consists of two basic portions; the roots which are normally underground, and the trunk, arms, and shoots which are usually aboveground (Fig. 2-1). The shoots consist mainly of stems, leaves, and flowers or fruit. The cells are the small structural compartments that make up tissues. Since they possess protoplasm (living material), they are the living units of vine structure and function. The vine consists of cells and the product of cells, and is an integrated collection of living and nonliving cells.

ROOTS

First Year Growth

Roots differ from stems because they lack nodes and internodes (regions that alternate throughout the length of the stem). The usual method of propagating vines is to use stem cuttings. Roots arise from meristematic regions near the surface of the cutting, and most develop near buds at the nodes. These roots, which do not arise from other roots, are termed *adventitious*. At the start of each growing season the overwintering roots develop new absorbing roots from many growing points (Pratt, 1974). As the root system develops and enlarges, root branches that arise from meristematic regions inside the root, may in turn produce new branch roots (Fig. 2-2). The finest roots, known as rootlets or feeder roots, are important because they greatly increase the absorption region of the roots.

At the apex of the root is the *root cap,* a mass of cells covering and protecting the apical meristems (undifferentiated tissues whose cells are capable of active cell division). Behind the root tip is a zone of elongation a few millimeters long (Pratt, 1974). Proximal to this is the zone of absorption of water and salts from the soil, about 3.4 in (10 cm) long, and many epidermal cells have elongated perpendicularly to the surface

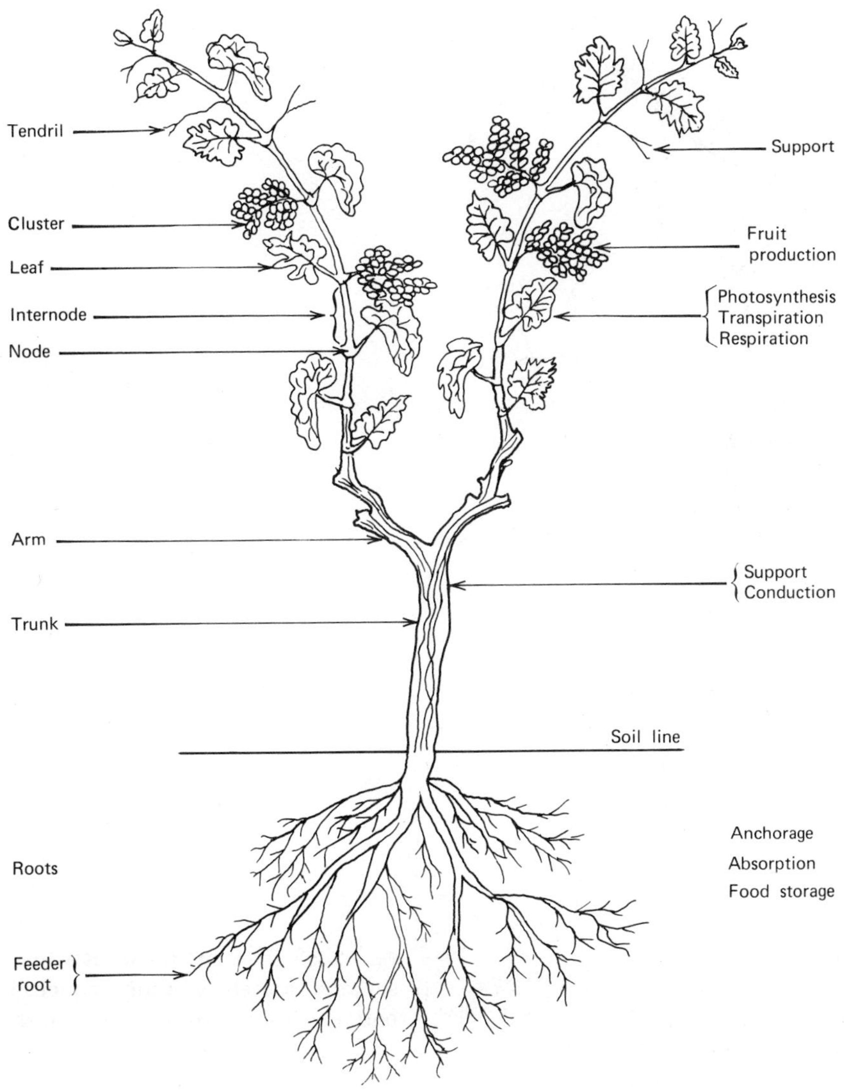

Fig. 2-1. Diagrammatic illustration showing important structures and functions of a grapevine.

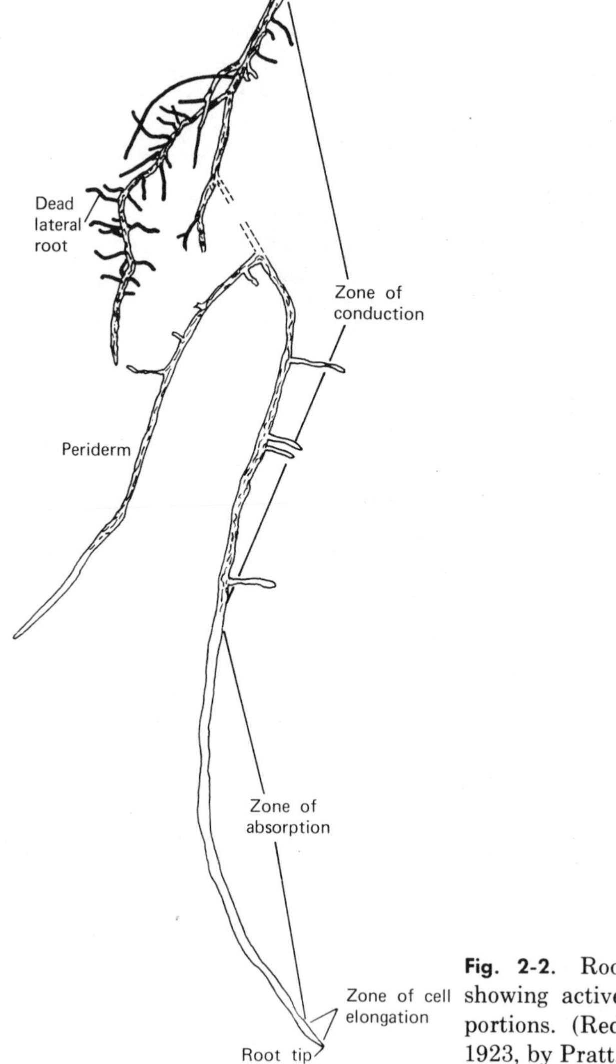

Fig. 2-2. Root of White Riesling, showing actively growing and dead portions. (Redrawn from Kroemer, 1923, by Pratt, 1974.)

to form root hairs that increase the absorbing area. This portion of the root, also known as the root hair region, is often yellowish in color. This zone, constantly replenished by new growth, and the root tip are very important regions of the root because growth in length, most absorption of water and nutrients from the soil, and development of primary tissues (first developed) occur here. Differentiation (development of tissues with many cells for performance of certain functions) begins in the upper cells of the meristematic region and extends upward through the zone of elongation into the lower zone of root hairs. In California, where soils are almost neutral to slightly alkaline, few root hairs are produced. Proximal to the zone of absorption is the zone of conduction, the mature stage.

Seeds are planted in breeding work to produce new varieties; a taproot develops and laterals arise from it. New roots then grow from the laterals and further branch roots develop.

Penetration and Spread

The root system of grapevines often penetrates deeply and spread laterally in the soil to a greater degree than the tops of the vines. It is a major component of the vine in terms of both absolute bulk and function. It often consists of one-third or more of the dry weight of the entire vine. Most of the roots are usually located in the upper 5 feet (1.5 m) of soil, but they can penetrate much deeper, often to 6 to 10 (1.8 to 3.0 m) or more (Kasimatis, 1967). In coarse sands or gravelly soils, roots can penetrate 25 feet (7.6 m) or deeper; penetration is usually less as the soil texture becomes finer.

Root depth may be limited by the occurrence of a water table, hardpan (a soil zone impervious to root growth), shallow soils, or by a zone of toxic materials in the soil. The root is structurally adapted for its major functions of absorption, anchorage, and storage.

THE ABOVEGROUND PART OF THE VINE

The aboveground part of the vine consists of the trunk and arms, shoots (including leaves, buds, and tendrils), flowers, and fruit.

Trunk and Arms

The trunk is the main stem of the vine that supports the canopy of leaves and other upper vine parts, and is the connecting link between the top of the vine and the roots. Water and mineral nutrients absorbed by the roots are transferred to the foliage, where food is manufactured for nourishment of the whole plant. A portion of these elaborated food materials are translocated downward through the food-conducting tissues (phloem) of the trunk to the roots.

The main branches of the trunk older than 1 year are called *arms*. They bear the spurs and canes kept at pruning for the production of the following year's crop of wood and fruit.

Shoots

The succulent stem with leaves that arise from a bud is termed a *shoot*. It is the current season's top growth. A *lateral shoot* is one growing out of the main shoot. The shoot is the portion of a vine that bears the flowers and fruit. The *shoot tip* refers to the apical end [about 6 in. (15.2 cm) long] where growth in length occurs as a result of cell division and cell elongation. A *cane* is a mature shoot after it has lost its leaves. Along the canes are slightly enlarged areas called *nodes*, where buds develop and from which leaves arise. The space between nodes is an *internode*, which may be short or long. The cane has a spongy *pith* at its center which, in most cultivated species (except rotundifolia and munsoniana), is interrupted at the node by a woody partition called a *diaphragm*.

The stems have rays that break up the secondary xylem (wood) into radial blocks (Fig. 2-3). The vessels (water-conducting tubes) are very large. There are also tangentially and radially defined blocks of phloem fibers as well as extensive food storage tissues (Pratt, 1974).

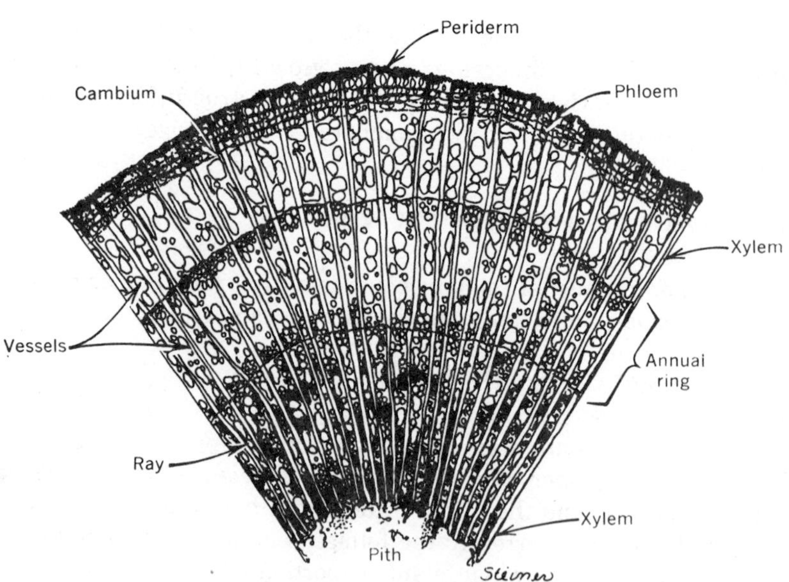

Fig. 2-3. Transverse section of part of 3-year-old arm of *vinifera*, showing periderm (dark outer tissue), phloem, cambium, distinct annual increments of xylem, rays, vessels, and annual ring resulting from one season's growth. ×8. (Redrawn from Esau, 1948.)

It is still unclear whether the stem is a single axis (a monopodium) or a sympodium. The latter theory states that at each node the main axis is carried on by the topmost lateral bud rather than by the terminal bud. In this case, the original terminal growth ceases and a lateral bud develops into a shoot.

Tendrils

Tendrils as well as the inflorescence can be considered to be lateral branches, each having its own specialized origin, structure, and function (Pratt, 1974). Leafless coiling tendrils occur opposite to or alternating with the leaves, and braces the vine by attaching to wires or other vine supports. Nearly all species have discontinuous tendrils; two adjacent leaves have tendrils but the third leaf has none. Usually the lower leaves of a shoot have no tendrils. *Vitis labrusca* has continuous tendrils, in which case there is a tendril or flower cluster opposite every leaf.

Buds

Buds develop from meristems axillary to a leaf. According to their subsequent behavior they can be classified as the summer lateral, the primary, secondary, and tertiary buds (Pratt, 1974) (Fig. 2-4). The primary, secondary, and tertiary are grouped together and appear as one bud. Hence, the three buds together are referred to as a *compound bud* or merely as a *bud*.

The *primary shoot* usually develops from a primary bud on the spur or cane. Before the bud goes into a rest period in late summer of autumn, it usually forms from 6 to 9 nodes (Pratt, 1974). Clusters are usually formed opposite the fourth to sixth leaves. In winter the buds are covered with dark hard scales and a soft tomentum. Each bud really consists of three buds: a central well-developed one, the *primary bud*, and two smaller ones known as *secondary* and *tertiary* buds (Fig. 2-5). The shoots usually arise from the primary bud, while the other two remain dormant. If the main bud has been killed, however, one of the secondary buds may begin to grow to replace the dead bud.

The *lateral shoot* or *summer lateral* arises from the primary shoot soon after the primary shoot begins active growth. The lateral shoot produces leaves and, in *vinifera* grapes, often several clusters referred to as the *second crop;* clusters on primary shoots are called the *first* or *primary crop*. Sometimes the second crop may make up 25 percent or

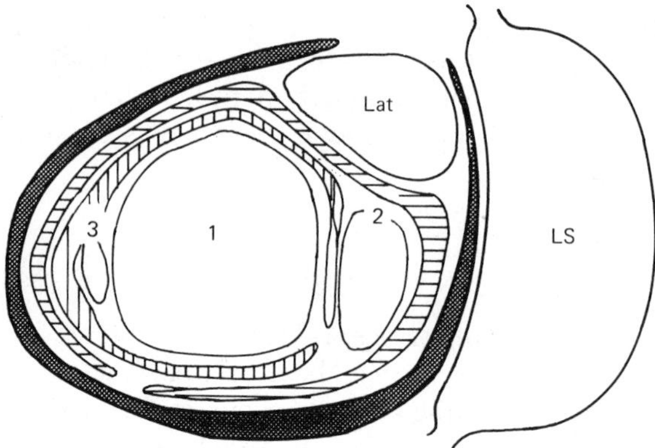

Fig. 2-4. Diagrammatic transverse section through a compound bud of Concord showing relative positions of leaf scar, lateral shoot, and three dormant buds. *Abbreviations for Figures 2-4 and 2-5: Lat,* lateral shoot, *LS,* leaf scar; *l,* primary bud in axil of bract (solid black) of lateral shoot; *2,* secondary bud in axil of basal bract (horizontally hatched) of primary shoot; *3,* tertiary bud in axis of next higher bract (vertically hatched) of primary shoot. ×17. (After Pratt, 1959.)

more of the total crop. Laterals may grow from leaf axils of the summer laterals in the same season.

Some buds, termed *adventitious buds* (arising from other places than leaf axils), remain imbedded in old wood. The weak growing shoots that occasionally arise from adventitious buds are called *water sprouts*.

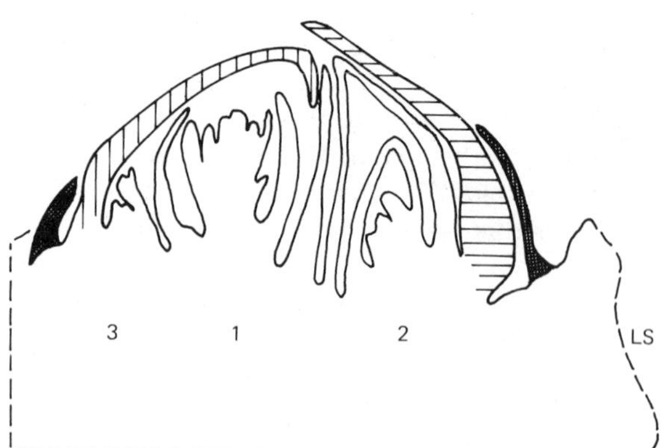

Fig. 2-5. Diagrammatic longitudinal section in plane of the cane axis showing the compound bud of Concord. Leaf scar is indicated by broken line at right of bud, ×17. (After Pratt, 1959.)

A *leaf bud* produces shoots that bear only leaves; *flower* or *fruit buds* contain a shoot possessing both rudimentary leaves and flower clusters. A fruit bud develops into a shoot that usually bears 1–4 clusters located opposite the leaves on the lower part of the shoot. In the dormant season the various types of buds can be identified only by microscopic techniques.

Leaf

The three parts of the leaf are the *blade,* the *petiole,* and two *stipules*. The latter are broad, short scales arising from the enlarged base of the petiole that partially encircle the stem. The pair of stipules can be seen on young leaves early in the growing season, but they soon dry up or fall off. About 30–40 days are required for full expansion of the blade, and senescence begins about 4–5 months after unfolding in full sunlight (Pratt, 1974). Leaves also thicken with age. There are few or no stomates (small openings on leaves) in the upper epidermis (outer layer of cells), but the lower epidermis has many. The cuticle (waxy layer on the outer walls of epidermal cells) consists of overlapping platelets of "soft wax" containing hydrocarbons, esters, aldehydes, alcohols, and unknown acids.

The palisade cells consist of one layer of cells containing many chloroplasts (Fig. 2-6). The spongy mesophyll cells are lobed cells containing many small chloroplasts and numerous air spaces.

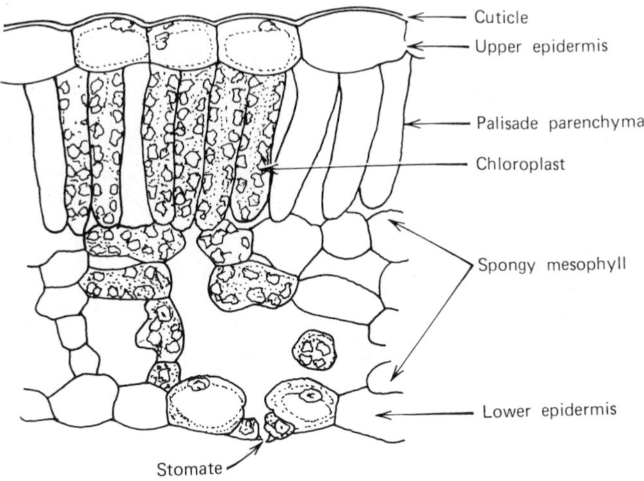

Fig. 2-6. Transverse section of mature leaf of *V. vulpina.* (After Pratt, 1974.)

The leaf blades are usually indented and most have five lobes. The shape of the leaves is sometimes helpful in varietal identification. Deep indentations between lobes are termed *sinuses*. The sinus located at the junction of the petiole and the leaf blade is known as the *petiolar sinus*. The margins of grape leaves are also often toothed. The tips of the lobe serrations end in hydathodes, structures which release liquid (p. 32). The primary functions of the leaf are photosynthesis and transpiration.

Flower

The *flower* and *fruit* comprise the reproductive parts of the vine. An inflorescence (flower cluster) is initiated during late spring and summer preceding the year in which flowering and fruiting occur. The cluster occurs opposite a foliage leaf in the same position as a tendril, to which it often shows transitional forms (Pratt, 1971). The flowers usually bloom about 6–10 weeks after the beginning of shoot growth, depending on climatic conditions. Flowers are born in clusters, and there may be several hundred flowers per cluster or bunch. The *rachis* is the main axis of the cluster, and the individual flowers are born on a *pedicel* or *cap stem*. The portion of the rachis from the shoot to the first branch of the cluster is termed the *peduncle*.

The main parts of a complete flower are the calyx, usually with five partly fused sepals; the *corolla* with five green petals united at the top to form a *cap* or *calyptra*, which falls at blooming; five *stamens* consisting of the *filament* and the pollen-producing *anther*, and a *pistil* (Fig. 2-7). The pistil consists of three parts: a *stigma*, a short *style*, and an *ovary* with two locules.

Most *vinifera* varieties have perfect or hermaphroditic flowers that have both a functional pistil and stamens (Fig. 2-8). Female or pistillate

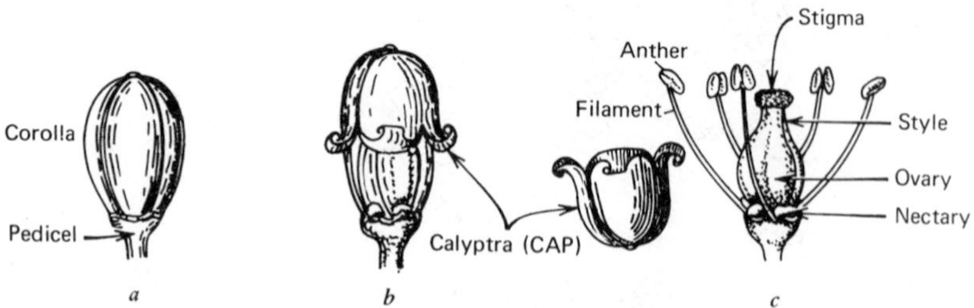

Fig. 2-7. Bloom sequence of grape flower: (*a*) calyptra attached, (*b*) calyptra separating, and (*c*) open flower. (After Babo and Mach, 1909.)

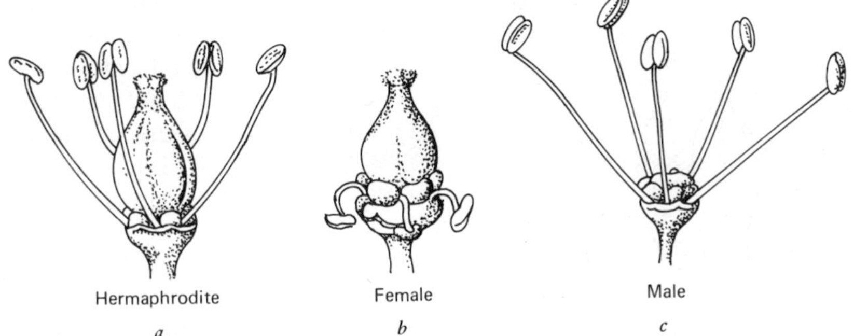

Fig. 2-8. Flower types: (*a*) hermaphrodite, (*b*) female, and (*c*) male (After Babo and Mach, 1909.)

flowers have stamens that are short and more or less reflexed, and produce pollen that is generally sterile. These types are found in many varieties of European grape as well as in certain American varieties. Male or staminate flowers have an undeveloped pistil that has neither a stigma nor a style, but contains only a small ovary which cannot be fertilized.

Many species of grapes such as some rotundifolia are dioecious. These have male flowers (functionally staminate) on one plant and female flowers (functionally pistillate) on another. To cultivate such varieties they must be interplanted; usually however, only plants having hermaphroditic flowers are selected for cultivation.

Pollination. During bloom, pollen grains fall upon the stigma where, under favorable conditions, they germinate. Anthesis occurs mainly between 6 and 9 A.M. with a rising air temperature, although much may also occur from 2 to 4 P.M. The pollen tube penetrates the tissue of the stigma, and grows down the style to the embryo sac in the ovary where fertilization occurs. Fertilization occurs 2 or 3 days after pollination (Pratt, 1971). The ovary then develops into the grape berry.

Fruit

Clusters consist of peduncle, cap stems, rachis, and berries. There are various types of bunch shapes such as cylindrical, (same thickness from top to bottom), conical or pyramidal, or globular or round branches (Fig. 2-10).

The *berry* consists of skin, pulp, and seeds (Fig. 2-9). The skins com-

Pedicel
Skin
Seed
Pulp
Stylar scar

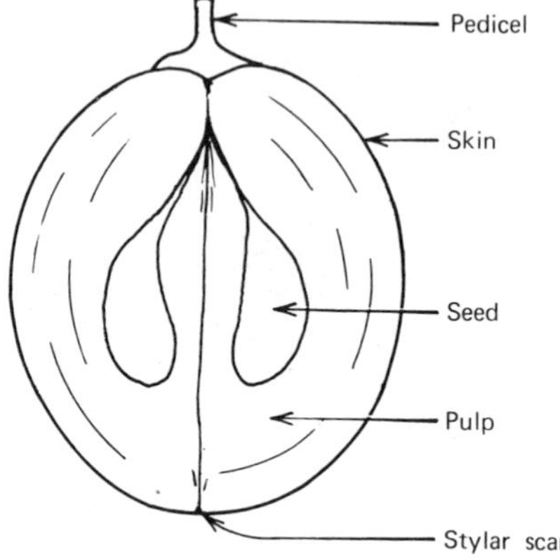

Fig. 2-9. Longitudinal cross section of grape berry.

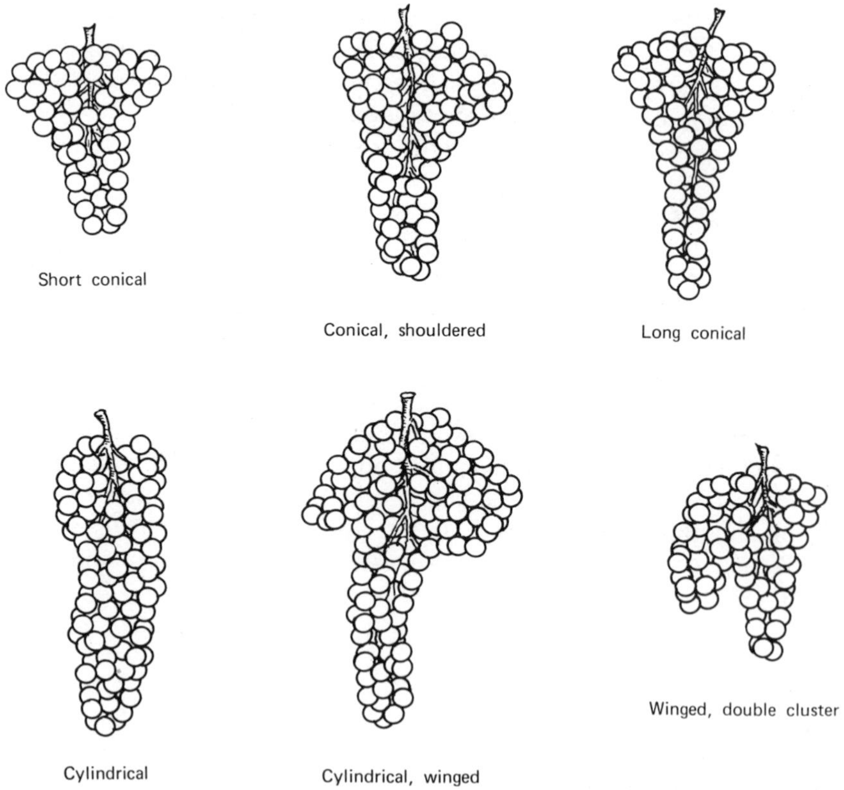

Short conical
Conical, shouldered
Long conical
Cylindrical
Cylindrical, winged
Winged, double cluster

Fig. 2-10. Diagrammatic representation of various cluster shapes (After Kasimatis et al., 1972.)

prise about 5 to 12% of the mature grape cluster (Amerine and Joslyn, 1970). The *bloom* is a thin, waxlike layer on the skin that enhances the appearance of the berry and prevents water loss and mechanical injury. The outer layers of the berry, mainly the skin, contain most of the aroma, coloring, and flavoring constituents. The skin-to-pulp ratio is greater in smaller berries than in larger ones. Thus, a ton of a small-berried variety of grapes would have more color and flavor than a ton of the same variety with larger berries (Singleton, 1972). This is one of the reasons that large-berried varieties such as Emperor, Ribier, and Malaga do not produce good wines. Different varieties of grapes often have berries of different shapes, which is useful in varietal identification (Fig. 2-11).

The *pulp* or fleshy pericarp is the portion surrounded by skin in which the seeds are embedded. The juice accounts for 80–90 percent of the crushed grapes (Amerine and Joslyn, 1970). The flesh of most grapes is translucent with colorless juice; in some varieties, however, the

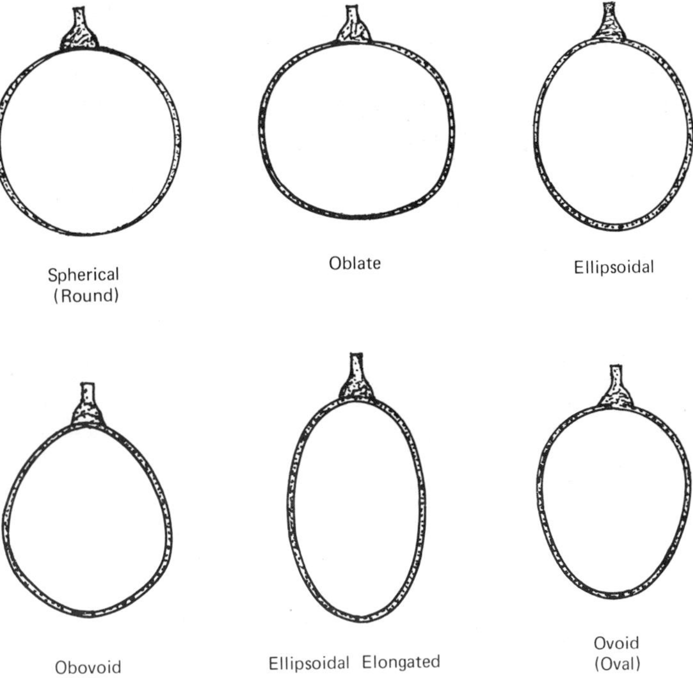

Fig. 2-11. Diagrammatic representation of various berry shapes. (Adapted from Kasimatis et al., 1972.)

pulp is light or dark red. White wines are made from varieties with colorless juice that may or may not have red pigment in the skin. Red wines are made from grapes with red pigment in the skin or pulp. Grapes with red juice are used only for making red wine.

The skins of the European-type grape adhere strongly to the flesh so that they are eaten together, but American varieties have *slipskins*.

Seeds consist of 0–5 percent of the weight of the crushed grapes (must) (Amerine and Joslyn, 1970). Seed number usually varies between zero and four per berry. Seeds are high in tannin (5–8 percent) and oil (10–20 percent of the weight of the seeds).

The stems of the cluster include the rachis and its branches, and pedicels, and make up 2–6 percent of the total weight of the cluster at maturity, depending on variety. There is much varietal difference in length of cluster parts, toughness, strength of pedicel attachment to berries, and rate of drying and browning of stems after harvest. These cluster characteristics are especially important with table grapes, which must be packed and shipped over long distances to market.

Freshly harvested stems sink in water, but after a drying period they float on water. As the cluster frameworks dry, pockets of air are probably retained in the tissues, reducing their specific density and causing them to float.

A ton of grapes can produce about 180–195 gal (681 to 738 liters) of wine depending on variety, type of processing, and other factors.

Chapter 3

GROWTH OF THE VINE

ANNUAL CYCLE OF GROWTH

Like most other plants, the grapevine has a fairly predictable cycle of growth. This chapter discusses the various growth stages beginning with the dormant season.

Dormant Season

This season begins in autumn in temperate regions when the vine sheds its leaves and enters the dormant period. There are some grape growing areas in the world, such as the subtropical climate of southern India, where grapes do not shed their leaves naturally. In such areas it is necessary to induce a type of dormancy by stopping growth for a time to get new shoots and good crops. Usually all leaves are removed by hand and often heavy root pruning and withholding of water are practiced. In some regions this must be done twice each year.

In the winter in temperate regions, much of the starch is converted into sugars that protect the vine against low temperature injury. Near the end of winter or the beginning of spring the vine may exhibit the phenomenon of *bleeding*. If a cane is cut, liquid flows from the xylem tissues. More than a gallon can be collected from one cut cane. If the cane tips of a vine are cut off every other day, from 5 to 7 gallons (18.9 to 26.5 liters) can be collected (page 32) (Winkler et al., 1974). Bleeding has no harmful effect on the vine.

Dormancy can be divided into periods of *quiescence* and *rest*. The first type of dormancy is under exogenous control, in which buds fail to grow because of unfavorable external conditions. *Rest* is under endogenous control, in which internal factors prevent growth despite favorable environmental conditions (Weaver, 1972). During the period of rest the inhibitor-promoter balance of hormones is weighed in favor of the inhibitor, but at termination of rest the balance is shifted in favor of the promoter. Bud rest can be broken by cytokinins, heat, cold, or desiccation. Rest is prolonged by gibberellin.

Grape seeds are in a resting condition at fruit maturity. The usual method of terminating rest is to stratify the seeds at about 40°F (4°C) for 3 months. Gibberellin can terminate rest.

Bud Break to Bloom

In the spring, when the mean daily temperature reaches about 50°F (10°C), the buds begin to swell and the green shoots emerge from them. This is commonly known as *bud break*. The shoots grow rapidly in length and thickness, and leaves, tendrils, clusters, and new buds in the leaf axils develop rapidly. As the daily temperature rises, shoots often attain a growth rate of an inch per day. Blooming usually occurs around 8 weeks after bud break, but the interval depends upon the weather. Bright warm weather brings on blooming earlier than cool rainy weather and reduces its duration. Rapid shoot growth in length usually begins to slow down by bloom time (Fig. 3-1).

The process of flower initiation for the following year's crop begins

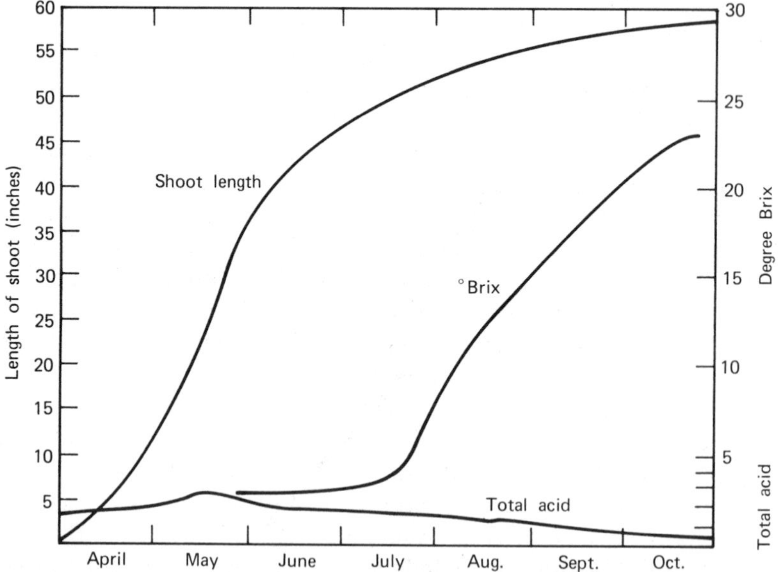

Fig. 3-1. Diagrammatic presentation of shoot growth, level of °Brix, and total acids of berries during the growth season of a *vinifera* grape.

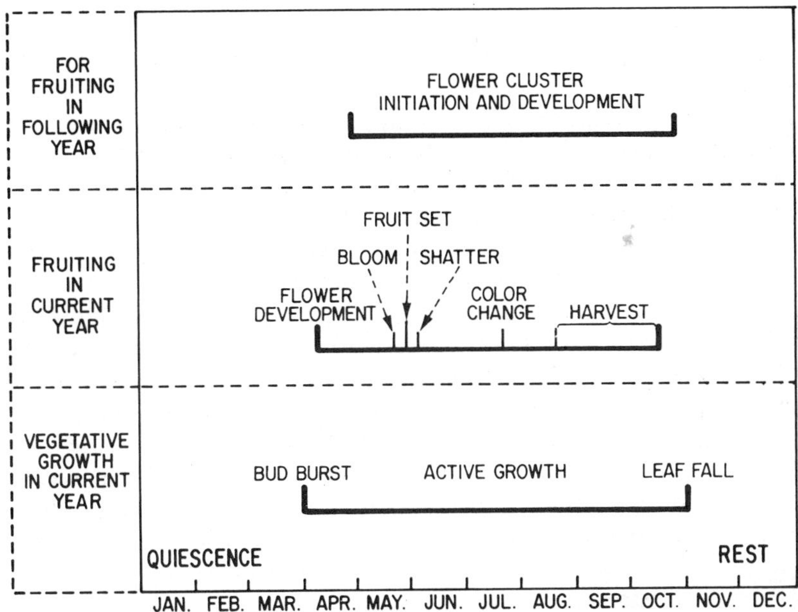

Fig. 3-2. Calendar showing when the stages in the growth and fruiting of a *vinifera* grapevine occur in an average year. Heavy lines indicate the periods of active growth. For emphasis, the events of shoot growth, flowering, and fruiting for the current year, and flower cluster initiation for the following year are shown on separate lines; actually many occur simultaneously in a mature vine. (Adapted from Shaulis and Pratt, 1965.)

before bloom, and the development continues until about harvest time (Fig. 3-2).

Bloom and Fruit-set

Bloom is the period when the caps (calyptras) fall from the flowers. Since bloom may progress for several days over a vine and an individual cluster, one must estimate the percentage of cap fall to designate the stage of development. Many grape growers consider full bloom that time when an average of 50 percent of the calyptras have fallen from the flowers.

When the calyptras fall from the flower a cloud of pollen is released from the anthers of the stamens which move away from the pistil. Pollen grains fall on the stigma and germinate if conditions are favorable. A pollen tube grows down the style to the embryo sac and serves as a pathway by which two sperm reach the embryo sac. One sperm then unites with the egg cell to form the zygote from which the embryo plant develops.

In cold rainy weather calyptras may not fall from the flowers. These persistent calyptras often reduce the amount of fruit-setting.

Several days after the bloom period, pistils and impotent berries—often 50 to 60 percent or more—shatter from the cluster. Many of these pistils have not been fertilized. This shatter of berries is important because it keeps clusters from becoming too compact. On the other hand, some varieties of clusters may set poorly and have many small, seedless berries that fail to enlarge (shot berries), a condition known as *millerandage*. Frost or rainy weather at bloom time may cause shot berry formation in clusters that would otherwise set well. The amount of set varies from season to season. A normal setting cluster of a seeded variety often has a mixture of berries containing from 0 to 4 seeds. A direct relationship exists between number of seeds and berry size: the greater the number of seeds, the larger the berry. This is because seeds produce gibberellins and other hormones that diffuse into the flesh of the berry and stimulate growth.

Some varieties set fruit without fertilization, a process known as *parthenocarpy.* Black Corinth usually require only the stimulus of pollination. In other seedless varieties (Perlette, Thompson Seedless, Black Monukka) fertilization occurs but the embryo subsequently aborts, a process termed *stenospermocarpy*. Other varieties, such as Chaouch, produce hard seeds that are hollow or empty.

The berries that do not fall from the cluster after bloom are said to have *set*. This is referred to as the fruit-set stage. Fruit-set and development are probably controlled by an interaction of naturally occurring hormones including auxins, gibberellins, cytokinins, ethylene, and inhibitors (Weaver, 1972).

Double-Sigmoid Curve of Berry Growth

When the increase in such variables as volume, fresh weight, dry weight, and diameter of fruit is plotted as a function of time after anthesis, seeded grapes are characterized by a double-sigmoid curve (Fig. 3-3). Two periods of rapid growth are separated by an intermediate period when either less growth or no growth in volume occurs. There are three clearly defined stages of growth. In the first (cell-division) stage, the ovary and its contents grow rapidly except for the embryo and

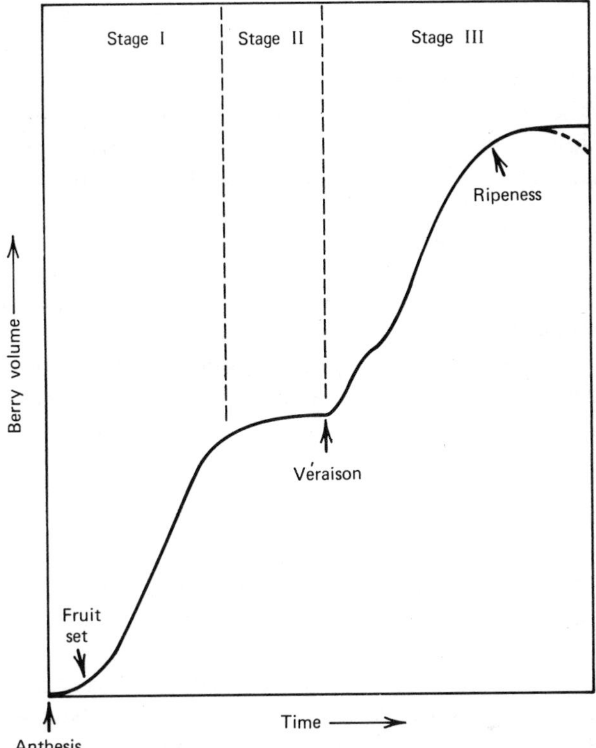

Fig. 3-3. Diagram of the growth curve of a grape berry, showing the extent of the three stages and the location of véraison (After Coombe and Hale, 1973.)

endosperm. Stage II is characterized by rapid growth of the embryo and endosperm, lignification of the endocarp, and slight growth of the ovary wall. In Stage III, rapid growth of the mesocarp occurs, causing the final swell of the fruit followed by maturation.

The explanation for the slow-growth period (Stage II) is unknown at present. Perhaps the osmotic pressure resulting from accumulation of sugars in the berry effects the beginning of Stage III, since the movement of water into the fruit causes cell enlargement and growth (Coombe, 1960). Ethylene may initiate the growth that occurs in Stage III (Maxie and Crane, 1968).

Seedless (stenospermocarpic or parthenocarpic) berries usually show less distinctive growth periods than do seeded berries. Generally they exhibit little or none of the double-sigmoid growth curve.

Green Stage of Berry Growth

This stage lasts from the setting of the berries up to véraison, the time that berries begin to color and soften (beginning of Stage III). Berry size increases rapidly (except for Stage II in seeded berries). The amount of sugars remains low and constant and the acidity is high.

Ripening Stage of Berry Growth

This stage begins when fruit starts to color and soften. In white varieties it is noted when the green color of the berries turn white or yellow. As ripening progresses the color of red or black varieties becomes more intense, the amount of sugar increases, and acidity decreases. Berries on the basal portions of clusters approach maturity more rapidly than berries on the apical portion.

Overripe Stage

This stage begins after the grape has passed the peak quality for its intended use. Due to the evaporation of water from the berries, the concentration of sugar increases while the amount of acid continues to decrease. Overripe grapes are easily attacked by fungi and insects. Berries shrivel and, in some varieties, the shatter of berries increases. Finally the grapes dry and raisin, and in many varieties unpicked grapes remain on the vine until they are removed by pruning.

Accumulation of Food Reserves

The stored foods in the roots and vine are rapidly utilized when the vine begins to grow in the spring, and there is a quick decline in reserve foods until about fruit-set. At this time shoot growth has slowed down and carbohydrates begin to collect rapidly in the shoots. After berry growth slows down and rate of shoot growth almost stops, the rate of accumulation of carbohydrate increases. By late autumn the food reserves in the shoots and roots are usually replenished. Collectively, the roots represent the main food storage organ of the vine, but carbohydrate is also found in the trunk, arms, and fruit. Starch and sugars are the main food reserves. In the dormant season available carbohydrate is stored mainly as sugar.

Sugars produced by photosynthesis can also be transformed into proteins and fats and other carbohydrates, each of which is essential to the life processes of the vine.

VINE GROWTH PROCESSES

The growing vine increases in bulk and complexity. Vine growth can be defined as an irreversible increase in size as measured by an increase in its dry weight. It reflects a net increase in protoplasm, the living substance in the vine. Vine development refers to differentiation and anatomical and physiological organization and specialization.

Photosynthesis

This process involves the use of light energy by the green leaves to convert carbon dioxide and water into energy-rich compounds. All organic matter in the vine is ultimately provided by photosynthesis, one of the most significant of all life processes. Photosynthetic reactions occur in the chlorophyll-containing plastids found in green leaves, and the reaction can be summarized as follows:

$$6\ CO_2 + 12\ H_2O + \begin{array}{l}\text{673,000 calories of light energy}\\ \text{in presence of chlorophyll, enzymes,}\\ \text{and cofactors}\end{array} \longrightarrow$$

$$C_6H_{12}O_6 + 6\ O_2 + 6\ H_2O$$

This chemical equation shows that 6 moles of carbon dioxide, 12 moles of water, and 673,000 calories of energy yield 1 mole of glucose, 6 moles of molecular oxygen, and 6 moles of water. The process is, of course, far more complex than this equation indicates. From the sugar produced, other compounds found in the vine are metabolized.

Carbon dioxide is normally found in the air at a concentration of 0.03 percent, and it enters the leaves through small holes called stomata (pp. 17). In confined areas it is possible to increase vine production by CO_2 enrichment of the air (Kriedemann, private communication). The rate of photosynthesis increases with light only to a certain intensity. At this point the leaf is said to be *light saturated*. In grapes this figure ranges from about 2500 to 5000 footcandles (ft-c) when grown under favorable conditions (Kriedemann and Smart, 1971). As the amount of light is reduced to about 125 ft-c the *compensation point* is reached, where the amount of food manufactured in photosynthesis is just equal to that lost by respiration (Kriedemann, 1968). With further reduction in light, the leaves can be said to be parasitic on the rest of the vine. Usually the outer tier of leaves absorbs around 90 percent of the light used for photosynthesis, leaving only 10 percent for use by the other leaves.

On a clear day, when leaves in the sun are at light saturation, there can be 12,000 ft-c of light. Since leaves shaded by other leaves will be below light saturation and may thus not be working at maximum capacity, as many leaves as possible should be in direct sunlight for greatest efficiency. This can best be attained by using proper training and trellising to give the most leaf exposure.

A high light intensity is also of value because more fruitful buds are produced. Vines grown in shade often produce little fruit in the subsequent year (May, 1965).

Grape leaves reach maximum photosynthetic activity when they attain full size (Kriedemann et al., 1970), about 30 to 40 days after unfolding; afterward the photosynthetic rate gradually declines. High temperatures inhibit photosynthesis. Above 86°F (30°C) the rate decreases and practically ceases at 113°F (45°C) (Buttrose, 1968; Kliewer et al., 1972). Optimum leaf temperature for maximum photosynthesis ranges from 77°F (25°C) to 86°F (30°C).

The stomata open in the morning when light strikes the leaf. This brings about an increase in sugar concentration that causes an increase in the osmotic pressure within the guard cells, and water then flows into the guard cells causing them to swell and become turgid. The thin peripheral wall bulges outward, pulling the thicker elastic wall along with it. Stomatal closure can result from lack of water in the soil, a high rate of transpiration, or low light intensities, or darkness. The growth inhibitor, abscisic acid, may play an important role in stomatal closure. Vines that wilt close their stomata; as a result CO_2 uptake is reduced, causing a reduction in photosynthesis. Leaf water potentials less than about −15 bars decrease the photosynthetic rate (Kriedemann and Smart, 1971).

Respiration

Energy from the photosynthetic reaction is stored in the chemical bonds of the sugars produced. Sugar is translocated to the various parts of the vine where its energy is used to run the cell machinery. Respiration is the process of obtaining energy from organic material. It is done at low and constant temperatures in controlled reactions. The following general equation is a simple expression of the very complex reactions involved in the biological combustion of sugar:

$$C_6H_{12}O_6 + 6\,O_2 \rightarrow 6\,H_2O + 6\,CO_2 + \text{energy}$$

The energy captured from light is released by the low temperature oxidation (removal of hydrogen) of sugars. A small amount of the energy is lost as heat, but most is channeled into chemical work. It is first stored as high energy phosphates and later used for the synthesis of organic materials needed for growth and development of the vine, and for other vital activities such as nutrient absorption.

Metabolism

All materials in the vine were derived from sugars produced by photosynthesis and nutrients and water absorbed from the soil. Metabolism refers to the synthesis and degradation of the organic materials formed. Plants are cultivated for the molecules they synthesize. Carbohydrates, proteins, and lipids are important compounds because they are the major constituents of our foods. In grapes the food used by man is mainly in the form of sugars.

Carbohydrate nutrition is the study of metabolism of sugars and starch, and mineral nutrition is the study of metabolism of nutrients absorbed from the soil. The study of plant nutrition is intimately concerned with soil chemistry and biology.

Water is the most abundant compound in an active grapevine cell, and in actively growing tissues water comprises from 85 to 95 percent of the total. Tissue from which the water has been removed consists mainly of organic matter derived from photosynthesis, and from inorganic nitrogen such as that found in ammonium or nitrates. The amount of mineral matter taken up from the soil is relatively small but nontheless essential to the life of the grapevine. The absorbed minerals are not lost in large quantities except at leaf fall in autumn. Water loss occurs constantly from the aerial parts of the plant, and this affects the rate of water absorption and movement.

Transpiration

The loss of water in vapor form from a living vine is called transpiration. Although most diffuses out through the stomata, some may be lost directly through the leaf cuticle. As water molecules evaporate from the wet cell walls inside the leaf, the wall imbibes more water from inside the cell, and the cell then gains water by osmosis from adjacent water-saturated cells. These cells in turn gain water from adjacent cells and, finally, from the water-filled tracheid of a vein ending. Thus a water-potential gradient is established from the vein to the outside air. The rate of transpiration is affected by several factors including air movements, relative humidity of the air, air temperature, light intensity, and soil conditions.

Water moves upward from the roots through the xylem, through the trunk and shoots, and then to the leaves from which vapor escapes to the atmosphere. A plausible theory about the processes involved in the ascent of water is known as the *transpiration pull* and *water cohesion theory*. The columns of water in the xylem are pulled up by the force of water evaporation from the leaf cells and the imbibitional forces that develop there. The columns of water maintain their continuity because of the strong cohesive forces between water molecules and adhesive forces between the cell walls and water molecules.

Absorption of Water

Large amounts of water are removed by transpiration and must be replaced by water absorption through the roots. Water may be absorbed by the roots by two methods. One type occurs when transpiration is low and absorption of water exceeds transpiration. Water is absorbed by osmosis (diffusion of a solvent through a differentially permeable membrane), leading to a pressure in the xylem sap, and *guttation* may then occur. This is the loss of liquid from the vine that occurs at the tips of the leaf lobe serrations. There are specialized openings called hydathodes through which the liquid emerges. Guttation liquid is not pure water but a dilute salt solution. It can occur in actively growing vines in the spring. If canes are cut in dormant season in late winter or spring, exudation of sap may occur from the cut ends of the vessels of the xylem. This condition known as *bleeding* depends, as does guttation, on the activity of living root cells, and the pressure generated in the xylem vessels is termed *root pressure*. Thus the time of cane bleeding in the spring probably coincides with renewed root activity and/or growth.

The usual mechanism of water absorption by roots results from osmotic movement of water into the roots. When soil solutes accumulate in the cell sap, water moves into the cells from the water in the soil solution. Under a high transpiration rate there is a water deficit in the plant tissues, and water moves into the root passively. Given these conditions, even dead roots can absorb water. When vines are subjected to conditions that cause rapid transpiration, the vine may show some temporary signs of water deficit and wilting since water absorption lags behind water loss despite available water in the soil.

Absorption of Nutrients

Nutrients or solutes may enter the root cells from the soil by diffusion, which is movement from a location of high solute activity to one of lower activity. Root cells can, however, accumulate nutrients in their

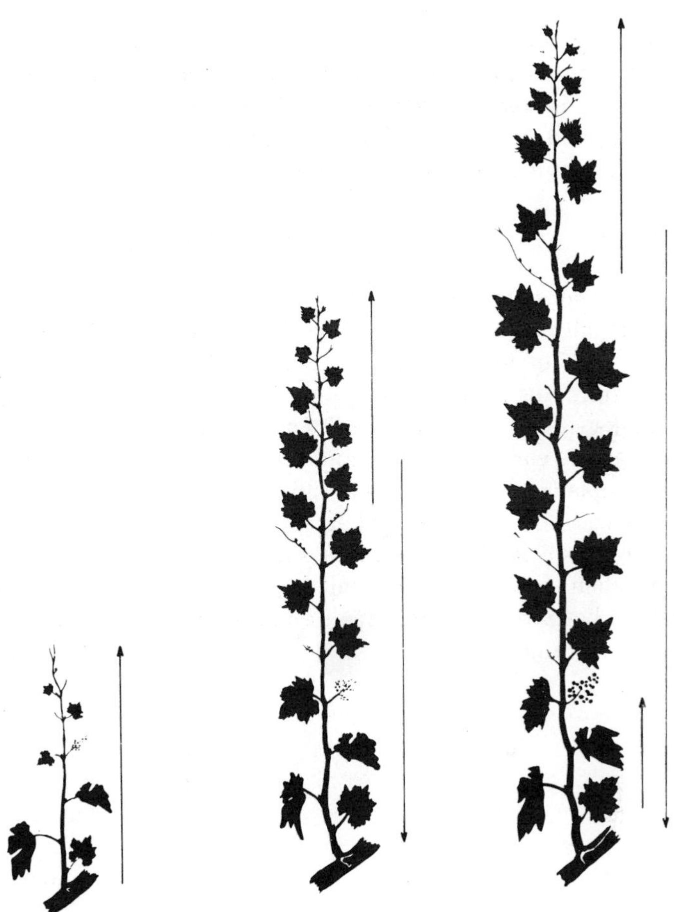

Fig. 3-4. Diagram of a rapidly growing grape shoot at three different developmental stages, showing main direction of movement of photosynthate. (*A*) Apical movement in a very young shoot tip. (*B*) Prebloom or bloom stage bidirectional export from two or three leaves below the shoot tip. Below this region movement is basal. (*C*) Apical movement of photosynthate into the cluster from leaves below the cluster, after the set of fruit. Several weeks following set stage, after the rate of shoot growth his decreased photosynthate moves basally from the tip. (After Hale and Weaver, 1962.)

vacuoles in concentrations far higher than that in the soil solution. This movement of nutrients from a region of low concentration to one of high concentration is termed *active solute absorption* or *accumulation*. Energy supplied by respiration is necessary for the movement of nutrients against a concentration gradient.

Translocation

Movement of most mineral nutrients occurs in the vessels of the xylem and it is usually upward. Salts may also be conducted in the phloem. The main path of food movement is through the phloem. The movement of assimilates is usually from a source (region of manufacture) to various sinks (regions of utilization). In the grapevine mature leaves are the main source for assimilates, and shoot tips, roots, and clusters are the major sinks (Hale and Weaver, 1962).

The sources and sinks and their power change during the growth season (Fig. 3-4). A young leaf is a sink since it only imports assimilates, but when the leaf is about half its full size it begins to export food materials to both the tip of the shoot and also downward. After fruit-set, berries become a powerful sink.

Chapter 4
VINE IMPROVEMENT

Vines grown from seeds usually differ markedly from each other and from the parent vine. Seeds are not used for propagation because the seedlings are usually of lesser quality than the parent vine in regard to vigor, productivity, quality of fruit, and wine produced from the fruit. However, seeds are valuable in breeding work and development of new varieties.

The *vinifera* species is very variable in its natural state, and accidental discovery of choice seedlings in the wild is the origin of most important grape varieties. Many varieties have also been established by man's breeding programs (Olmo, 1939).

BREEDING OF GRAPES

In the past 200 years man has attempted to produce new grapes that better suit his needs. Both selection and crossing have been used. Grape breeding work begins just before the vine flowers. The grape flower is green and about ⅛ to $^{3}/_{16}$ (3.0 to 4.8 mm) inches long. On large clusters there are often more than a thousand flowers. Practically all grape *vinifera* varieties in California have perfect flowers, that is, the flowers contain both male organs (stamens) and female organs (pistils) (Fig. 2-8). Plants grown from self-fertilized seeds may somewhat resemble the parent, but they are almost always inferior. Many are weak, have smaller fruit and other undesirable features.

CROSSING OF VARIETIES

Most grape breeders have obtained best results by crossing two varieties (Olmo, 1939). If a cross is made between Carignane and Zinfandel, the following steps must be taken. Self-fertilization of the Carignane must be prevented by removing the stamens with a pair of forceps before the pollen has been shed, several days before the flowers begin to open. This procedure is called emasculation (Fig. 4-1). After this operation is performed on many flowers of the cluster, the others, not emasculated, are plucked off. Pollen from the Zinfandel is then dusted over the pistils. It is necessary to have the Zinfandel cluster to be used for pollen bagged, especially if other varieties are growing in the vicinity. The blossoming cluster can be cut off and used for dusting the emasculated clusters, or the pollen may be collected in a vial. The treated cluster is then enclosed in a paper bag and tightly secured to prevent contamination of the cross by other pollen carried by wind or insects from neighboring vines.

The essential part of the fertilization process is the combination of the germ cells, one from the Carignane and one from the Zinfandel. It makes no difference which way the cross is made. One could have emasculated the Zinfandel flowers and used Carignane pollen.

The berries develop and their seeds contain embryo plants that have characteristics of both parent plants. These new plants are hybrids and will be unlike either parent; in a thousand or more a few might be superior to either parent. The production of desirable new grape varieties depends largely on the choice of proper parents and the growing of a sufficiently large number of hybrid plants. One must also know which

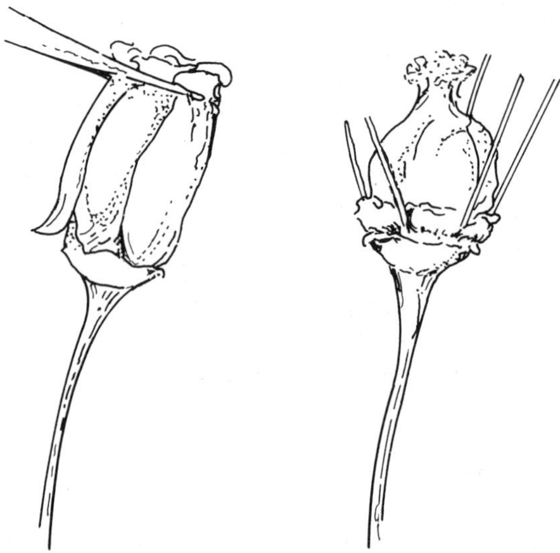

Fig. 4-1. Emasculating a Tokay grape flower by removing calyptra and stamens with forceps.

varieties best transmit superior qualities, since some very vigorous varieties yield sick and weakly offspring. Thus knowledge about the breeding behavior of many varieties is required in a successful grape breeding program.

GROWING THE SEEDLINGS

At maturity the bagged clusters are collected and the berries of each cluster counted. The seeds are extracted from the pulp, washed, dried, and then filed in packets until planting time (Olmo, 1939). In December or January the seeds are planted in boxes containing a mixture of peat moss, sand, loam, and manure. The compost must first be sterilized with steam, or soil organisms may kill the tender seedlings. The boxes are stored out-of-doors for eight to nine weeks. This chilling is very beneficial and breaks the seed rest, resulting in a high and uniform germination. Early in March the boxes are moved to a warm greenhouse and within three weeks the seeds germinate. Three weeks later, after the danger of frost is over, the young seedlings are transplanted from the boxes into cloth-covered frames out-of-doors. Here young vines are spaced at appropriate intervals to allow them to grow as much as possible during their first year. By November, shoots are 2 ft (0.6 m) or more in length, and the vines are dug, bundled, and stored in moist sand until time for planting in the vineyard the second year.

In the vineyard, the vines are set 2 ft apart in the row to get the vines to the fruiting stage in as little space as possible. At the beginning of the third year all brush is removed except the strongest cane, which is cut back to 3 or 4 in. in length. During the spring all shoots are removed except the strongest one which is trained to a stake. At the end of the growing season there is often a 5 or 6 ft (1.5 to 1.8 m) trunk.

In the fourth growing season some vines (about 30 percent) have a partial crop; in the fifth, all vines bear a crop and the selection process can then procede (Olmo, 1939). A very few promising vines are grafted to healthy rootstocks at commercial spacings in 5 or 10 vine lots. Another evaluation and selection is made from these trial blocks. Vines that pass the trial block selections are given to various growers for trials and, after they had been shown to be entirely satisfactory, are then released for commercial use.

If seeds are planted immediately after harvest, the breeding process can be shortened by one year (Olmo, private communication). After the seeds are removed from the clusters, they are planted in beds of soil on greenhouse benches in December. The greenhouse heat is turned off so the seeds will stratify in the cold. In February the heat is turned on and the seeds germinate and begin to grow. By May they have developed a good root system and shoots are often 8 to 10 in. (20.3 to 25.4 cm) long. In early June the plants are transplanted into vineyard rows at intervals of 2 ft (0.61 m), planted in broad furrows, and carefully irrigated. By the fourth season all vines bear a good crop and the selection process can begin.

Wine Grapes. The procedure for breeding new wine grapes is the same as for table grapes except that wines must be made and their quality judged.

CLONAL SELECTION

In clonal selection the best vines of a variety are selected from the best material or vineyards available. In many countries, such as Germany and Australia, there are large research programs in clonal selection. The better strains collected may merely be those that are virusfree, although there may also be genetic changes, known as mutations, that occur in plants.

DEGENERATION OF VARIETIES

Historically each grape variety arose as a seedling from a single seed. When a single plant is propagated continually by cuttings or buds, the resulting progeny in called a clone, and presumably each daughter plant contains the same genes or inherited units (Olmo, 1951). When the vine has been propagated for many generations, however, the grape clone can change because of virus infection or mutations within the plant cells.

Viruses are minute particles that reproduce within the host and parasitize the plant. The vine can be weakened or even killed by a virus. When cuttings and buds are used for propagation the virus may be carried indefinitely. When a rootstock is infected with a virus, it infects the scion portion of the vine, although the rootstock plant itself may show no visible effects of the virus it carries.

Mutations arise when the heredity units of the cells radically change their function. They arise at random and usually affect the plant adversely. Mutations are also commonly known as *bud sports,* and these continue in their new state about as stable as the type from which they arose. A positive selection process takes advantage of the rarely occurring beneficial changes. Varieties should be maintained and improved whenever possible by a selection of the best individual plants.

PRODUCTION OF VIRUSFREE WOOD

With the high incidence of virus disease in vines, it is important to provide growers with clean wood (free of virus). In California this is

Table 4-1 Indexing of virus and viruslike diseases of grapevines at Davis, California, 1974*

Disease	Index Method†	Indicator Plant	Minimum Time for Symptom Expression (mo)
Leafroll	Chip-bud graft	Mission LN-33 Baco-22A	6–18 months 6–18 months 5 months
Corky bark	Chip-bud graft	LN-33	6–18 months
Fanleaf (yellow mosaic, vein-banding)	Chip-bud graft	Rupestris du Lot (St. George)	15 days–15 months
	Mechanical inoculation of pressed sap	Chenopodium amaranticolor or C. quinoa	8–14 days
Yellow vein (tomato ringspot)	Mechanical inoculation of pressed sap	C. amaranticolor or C. quinoa	5–15 days
Asteroid mosaic	Chip-bud graft	St. George	15 days–15 months
Fleck	Chip-bud graft	St. George	15 days–15 months
Pierce's disease	Leafhopper transmission	Any vinifera variety	2 months
	Electron microscopy		2–3 days
Yellow speckle	Unknown		

* Summary tabulation presented by A. C. Goheen, at first meeting of Interregional Grape Virus Workers, Geneva, New York, September 15, 1974.

† All graft-inoculation tests of indicators are set in the field nursery and held for 18 months. All mechanical inoculation tests of herbaceous indicators are made in the greenhouse and held for approximately 30 days.

done by the nonprofit Grapevine Registration and Certification Program. This program is a cooperative effort of the United States Department of Agriculture, the University of California, the California Department of Food and Agriculture, and nurserymen.

The research organizations use transmission tests, known as indexing, or electron microscopy (Table 4-1) to identify the individual diseases or disease organisms. For each virus disease, indexing involves the use of a variety called an indicator that is especially susceptible and shows the symptoms clearly within a definite time period. Some varieties will show one virus disease clearly but not others, and rootstock behave in a similar fashion. Rupestris du Lot (St. George) shows the symptoms of fanleaf and other minor diseases very well; Mission and LN-33 illustrate leafroll; and LN-33, those of corky bark.

The present indexing at the University of California at Davis involves graft inoculation of indicator plants with tissue from the plant to be tested. The particular virus disease will produce definite symptoms of the virus in the leaves of the indicator if the plant being tested is affected. If unaffected the indexed plants become the source of wood for foundation plants, which are in turn the source of wood for nursery stock free of known viruses. The indexing program handles commercial varieties and rootstocks, new introductions from overseas, and new varieties from the breeding program.

HEAT INACTIVATION OF VIRUSES IN GRAPEVINES

Heat treatments of vines or vine parts can inactivate viruses. Goheen and Luhn (1973) grew virusfree rootstocks in containers. Then single, dormant buds from the infected mother plant were chipbud grafted each to a single rootstock. After 8 days the budded plants were placed in a heat chamber at 100°F (38°C) for 60 days. After removal from the heat chamber, plants were decapitated to force the surviving buds into growth. Over 77 percent of the buds that grew were free of the viruses present in the source vines.

PART 2
STRATEGY FOR GRAPE PRODUCTION

Chapter 5
SITE SELECTION AND WEATHER MODIFICATION

The two most important characteristics of a grapevine site are the climate and the soil. Temperature is usually the most important climatic factor.

TEMPERATURE

American Varieties. The winter minimum is especially important in areas such as the middle western and eastern parts of the United States. In New York State length of growing season (from bud burst to harvest) is important, and a growing season of at least 170 days or longer is optimum for the fruit to ripen properly (Shaulis and Dethier, 1970). In New York, temperatures of −18°F (−28°C) or below usually result in commercially important damage to Concord buds or trunks (Shaulis et al., 1964, 1973).

Concord grapes withstand humid summers and cold winters better than pure *vinifera*. They grow better in regions of moderate summer humidity than in the very dry climate of California's interior valleys. Grapes usually grow poorly in a hot tropical climate with high humidity.

Vinifera Varieties. Most vinifera grapes require long, warm-to-hot, dry summers and cool winters for best development. They are not adapted to humid summers because they are susceptible to certain fungus diseases that flourish under such conditions. They cannot withstand winter cold below −8°F (−22°C) to −15°F (−26°C) without protection. Frosts that occur after vine growth starts in the spring can kill most of the fruitful shoots and reduce the crop.

Rain is desirable in winter although irrigation can make up for deficiencies. When rains occur early in the growing season it is difficult to control disease, although vine growth is not otherwise hindered. Poor berry-set may occur as a result of rains, cold, or cloudy weather during the blooming period. During the ripening and harvesting, rains can result in severe rotting of fruit. Grapes can tolerate a higher humidity in cool regions than in warmer regions. For the sundrying of grapes for raisins a month of clear, warm, rainless weather after the grapes ripen is necessary.

Vinifera grapes usually require a winter rest period of about 2 months, with an average daily mean temperature below 50°F (10°C), and some freezing but no temperatures below 10°F (−12°C). Shoot growth in spring begins soon after the daily mean temperature reaches 50°F (10°C).

Heat Summation. For proper vine development and maturation, most varieties require a daily mean temperatures of at least 65°F (18°C); some require temperatures of 70°F (24°C) to 85°F (29°C). The time required for grapes to reach maturity is determined mainly by the total amount of heat received, which can be expressed in terms of temperature-time values called *degree days* or heat units. The effective heat summation for a given location largely determines the length of time from bloom to ripening for a given variety. Heat units are usually determined for the months from April through October, but sometimes are calculated from full bloom.

The number of heat units required for a growing area can be estimated as follows (Jacob, 1950):

Determine the average daily temperature by averaging the lowest and highest temperature of each day. Substract 50°F from the mean daily temperature and add up the mean daily temperature (−50°F per day) for the months from April through October to determine the degree-days for the season. It is easier to use the mean monthly weather data. Multiply this figure (−50°F) by the number of days in the month, which gives the degree-days for that month. For example, if 59.4°F is the mean temperature at Fresno for April, (59.4°–50°F) × 30 days = 282 degree-days for the month of April.

Early grapes require about 1600 degree-days to mature, and late ones at least 3500 degree-days (Jacob, 1950). If the temperature measurements are begun at full bloom, Thompson Seedless will be mature for table use (18° Brix) when the heat summation above 50°F reaches about 2000 degree-days. This variety will be ripe for drying of raisins when the summation reaches 3000. Similarly Tokay will be ripe for table use at about 2000, and Emperor at about 3300 degree-days. Temperature, especially during the ripening period, greatly influences the sugar and/or acid content of grapes and thus affects their quality for various uses.

Climatic Regions for Grapes. Based on the summation of heat as degree-days above 50°F for the period from April 1 to October 31, grape-producing areas in California fit into one of five temperature groups or regions (Jacob, 1950). These regions with representative locations are listed below:

1. Cool regions—less than 2500 degree-days of heat. Napa, Hollister, Mission San Jose, Saratoga, Bonny Doon, Guerneville, Santa Rosa, Sonoma, Lompoc, Watsonville, Campbell, Aptos, Santa Cruz, Gonzales, Hayward, Peachland, and Santa Maria.
2. Moderately cool regions—2501 to 3000 degree-days of heat. Rutherford, St. Helena, Glen Ellen, Healdsburg, San Jose, Los Gatos, Santa Barbara, Gilroy, Sebastapol, San Luis Obispo, Soledad, San Jose, Grass Valley, Napa, Sonoma, and Placerville.
3. Warm regions—3001 to 3500 degree-days of heat. Calistoga, Hopland, Cloverdale, Livermore, Alpine (in San Diego County), Ukiah, Paso Robles, Pinnacles, Cuyama, Santa Ana, King City, St. Helena, Healdsburg, Clear Lake Park, Jamestown, Camino, Mokelumne Hill, Potter Valley, Ramona, Mandeville Island, and Lodi.
4. Moderately hot regions—3501 to 4000 degree-days of heat. Davis, Manteca, Modesto, Ontario, Martinez, Escondido, Upland, Suisun, Colfax, Turlock, Linden, Vacaville, Sacramento, Clarksburg, Sonora, San Miguel, Fontana, Pomona, Stockton, and Auburn.
5. Hot regions—more than 4000 degree-days of heat. Madera, Fresno, Delano, Visalia, Bakersfield, Chico, Red Bluff, Redding, Ojai, Oakdale, Brentwood, Antioch, Woodland, and Reedley.

The best table wines are produced in the cool and moderately cool regions, the best natural sweet wines in the warm regions, and the best dessert wines and the commercial table and raisin grapes come from the moderately hot and hot regions.

The five climatic grape-growing regions in California are shown in Figure 5-1.

The heat summation method described above is the most widely used technique for forecasting the maturation date for various varieties. In Michigan, a modified heat-unit method has been successful for predicting the harvest date of the Concord grape (Brink, 1974). Heat units were counted from April 1 at a base temperature of 50°F (10°C). Optimum maturity was reached 85 days after the date when 1000 units were accumulated for the season. Maturation was largely determined by temperatures during the first 100 days of the season.

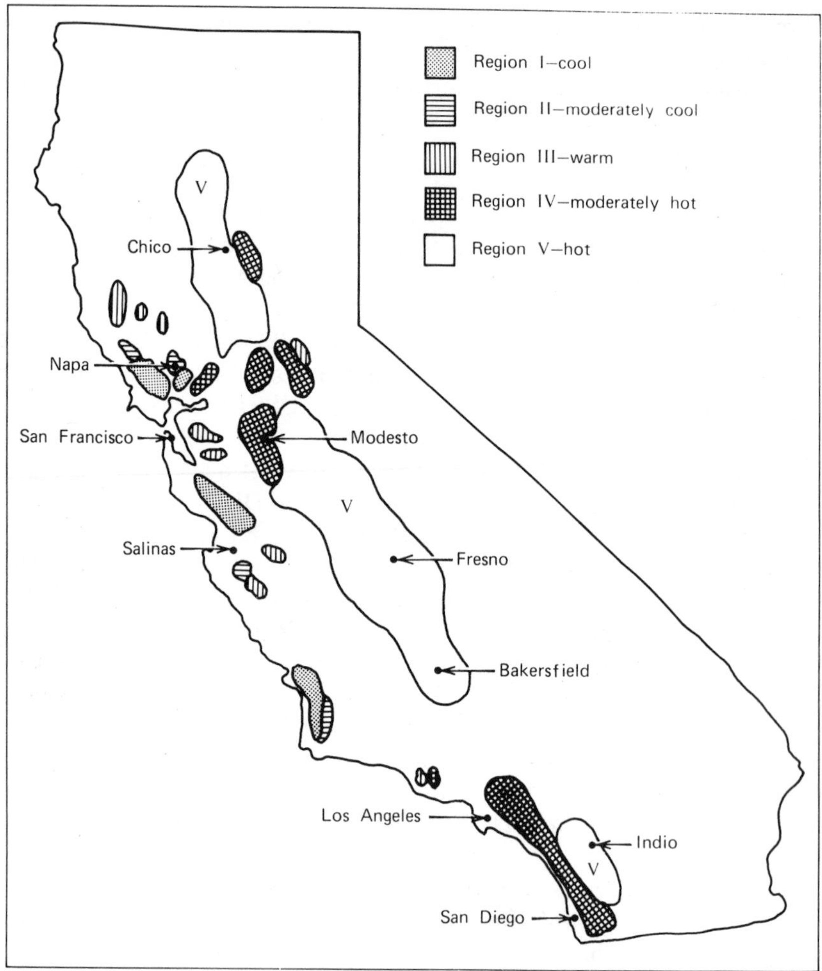

Fig. 5-1. The climatic regions for grape growing in California based on heat summation above 50°F (10°C).

SOILS FOR GRAPES

Grapes grow fairly well in many different types of soil. They are commercially grown throughout the world in almost every type of soil from gravelly sand to clay loam, shallow to very deep soil, and in soils ranging from high to low in fertility. It is preferable, however, to avoid very heavy clays, very shallow soil, poorly drained soil, and soil with relatively high concentrations of alkali salts, boron, and other toxic materials.

The deeper and more fertile soils usually produce the heaviest crops and are therefore preferred for raisins, standard wine grapes, and table grapes such as Tokay and Thompson Seedless. Certain varieties, such as Malaga and Emperor, attain higher quality on soils of limited depth.

AIDS FOR VINEYARD SITE SELECTION

In selecting a site for grape growing one should consult the local County Farm Advisor and successful growers in the area. Information on climatic conditions, temperature, frost, wind, topography, rainfall, soil depth and fertility, availability of water, labor sources, and distance to market is valuable. It is safest to plant grapes in a new area on a trial basis only, and also to plant a variety that has already been grown successfully in an area. Although site selection refers mainly to choosing new vineyard locations, site selection and rejection is a continuous problem for the life of the vineyard. A vineyard site should be constantly on trial (Shaulis and Dethier, 1970).

TECHNIQUES USED TO MODIFY THE WEATHER

Frost protection during the spring months after shoot growth begins is desirable in many areas where air temperatures in grape vineyards drop down to 31°F (−0.5°C).

Frost Protection with Overhead Sprinklers

Mechanism. Water releases heat as it freezes. When it is sprinkled on vines, heat is produced in the brief time that droplets are freezing on the green shoot and leaf surfaces. The release of heat maintains the temperature of vine parts covered by the ice-water mixture to near 32°F (0°C) even when the air temperature and plant parts not in reach of sprinklers is as low as 15°F (9°C). To prevent freezing damage the vines must be kept wet constantly. Sprinkler installations provide sufficient water application rates to maintain plants at a minimum of 31°F (0.5°C), a safe temperature for any developmental stage of the vine.

The ice film itself provides almost no protection because it is a good heat conductor. This explains why heat produced at the surface is readily conducted to the plant.

Precipitation rates required for protection under frost conditions have been fairly well determined (Marsh and Meyer, 1973). Water must be sprinkled on all parts of the vineyard throughout the frost period. However, rates vary with the variety, stage of cluster growth at time of frost, wind conditions and relative humidity during the frost period, and the type of frost.

The precipitation rate should be kept at the minimum required to provide frost protection to avoid waterlogging of the soil. Sprinklers must be operated continuously during frost periods to maintain a water-ice interface. A precipitation rate of 0.11 to 0.12 in. (2.8–3 mm) per hour or more is usually sufficient for over-vine sprinklers. The amounts of water required in gallons per minute per acre (gpm/acre) for various precipitation rates are presented in Table 5-1 (Meyer and Marsh, 1972).

When to Begin Sprinkling. A drop in temperature due to evaporative cooling usually occurs when the sprinklers are first turned on. When a frost is expected, the usual practice is to start the sprinklers when the temperature drops to 34°F (1°C), thus providing a margin of safety (Marsh and Meyer, 1973). Since grapevines at the bud break stage of development are more sensitive to frost when wet than when dry, a starting temperature for sprinklers lower than 34°F (1°C) should not be risked.

When very dry atmospheric conditions prevail (low dewpoint temperature), it may be necessary to start the sprinklers at a higher temperature than 34°F (1°C) to increase the relative humidity before

Table 5-1 Water Requirements for Frost Protection*

Precipitation Rate		Amount of Water	(liters pm
(inches per hour)	(mm per hr)	(gpm per acre)	per hectare)
0.06	1.52	27	252.4
0.08	2.03	36	336.6
0.09	2.29	40	377.0
0.10	2.54	45	420.7
0.12	3.05	54	504.9
0.14	3.56	63	579.7
0.16	4.06	72	673.2
0.18	4.57	81	757.3

* After Meyer and Marsh, 1970.

Table 5-2 Air Temperature to Turn on Sprinklers to Maintain 31°F (−0.6°C)*

Dewpoint Temperature	Air Temperature at Which Sprinklers Should be Started When Barometric Pressure is 30 Inches (76.2 cm) (0–500 ft (0 to 152.4 m) elevations)
15°–16°F (−9.4 to −8.9°C)	39°F (3.9°C)
17°–18°F (−8.3 to −7.8°C)	38°F (3.3°C)
19°–21°F (−7.2 to −6.1°C)	37°F (2.8°C)
22°–23°F (−5.6 to −5.0°C)	36°F (2.2°C)
24°–25°F (−4.4 to −3.9°C)	35°F (1.7°C)
26°–27°F (−3.3 to −2.8°C)	34°F (1.1°C)
28°F (−2.2°C)	33°F (0.6°C)

* After Meyer and Marsh, 1973.

the frost occurs. The dewpoint is the temperature at which the relative humidity reaches 100 percent. Table 5-2 shows the proper temperatures to begin sprinkling at various predicted dewpoints to prevent evaporative cooling below 31°F (−0.5°C).

A larger amount of water must be immediately available for frost protection as compared to that required for irrigation, since during a freeze all parts of the vineyard must be sprinkled at the same time. For example, a precipitation rate of 0.12 in. (3.0 mm) per hour to protect 40 acres (16.2 hectares) of vines requires a sprinkler system with a total capacity of 2160 gpm (8176 liters).

An over-the-vine continuous sprinkling at a rate of 5 gpm (18.9 liters pm) provides protection of 6°–8°F (3°–4°C).

Sprinklers can be turned off when the air temperature outside the treated area has risen to 32°F (0°C), if there is no wind. If it is windy, it is best to wait until the air temperature has risen to 34°F (1°C), although it is not necessary to wait until all the ice has melted.

Cultural Practices to Prevent Frost Damage

Cultivation and Training. Table 5-3 shows data collected during radiant frosts in the central San Joaquin Valley concerning expected protection for vines under different soil surface conditions (Swanson et al., 1974).

For warmest temperature the soil must be bare. Trash, weeds, and

Table 5-3 Effect of Different Soil Surfaces on Protection from Frost—Temperature Measured at 4 Feet (1.2 m) above Soil Surface*

Soil surface	Amount protection
1. Bare, firm, moist soil	warmest
2. Moist soil, shredded covercrop	½°F (0.6°C) colder than 1
3. Moist soil, low covercrop	1°–3°F (0.6°–1.7°C) colder than 1
4. Dry, firm soil	2°F (1.1°C) colder than 1
5. Freshly disced ground	2°F (1.1°C) colder than 1
6. High covercrop	2–4°F (1.1°–2.2°C) colder than 1 (occasionally 6°–8°F (3.3°–4.4°C) colder than 1)

* After Swanson et al., 1974.

cover crops are detrimental because they insulate the soil surface so it cannot absorb or release heat readily. Any ground cover that is kept should be mowed as low as possible.

Proper training may provide some protection against frost. The air is cooled at the ground surface and gradually builds up a cold layer; thus the lower shoots are usually frozen first. This advantage is utilized in the Orange River Valley of South Africa, where Thompson Seedless vines are trained to a high trellis with a horizontal cross arm at the top.

A compact, bare soil can be made by turning under cover crops or heavy weed growth. However, a loosened soil is as hazardous in frosts as soil with vegetation.

Late or Double Pruning. Late pruning after the buds on the apical parts of the canes have started to grow will delay the leafing out of the buds on the retained spurs. This delay may vary from 7 to 10 days depending on the temperature (Schultz, 1962): when it is very warm the delay will be short, in cold weather the delay will be longer. Shoots on the higher parts of the vines should not be permitted to grow more than 3 or 4 in. (7.6 or 10.1 cm) before pruning, lest the vines be weakened (Fig. 5-2).

To delay pruning in large vineyards until the apical buds have grown 3 or 4 in. (7.6 or 10.1 cm), however, may result in difficulties when labor is scarce, especially in seasons when the shoots grow rapidly. Thus if one desires to delay the growth of the buds on the spurs, double pruning may be used. During the winter, all canes except those to be used for spurs should be removed and the retained canes should be pruned to rods or half-long canes 15 to 20 in. (38.1–50.8 cm) long. In spring, after the apical buds on these half-long canes have produced shoots 3 to 4 in. long, the canes should be cut back to spurs one to four buds long depending on their diameter. Double pruning will delay leafing out of the lower buds just as effectively as late pruning and it facilitates work in the dormant season, such as brush disposal and cultivation. This method also avoids the possibility of a labor crisis or of working under adverse weather conditions at the critical stage in growth, since cutting back of half-long canes can be done very rapidly. Late pruning or double pruning is only practical for small acreages.

Enhancement of Air Flow

Elimination of cold air pockets is another technique to prevent local spring frosts (Schultz et al., 1962). On clear and calm nights, the coldest air settles near the ground. The influence of the terrain often brings this cold air into motion down the slopes and out of the valleys. Cold air drainage flow is very beneficial because it always draws warm air down

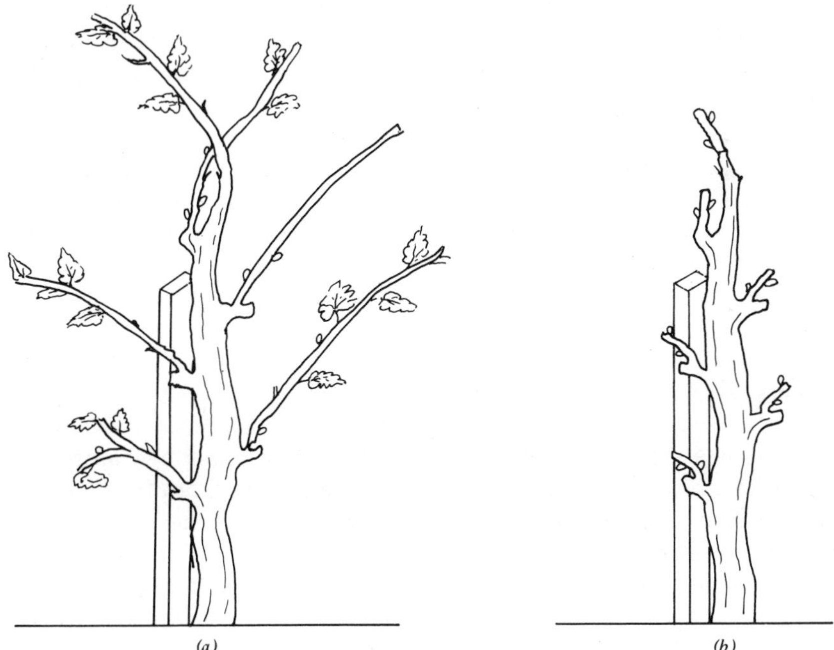

Fig. 5-2. Double pruning of wine grapes near Ukiah, California. (*A*) Vine pruned to rods. (*B*) Vine after being pruned to normal spurs.

from overhead. Wherever the cold air flow is hindered by embankments, a row of brush, low-branched trees, or barns, frost danger areas will form. Vegetation and buildings that stop the air flow and dam up the cold air should be removed.

Wind Machines

During the day the sun warms the earth's surface—soil and vines—which in turn warms the air in contact with it. A layer of warm air is formed over the vineyard as the air at ground level warms and rises.

At night the process is reversed and the soil surface loses heat through outward radiation to the sky. The air in contact with the earth's surface is cooled and, since cold air is heavier than warm air, it remains at ground level or flows into low areas. This nighttime condition is called *inversion.*

Frosts that develop under these conditions are called radiation frosts, and are characterized by colder air below and a warm air layer ranging from 20 to 100 and more feet (6.1–30.5 m) above the ground. Both wind machines and heaters work better when the warm air or ceiling is near the ground, a condition called strong inversion by meteorologists.

The main purpose of wind machines is to mix the upper warm air layer with the cold air layer near the vines. The difference between the air temperature at 5 ft and 40 or 50 ft (12.2 or 15.2 m) above the ground is a common method used to measure the strength of an inversion. Well-designed wind machines require about 8 to 10 hp per acre. This is almost equivalent to two tower-mounted dual machines or four movable or ground-level machines for 30 to 40 acres (12.1–16.2 hectares). The two kinds of wind machines on the market are the permanently installed tower-mounted type and the movable type mounted on a short tower.

Heaters

Orchard heaters provide heat by direct radiation and by convection. Hot-stack heaters, like the return-stack heater, give out 25 to 30 percent radiant heat (Fig. 5-3), which travels directly from the heater to the vine or any other object in the vineyard.

The hot air and gases created by the heater tend to rise out of the vineyard. If there is a strong inversion creating a low ceiling of warm air above the vineyard, the heated air rises less. In this case, all types of heaters burning at equivalent rates work equally well. When there is little effective ceiling, the hot-stack heaters provide greater protection because their radiant heat output is not affected by the small inversion.

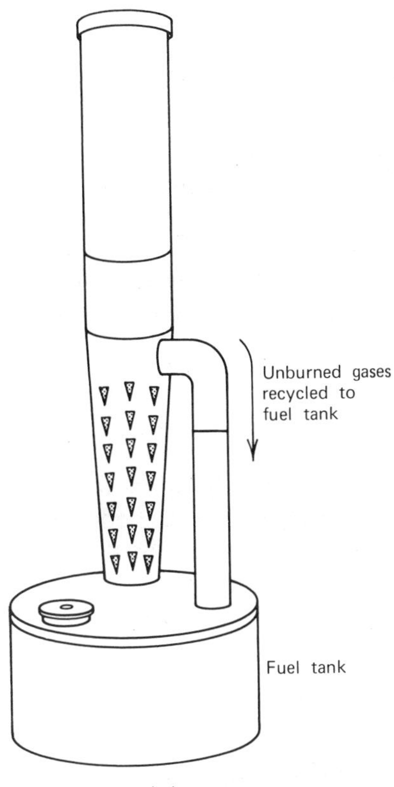

Fig. 5-3. Diagram of a return stack heater, often used to prevent frost damage in vineyards.

An installation of 40 to 50 heaters per acre (99–124 per hectare) is recommended for coastal grape vine districts, although they would seldom have to burn all at the same time. This provides 5°F (3°C) protection or more depending on the burning rate and type of inversion. Twenty-five heaters would give 3°F (2°C) protection or more, depending on the same variables. At the starting time which for economic reasons might not be above 31°F (minus 1°C) air temperature, it is necessary to light only one-third of the total number of heaters although additional heaters should be lit if the temperature drops. Distribution of the burning heaters should be even except along the edges, where a greater concentration is needed for border heating. This is especially important at borders where the nightly drainage flow of cold air occurs or at those exposed to adjacent grassland or alfalfa fields (Schultz et al., 1962).

Growers who use heaters should carefully abide by the regulations of the Air Pollution Control District and do everything possible to hold smoke nuisance to an absolute minimum.

Wind machines and heaters working together provide more frost protection than either used alone. This increased efficiency results from mixing the hot air and gases generated by the heaters directly with the air in the vineyard. The most beneficial effect from the heater and wind machine combination occurs on nights when there is only a small inversion (little warm air above the vineyard for the wind machine to pull down, and no ceiling to stop the upward progress of the warm air generated by the heaters). For example, in the North Coastal region of California 20 to 25 heaters per acre (49–62 per hectare) with a wind machine provide up to 4°–5°F (2°–3°C) protection (Burlingame et al., 1971).

Helicopters

An inversion layer of upper warm air is needed by helicopters for beneficial effects similar to that for wind machines; with helicopters, however, the operator has the advantage of selecting the flight pattern that will give maximum heat at the vine level. Depending on the inversion, one helicopter can protect from 40 to 100 acres (16–40 hectares) (Swanson et al., 1974). The helicopter should carry a rapid response thermometer to determine temperature in the upper warm layers. It is also a good idea to monitor the helicopter from the ground so the pilot can be guided to colder spots in the vineyard.

Crop duster planes can also be useful for mixing upper warm air.

Fog Machines

Fog machines can produce a blanket of fog over the vines that reduces heat loss by radiation. Fog can keep air temperatures 1–2°F (0.6 to 1.1°C) higher and leaf temperatures 2–3°F (1.1 to 1.7°C) higher (Swanson et al., 1974). The main problem with using fog is to keep it over the vineyard and not let it drift away.

Heat Blocks

Blocks such as Tree Heat can be used; at least 100 or 200 blocks per acre (247 or 494 per hectare) should be used. They can also be used to supplement other heating devices in more severe frosts.

Heat blocks are also used in New York State. These petroleum blocks will burn for approximately 5 hours, and 100 blocks per acre (247 per hectare) will raise the temperature about 2°F (1°C).

Covers

In Japan, Holland, and other northern European countries where vines cannot be grown out-of-doors, high quality table grapes are grown in greenhouses. Both glass and plastic covers are used in these greenhouses, which are usually steam heated in winter.

Covers of polyethylene strips or other material over the vines can give 3°F (2°C) protection against frost (Schultz, 1961), and this amount can be increased by extending the plastic to the ground at night.

In some countries such as Israel, Italy, and South Africa, there has been experimental and small-scale use of plastic to hasten ripening of grapes. When plastic is placed over the vines before bud break, shoot growth is much more rapid and fruit ripens a couple of weeks earlier than on the control. However, the plastic must be removed before the weather gets too hot, usually soon after flowering, to prevent burning of the shoots.

LABRUSCA VINES IN THE EASTERN UNITED STATES

In New York the winter minimum temperature often falls low enough (−12°F to −18°F) (−24 to −28°C) to kill grape tissues or vines. High position of canes can decrease injury from spring freeze, and there is usually less injury to shoots above 4 ft (1.2 m) from the soil than to those at heights of less than 3 ft. (0.91 m). In May 1963 the shoot kill was about 30 percent and 60 percent, respectively (Shaulis et al., 1964).

A major portion of injury from winter freeze in New York vineyards is to the trunk at heights of 1 to 3 ft (0.30–0.91 m) above the soil. Use of double trunks from the soil level can reduce loss of vines from freezing injury in winter. Loss may be reduced by 80 percent (Shaulis et al., 1964).

A vigorous vine with good foliage cover until the crop is harvested is essential to reduce damage from a preharvest freeze. Exterior foliage protects the partially exposed foliage beneath. Foliage with potassium deficiency symptoms is more susceptible to freezes than is healthy foliage.

TREATMENT OF FROST DAMAGED VINES

After a freeze, damage to the grapevines becomes apparent after a few hours although the full extent of cluster damage is often not clear until after fruit-set (Kasimatis and Kissler, 1974; Swanson et al., 1974). In most vineyards the best treatment may be to do nothing, but some recovery may be obtained by shoot removal treatments.

After vines have been frozen some recovery of crop might be obtained from secondary or tertiary buds or from dormant and latent buds. If the shoots produced from the primary bud freeze, partial recovery can sometimes be obtained by breaking out shoots at the base, if the secondary and/or tertiary buds are fruitful. Usually however, shoot breakout treatments fail to enhance growth from secondary growing points (Kasimatis, 1974). Secondary and tertiary buds usually develop in about two weeks in cases where growth does occur. Shoot removal should be done immediately after a freeze, and best results occur when shoots are less than 6 in. (15.2 cm) long when frozen. When longer shoots are broken out the secondary and tertiary growing points are also removed. After a frost one should break off only those shoots on which clusters have frozen, omitting those shoots that have been killed back to the base. No benefit can be obtained by breaking off completely frozen shoots.

Since Thompson Seedless has no fruitful secondary or tertiary buds, no treatment is recommended. If the crop is lost by freezing, canes may be cut back to spurs to promote shoot growth and better canes for the following year. No treatment is beneficial on spur-pruned vines with nonfruitful secondary or tertiary growing points such as in the Emperor variety. However, in spur-pruned varieties with fruitful secondary and tertiary buds, such as Cardinal, Ribier, and most wine varieties, shoot removal may sometimes result in greater yields.

The growth of dormant and latent buds present at the bases of spurs and canes is of little benefit in cane-pruned Thompson Seedless but can sometimes increase yields in spur-pruned vineyards. Suckers or water sprouts from other parts of the vine can produce some fruit especially in wine varieties. If shoots are pruned off, only lateral buds develop near the base of the shoots that are not fruitful.

HEAT SUPPRESSION WITH OVERHEAD SPRINKLERS

From June to September the San Joaquin Valley of California has high daytime temperatures. Maximum daily air temperatures often consistently exceed 90°F (32°C) to 95°F (35°C) for extended periods, and the relative humidity drops below 25 percent during the day. Temperatures in excess of 112°F (44°C) have frequently been recorded in the San Joaquin Valley.

Air temperatures ranging from 77°F (25°C) to 86°F (30°C) are considered optimum for growth of most temperate-zone crops. Leaf temperatures higher than 86°F (30°C) can cause internal water deficiencies, sunburn damage, dehydration of fruit, and reduced growth rates. High temperatures can also decrease the acid content and increase the pH levels of wine grapes, inhibit color formation and increase raisining (sunburning) of table grapes. Low atmospheric relative humidity places additional water stress on the vines (Aljbury et al., 1973).

Preventing injury by evaporative cooling can be obtained by intermittent operation of sprinklers (Fig. 5-4), which can be turned on for 3 minutes, followed by an off period of 15 minutes or more. Decreases in leaf temperature can be as much as 25°F (14°C) during sprinkling at 15–20 percent relative humidity. Little rise in air temperature occurs between water applications.

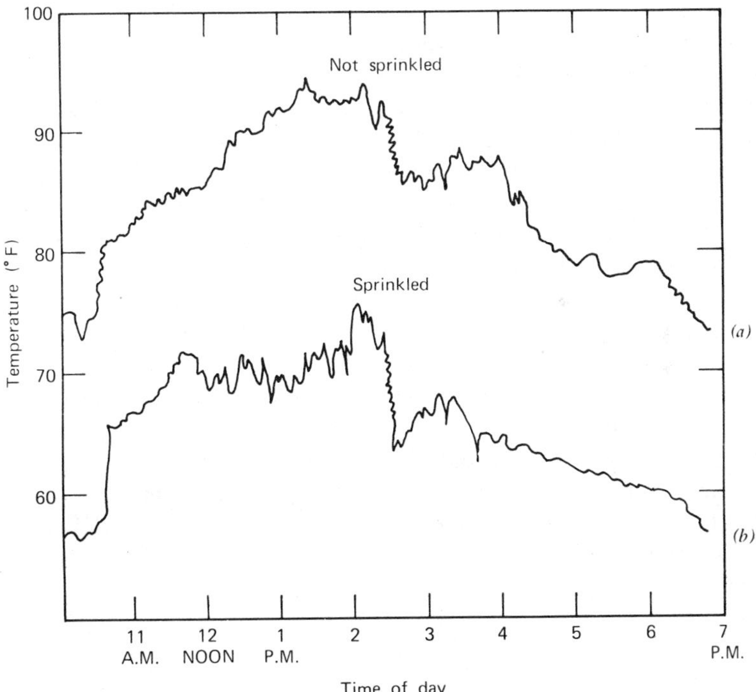

Fig. 5-4. Effect on berry temperature after one day of sequential sprinkling of berries in the sun at Lodi, California. (*a*) Not sprinkled; (*b*) sprinkled. (After Aljibury et al., 1973.)

Increases in relative humidity accompany cooling. Under conditions of 15–20 percent relative humidity, the humidity can be increased by approximately 10 percent. Precipitation rates during sprinkler operation need be only 0.08–0.10 in. (2.0–2.5 mm) per hour or less with sprinkler spacings similar to those used in irrigation.

Simple on-off switch panels for pumps with manual lateral line valves are commonly used.

Water Quality

Water quality may limit the usefulness of this technique for cooling vines. "Burning" of foliage may occur with waters that contain in excess of 7.5 meq/liter of total dissolved salts or 3 meq/liter of sodium. Waters containing 1.5 mg/liter of bicarbonates can cause deposits to form on the leaves and fruit (Aljibury et al., 1973).

Time of Day

Air and plant temperatures were reduced no matter what time of day sprinkling was done. Air temperatures were reduced 7–10°F (4–8°C) during the peak heat period of the day, and humidity was increased 10 to 20 percent. Temperatures of leaves and exposed berries were usually decreased about 7–20°F (4–11°C), and that of shaded berries were decreased about 7–12°F (4–6.7°C) (Aljibury, 1973). Fresh weight of berries of cooled fruit is increased over that of uncooled fruit (Kliewer and Schultz, 1973; Aljibury et al., 1975).

Bunch rot was increased by sprinkler cooling but avoidance of nighttime spraying and fungicide treatments might lessen these effects. Vine growth was stimulated by cooling, and vines that are cooled are greener, more lush, and more vigorous. This result may be partially due to an increase in soil moisture.

In a vineyard near Reedley, California, cooling of Chardonnay, Semillon, and Chenin blanc fruit by sprinkling sequentially from véraison to harvest reduced the levels of soluble solids and delayed fruit maturation by two or three weeks (Aljibury et al., 1975). However, sprinkling from bloom to shortly after véraison had little effect on soluble solids of Cardinal, Carignane, and White Riesling (Kliewer and Schultz, 1973).

Sprinkling increased coloration of Cardinal fruit (Kliewer and Schultz, 1973). The value of overhead sprinklers to cool commercial vineyards for improvement of quality is not yet known.

Wine Quality

Wines produced from sprinkler grapes are rather unsatisfactory and no better than wine made from the same varieties unsprinkled. Whether heat suppression can result in improved wine quality is still unknown (Aljibury et al., 1973).

Chapter 6 VARIETIES

About 25 million acres (10,121,000 hectares) of grapes are grown worldwide, and over the past several years plantings have increased sharply in some locations. The acreage of grapes grown and gallons of wine produced in various countries of the world in 1973 is shown in Table 6-1, and the total tonnage of grapes produced in California and several other states in 1972 is shown in Table 6-2.

In California the largest increase in grape plantings from 1969 to 1973 was in wine grapes, which increased from 7,064 acres (2,860 hectares) in 1969 to 57,385 acres (23,233 hectares) in 1973 (Table 6-3). In 1974 plantings of wine grapes decreased by over 50 percent. The bearing acreage of 645,981 (261,531 hectares) is made up of table (11%), wine (50%), and raisin (39%) grapes.

There are around 8000 varieties of grapes in the world that have been named and described, but only about 2000 of these are grown in California. Perhaps 60 of the total number are important varieties.

TABLE GRAPES IN CALIFORNIA

Table grapes are of far less importance than wine or raisin grapes on a worldwide basis. In California, however, table grapes constitute an important industry. The Republic of South Africa also has a considerable table grape industry. The following pages give descriptions of important kinds of table grapes grown in California (after Jacob, 1950). The number of acres at the end of each description refers to acres of the grapes grown in California, both bearing and nonbearing.

Almeria (Ohanez). A late, white, seeded variety with excellent shipping quality. Stems are tough and berries firmly attached. Clusters are of medium to medium-large size, straggly to compact. Berries are medium-large, cylindric, greenish-white, with neutral flavor and tough,

Table 6-1 Grape Acreages and Wine Production of Various Countries of the World in 1973*

	Area (acres)	Area (hectares)	Wine in Thousands of Gallons	Wine in Thousands of Hectoliters	Rank in Wine Production
Spain	4,124,900	1,670,000	1,301,549	49,269	3
Italy	3,381,430	1,369,000	2,026,622	76,716	2
France	3,255,460	1,318,000	2,204,145	83,436	1
USSR	2,677,480	1,084,000	596,844	22,593	4
Turkey	2,025,400	820,000	13,737	520†	23
Portugal	879,320	356,000	299,096	11,322	7
Romania	827,450	335,000	234,188	8,865	9
Argentina	770,640	312,000 (1972 data)	594,836	22,512	5
Algeria	615,030	249,000	166,957	6,320	11
Yugoslavia	615,030	249,000	203,439	7,701	10
California (U.S.A.)	607,620	246,000	338,140	12,800	6
Hungary	526,110	213,000	166,059	6,286	12
Greece	506,350	205,000	134,173	5,079	14
Bulgaria	494,000	200,000†	79,252	3,000†	15
South Africa	350,740	142,000 (1972 data)	142,309	5,387	13
Afghanistan	269,230	109,000	—	—	—
Iran	247,000	100,000†	106	4†	—
Germany (West)	237,120	96,000	282,585	10,697	8
Brazil	180,310	73,000†	60,760	2,300	18
Morocco	172,900	70,000†	24,357	922	22
Syria	167,960	68,000 (1972 data)	—	—	—
Australia	167,960	68,000 (1972 data)	72,145	2,731	16
Austria	116,090	47,000	63,507	2,404	17
Cyprus	121,030	49,000 (1972 data)	1,040 (1972 data)	—	—
Czechoslovakia	93,860	38,000	38,463	1,456	19
Tunisia	91,390	37,000	28,953	1,096	21
Mexico	81,510	33,000	3,963	150†	28
Japan	54,340	22,000†	4,095	155†	27
Lebanon	41,990	17,000†	1,057	40†	31
Egypt (U.A.R.)	37,050	15,000	1,585	60†	30
Albania	32,110	13,000†	1,849	70†	29
Switzerland	32,110	13,000	34,316	1,299	20
Canada	22,230	9,000†	13,473	510†	24
Bolivia	14,820	6,000 (1971 data)	159	6†	34

Table 6-1 (Continued)

	Area (acres)	Area (hectares)	Wine in Thousands of Gallons	Wine in Thousands of Hectoliters	Rank in Wine Production
Jordan	14,820	6,000†	396	15†	33
India	93,860	38,000	—	—	—
Libya	7,410	3,000†	528	20†	32
Pakistan	4,940	2,000 (1971 data)	—	—	—
Paraguay	4,940	2,000 (1971 data)	—	—	—
Luxembourg	2,470	1,000	4,914	186	26
Thailand	2,964	1,200 (1971 data)	—	—	—
New Zealand	2,470	1,000	6,604	250	25
Malta	2,470	1,000 (1971 data)	528	20	32
Taiwan	2,470	1,000 (1971 data)	—	—	—
Belgium	864.5	350 (1971 data)	132	5	35
Colombia	741	300 (1971 data)	—	—	—
World total (approximate)	24,611,080	9,964,000			

* Bulletin de International de la Vigne et du Vin, **47** (525): 916–950 (1974).
† Estimate.

thick skins. Susceptible to Ohanez spot, probably due to heat injury. Located mainly in Tulare county and mostly grown on arbors. The variety is self-sterile and requires cross pollination. Spain exports large quantities of Almeria packed in granulated cork. (1007 acres, 408 hectares.)

Calmeria. An open-pollinated seedling selection of Almeria with excellent storage and shipping quality. The vine resembles Almeria and has large, long conical, well-filled clusters. It is self-fertile. Berries are white, large, cylindrical, with tough skins and firm pulp. Introduced in 1950 at Fresno, California, by E. Snyder and F. Harmon (U.S. Department of Agriculture.) (4,053 acres; 1,641 hectares.)

Cardinal. A very early red, seeded, table grape which is a cross of Tokay and Ribier. Clusters are medium to large, conical, and loose to

Table 6-2 Tonnages of Grapes Produced in Several States in 1972*

State	Tons
California	2,266,000
New York	103,000
Washington	62,100
Michigan	53,000
Pennsylvania	37,600
Arizona	13,600
Ohio	12,000
Arkansas	9,500
South Carolina	6,400
Missouri	3,600
North Carolina	2,200

* Source: Fruit Situation TFS 190, February 1974, Economic Research Service, U.S. Department of Agriculture, Washington, D.C.

Table 6-3 Acreage of Various Types of Grapes Standing by Year Planted*

Type	1969	1971	1972	1973	1974†	Bearing	Non bearing‡	Total
Raisin	1,439	2,778	2,817	5,735	1,720	240,828	10,272	251,100
	(582)	(1,124)	(1,140)	(2,321)	(696)	(97,459)	(4,157)	(101,616)
Table	551	1,376	1,894	3,180	865	66,898	939	72,837
	(223)	(557)	(766)	(1,287)	(350)	(27,073)	(380)	(29,476)
Wine	7,064	34,616	56,829	57,385	25,990	181,840	140,204	322,044
	(2,859)	(14,009)	(22,998)	(23,223)	(10,518)	(73,588)	(56,734)	(130,326)
Grapes	9,054	38,966	61,540	66,330	28,575	489,556	156,415	645,981
(except rootstocks)	(3,664)	(15,769)	(24,904)	(26,843)	(11,564)	(198,116)	(63,299)	(261,419)
Rootstock	48	152	421	759	1,360	—		2,486
	(19)	(62)	(170)	(307)	(505)	—		(1,006)
All grapes	9,102	38,892	61,961	67,059	29,935	—		648,827
	(3,683)	(15,739)	(25,075)	(27,138)	(12,114)			(262,571)

* After California Grape Acreage 1973, California Crop and Livestock Reporting Service. Areas in hectares are shown in parentheses.

† Includes 337 acres (136 hectares) not yet planted when surveyed, but expected to be planted before January 1, 1975, and is comprised of the following types: raisin, 13 acres (5.3 hectares) wine, 324 acres (131 hectares).

‡ Nonbearing, includes plantings in 1971, 1972, 1973.

compact. Berries are very large, round to short oval, depressed at apex, and may have one or more shallow sutures. Berries are cherry-red, changing to reddish-black as maturation proceeds. Vines are vigorous and grow well with cordon pruning, and require flower-cluster or cluster thinning. The vine thrives is hot areas, and was introduced in 1950 at Fresno by E. Snyder and F. Harmon. (2,779 acres; 1,125 hectares.)

Emperor. A very popular late-ripening, seeded variety. The clusters are large, long conical, and loose to well filled. Berries are uniform, large, elongated obovoid or ellipsoidal, light red to reddish-purple. They are seeded, moderately firm, neutral in flavor, and have thick, tough skins. Stems are tough, and berries adhere very firmly. Vines are moderately vigorous, of medium productivity, and are cordon-pruned.

The Emperor attains a red color and a large berry near the foothills along the eastern side of the San Joaquin Valley in Tulare, Fresno, and Kern counties where about 90 percent of the Emperor is produced. It is an excellent storage and shipping grape and much of the crop is held in cold storage. Its origin is unknown. (23,974 acres; 9,706 hectares.)

Tokay (Flame Tokay). This midseason variety once ranked first in popularity, but is now surpassed by Thompson Seedless and Emperor. Clusters are large, shouldered, short conical, and compact. Berries are large to very large, avoid truncate, pink to red, seeded, very firm, neutral in flavor, and have thick, tough skins. Stems are large and tough, and the berries adhere firmly. The fruit sunburns easily. The vines are usually head-pruned, but also grow well when cordon-pruned.

The principal producing area in around Lodi in Region IV. In the hotter regions the variety does not color well and sunburns badly, whereas in the cooler coastal sections it does not ripen properly. Tokay has good storage and shipping qualities. It originated in Algeria. (19,895 acres, 8,055 hectares.)

Italia. Clusters are large to medium, conical, and well filled. Italia has very large, long oval berries, a heavy orange-white bloom and a mild muscat flavor. The vigorous vines respond well to cordon pruning. The shipping quality is fair. The source of Italia is Italy. (1,149 acres; 465 hectares.)

Malaga. Malaga is a white, seeded grape that ripens in midseason and has good storage and shipping qualities. It was once the leading table grape variety in California but has been largely replaced in the market by the Thompson Seedless from girdled vines. Most of the grapes are used for distilling material or low-grade wines.

Clusters are large to very large, conical, and well filled. Berries are uniform, large, ellipsoidal, whitish-green to whitish-yellow, normally seeded, firm, neutral in flavor, and have thick, tough skins. Stems are tough and the berries adhere firmly. Vines are vigorous and very productive. Cordon pruning is most suitable, but head pruning is satisfactory. It is grown mainly in the warm regions of the San Joaquin Valley. (4,213 acres; 1,706 hectares.)

Perlette. Perlette is a very early ripening, seedless white grape. Clusters are large to medium in size, conical, shouldered, and very compact. Berries are round, of medium size, and have a white, waxy color. This vigorous variety responds well to cordon pruning, but requires heavy berry thinning. Introduced in 1946 by H. P. Olmo, University of California at Davis. (3,790 acres; 1,534 hectares.)

Red Malaga (Molinera). This red, early midseason, seeded grape with excellent shipping and storage qualities is harvested before the Tokay. Clusters are very large, irregular in shape, and loose to well filled. Berries are large, spherical to short ellipsoidal, pink to reddish-purple, often faintly striated, very crisp and hard, neutral in flavor, low in acidity, and have tender skins. Berries are firmly attached and stems are tough. The vines are very productive when cordon-pruned. Red Malaga flourishes in most of the San Joaquin Valley. (1,216 acres; 492 hectares.)

Ribier (Alphonse Lavallée). This black, seeded grape with good keeping and shipping qualities ripens in early midseason. The grape was misnamed Ribier in California. It is not the Gros Ribier grown in Europe. It is one of the best greenhouse varieties in Europe. Clusters are of medium size, short conical, often heavily shouldered, and vary from loose to compact. Berries are very large, oblate to ellipsoidal, jet black, normally seeded, firm, neutral in flavor but mildly astringent, low in acid, and moderately tough-skinned. Stems are tough, and the berries firmly attached. The vines are moderately vigorous, very productive, and are cordon-pruned.

Ribier is best suited to the warm middle and upper San Joaquin Valley. (7,496 acres; 3,035 hectares.)

Thompson Seedless. See raisin grapes (p. 87).

Table Grapes of Minor Importance in California

Barlinka. A black, late maturing grape with large, short oval berries. It is the leading table grape in the Republic of South Africa.

Beauty Seedless. A very early black seedless variety with small to medium oval berries with a neutral taste. Introduced in 1954 by H. P. Olmo, Davis, California. (193 acres; 78 hectares.)

Black Hamburg. A black seeded variety with large spherical berries.

Black Monukka. A black, seedless, midseason grape with large berries. (348 acres; 141 hectares.)

Black Prince. A black, seeded variety with large, spherical berries and a crisp texture. (147 acres; 60 hectares.)

Blackrose. A jet black grape that ripens in early midseason. It has a very attractive appearance. Introduced in Fresno, California by E. Snyder and F. Harmon. (68 acres; 28 hectares.)

Concord. A black, seeded American variety (118 acres; 48 hectares.)

Cornichon. Berries are purplish black and short oval. Vines rather tolerant of powdery mildew. (88 acres; 36 hectares.)

Delight. An early white seedless grape with medium-size ellipsoidal berries having a slight muscat flavor. Introduced by H. P. Olmo in 1947 at Davis, California.

Early Muscat. A very early yellow grape with round medium-size berries and a mild muscat flavor. Introduced in California by H. P. Olmo in 1958.

Dattier. A white seeded grape with large ellipsoidal berries. Fruit has pleasing flavor and ripens in midseason.

Early Niabel. A black, seeded, tetraploid, midseason grape, suitable for sweet juice and semisweet wines. Colors two weeks earlier than Niabell. Introduced by H. P. Olmo in 1958 at Davis, California.

Exotic. A black, seeded, early maturing grape with large oval berries. It has large clusters and berries crack easily. Introduced by E. Snyder and F. Harmon in 1958 at Fresno, California. (740 acres; 300 hectares.)

Gold. An early midseason grape with large oval berries. The gold-colored berries have a mild muscat flavor. Introduced by H. P. Olmo in 1958 at Davis, California.

Golden Muscat. An American variety with a muscat flavor that grows well in California. (101 acres; 41 hectares.)

Kandahar. A white, seeded grape that ripens in midseason. It has very large cylindrical berries and brittle stems (62 acres; 25 hectares.)

Muscat Hamburg. A black midseason variety with berries of medium size and ellipsoidal shape that have a muscat flavor.

Niabell. A black, seeded, tetraploid, midseason grape with very large, spherical berries, possessing a distinctive *labrusca*-like flavor. Vines are rather tolerant to powdery mildew. Introduced by H. P. Olmo in 1958 at Davis, California. (111 acres; 45 hectares.)

Olivette blanche. A late season, seeded white grape with very large clusters and berries. Vines are vigorous and moderately productive if cane-pruned. Clusters are usually straggly to loose. Grows well in all grape-growing areas of the San Joaquin and intermediate central valley regions. (115 acres; 47 hectares.)

Pearl of Csaba. A very early, white seeded variety with medium-sized berries of spherical shape. It has a muscat flavor.

Queen. A red, midseason grape with large oval berries and a neutral flavor. It is a good shipping variety. Introduced by H. P. Olmo in 1954 at Davis, California. (394 acres; 160 hectares.)

Rish Baba. A white seeded grape that ripens in early midseason. Berries are large and elongated, with one side nearly straight and the other bulging near the middle. It has a neutral flavor, and is thin-skinned and easily bruised. Stems are brittle. Both Rish Baba and Olivette blanche are marketed as "Lady Fingers." Originated in Iran.

Superior Seedless. A white seedless with large, firm berries that ripens earlier than Thompson Seedless. Introduced recently by the Superior Farming Company (121 acres; 49 hectares.)

Robin Cardinal (Thornburgs Robin). A bud mutation of Cardinal that ripens about 5 to 10 days before Cardinal. Introduced in Arizona by W. Thornburg. (75 acres; 30 hectares.)

Ruby Seedless. This variety has large reddish-black to dark red clusters that ripen in late midseason. Has fair eating quality. Originated at Davis, California, by H. P. Olmo in 1968. (131 acres; 53 hectares.)

WINE GRAPES IN CALIFORNIA

Description of some of the important wine varieties are listed below (after Jacob, 1950; Amerine and Winkler, 1963). The total acreage of each variety in California is also included.

Aleatico. A black grape with a fragrant muscat aroma, important in the Tuscany region of Italy. In California the vine is not popular because of its orange-red color, early ripening, poor vigor and productivity, and tendency to sunburn. (249 acres; 101 hectares.)

Alicante Bouschet. A black variety with red juice. It makes poor wine with intense color which fades with age. The grapes have fair shipping quality and many are sent to eastern markets. Clusters are medium sized, shouldered, conical, and well filled to compact. Berries are medium sized, spherical, brilliant black with a blue-gray bloom. Ripens in late midseason. Best suited to fertile soils in the warmer parts of the coastal valleys and in the intermediate central valley region. (6,820 acres; 2,760 hectares.)

Almission. A black grape that is a cross between Mission and Carignane. It makes poor wine. (97 acres; 39 hectares.)

Aramon. A late season red grape having poor color. It is used for dry table wine. (75 acres; 30 hectares.)

Barbera. A red grape having very high acid content that makes it valuable for blending with other grapes for table wine production in moderately warm regions. These grapes can make a good high-acid wine. Clusters are medium sized, conical, winged, well filled. Berries are medium sized, ellipsoidal, black, with colorful skin, neutral flavor, astringent. They ripen in midseason, and are best suited to the warm areas of the coastal valleys and the intermediate central valley region. (20,576 acres; 8,330 hectares.)

Beclan. A red midseason variety with medium productivity. It has low acid, medium color, and is used for dry table wine. (110 acres; 45 hectares.)

Black Malvoisie (*Cinsaut*). A black grape, probably imported from southern France. Clusters are medium sized, winged cylindrical, and loose to well filled. Berries are medium large, ellipsoidal, reddish-black to black. They ripen in early midseason, and become soft soon after harvest. Vines are vigorous and productive. Its principal use in California is for blending with other varieties to make dessert wines. The grapes are low in acidity and color and attain a high sugar content. (738 acres; 299 hectares.)

Burger. A very productive white grape that is susceptible to damage from bunch rot in cool regions. It is best suited to warm locations. Clusters are large to medium, shouldered to winged cylindrical, and compact. Berries are medium sized and ripen late. (2,083 acres; 843 hectares.)

Cabernet Sauvignon. Important in the production of the famous claret wines of France's Gironde region. In suitable locations in California this grape produces a wine of pronounced varietal flavor, high acidity, and good color. It is one of the finest red table wine varieties in California.

Clusters are small to medium and irregular in shape, but often long conical. They are loose to well filled. Berries are small, very seedy, nearly spherical, and black with a gray bloom. They ripen in midseason. Skin is tough, and the flavor is pronounced and characteristic. Vines are very vigorous and productive with cane pruning. The grapes attain their highest quality in the cooler parts of the coastal valleys. (24,539 acres; 9,935 hectares.)

Carignane. Of Spanish origin, and also important in southern France and Algeria. In California it is utilized mainly for making bulk red wines of medium acidity and color that usually have no striking varietal characteristic. It is very susceptible to powdery mildew.

Clusters are medium sized, shouldered, cylindrical, and well filled to compact. Berries are medium sized, ellipsoidal, and black with a heavy blue-gray bloom. Berries ripen in late midseason. Vines are very vigorous and productive. Cane is large, semierect to erect, and the vine grows well with head pruning. (30,710 acres; 12,433 hectares.)

Chardonnay (*Pinot Chardonnay*). Produces fine wines in the Cote d'Or region of France and other European countries. It can also produce excellent wines in California, but is a shy bearer. It is best adapted to Regions I and II. Clusters are small, loose to well filled, cylindrical, and winged. Berries are small, round, and usually have one seed. Leaves are large and have a rough texture. At basal edge of the leaf there is often a vein adjacent to the leaf margin. (10,037 acres; 4,064 hectares.)

Chenin blanc. A regular producer, susceptible to bunch rot. The wine, which has a fresh fruity flavor, is recommended for cooler regions as well as for Regions IV and V. Clusters are large, long conical, and compact. Berries are medium sized, oval, and have tough skins. Canes are semierect. (19,826 acres; 8,027 hectares.)

French Colombard. A vigorous productive variety that produces a standard wine of good quality. It is best suited for Regions III, IV, and V. Clusters are medium, conical, and compact. Berries are medium sized, and canes are usually upright. (26,666 acres; 10,796 hectares.)

Gamay. Important in the French Beaujolais region. In California this variety, or a similar one, is called *Napa Gamay* and is a very productive. Berries are medium sized, round, and have tough skins. It is a late midseason grape and clusters often contain small, green shot berries. Gamay produces good red and rosé wines in Regions II and III. (4,760 acres; 1,927 hectares.)

Gamay Beaujolais. Probably a clone of Pinot noir that was introduced into California from the Beaujolais region of France. It is best adapted to Regions I, II, and the cooler parts of III, where excellent wines similar to Pinot noir can be produced. Clusters are medium sized, compact, and shouldered to winged. Berries are medium sized, black, short-oval, and seeds are small and light brown. (4,490 acres; 1,818 hectares.)

Grenache. A Spanish variety grown in California mainly for the production of rosé and port wines. It thrives in the hot regions, bearing excellent crops. Its wines are medium to low in acidity. Sometimes the grapes are dificient in color and must be blended with other well-colored varieties. High quality pink or rosé wines are produced in the cooler coastal regions. The vines are susceptible to powdery mildew.

Clusters are medium sized, short conical, sometimes shouldered or winged, and loose to well filled, stems are thick. Berries are small-medium, short-ellipsoidal, nearly spherical, reddish-purple to black, and ripen in late midseason. Vines are unusually vigorous, erect, and very productive. (20,244 acres; 8,196 hectares.)

Mission. Jesuit missionaries planted the first *vinifera* grapes in California at the San Diego Mission in the latter part of the eighteenth century. Until about 1870 it was the principal variety grown in California. It is often used for white dessert wines such an Angelica. It is low in acidity, deficient in color, and best adapted to the central valley and southern coastal regions.

Clusters are large, conical, heavily shouldered, and loose. The stems are rigid. Berries are medium sized, oblate, reddish-purple to black, and ripen in late midseason. The pulp is firm but juicy. Vines are very vigorous and occasionally attain huge size, if given a sufficient area in which to develop. It is a heavy bearer. (6,356 acres; 2,573 hectares.)

Palomino. The principal sherry grape of Jerez, Spain. It makes good sherry, but inferior dry wine. Clusters are large-medium, shouldered, widely branched with stiff stems, and are loose to well filled. Berries are medium sized; oblate, greenish-yellow, firm to tough, and ripen in late midseason. Vines are very productive. Leaves are dull, dark bluish-green, rough on the upper surface, with a heavy tufted pubescence on the lower surface. Either head or cordon pruning is satisfactory. Palomino grows well in the San Joaquin, Sacramento, and intermediate central valley regions, and in warm parts of the coastal valleys. (5,992 acres; 2,398 hectares.)

Petite Sirah. Yields well and produces good red table wines. The skins are highly colored. In hot regions or hot seasons the fruit may sunburn. It is best adapted to moderately cool locations such as valleys of the northern coastal region. Clusters are medium sized, winged, cylindrical, and compact. Berries are of medium size, slightly ellipsoidal, black, with a dull bluish-gray bloom, and ripen in early midseason. Vines are of moderate vigor and productivity. (13,074 acres; 5,293 hectares.)

Pinot noir. A black grape used for making the famous wines of Burgundy in France. It ripens early, is moderately vigorous, and is best adapted to the cooler locations of Region I. Clusters are small, cylindrical, winged, and well filled to compact. Berries are small-medium, black, oval. Seeds are large, plump, and light brown. (10,098 acres; 4,088 hectares.)

Rubired. A hybrid of Alicante Ganzin X Tinta Caõ. Vines are productive and can be used to increase color in port-type wine production and also in blending. Clusters are loose to well filled and medium sized. The small berries are ellipsoidal and ripen in midseason. The grapes are resistant to spoilage and best suited for Regions IV and V. Introduced by H. P. Olmo, University of California at Davis, in 1958. (13,112 acres; 5,309 hectares.)

Ruby Cabernet. A cross of Cabernet Sauvignon and Carignane. It is highly productive, and the red wine has a distinctive Cabernet aroma. It is best suited for Regions IV and V. Clusters are medium-large, long conical, and loose to well filled. Berries are medium sized and round. Vines are large and canes upright. Introduced by H. P. Olmo, University of California at Davis, in 1968. (17,583 acres; 7,119 hectares.)

Salvador. A direct-producing hybrid of *Vitis rupestris* and *V. vinifera.* It is highly colored and can be used for blending to increase color in other types of wine. Salvador is inferior to Ruby Red and Royalty in wine quality. Clusters are small and may be shouldered. Berries are small to medium, short oval, and have a gelatinous pink pulp. Leaves are similar to those of *Vitis rupestris*. It is best adapted to Regions IV and V. (3,795 acres; 1,536 hectares.)

Sémillon. The famous sauternes of France owe much of their character to the Sémillon grape. In California, the dryness of the climate prevents "noble rot" (*Botrytis cinerea*) from growing on the grapes as they ripen, so that the finished wines differ from the French Sauternes in flavor and aroma. Sémillon is best suited to Region III.

Clusters are small to medium, short conical, and well filled. Berries are medium sized, spherical, golden yellow, have a figlike flavor, and ripen in early midseason. Vines are moderately productive. (3,356 acres; 1,359 hectares.)

White Riesling (Johannisberger Riesling). The main variety used in producing the Rhine wines of Germany. Its wines have a strong varietal flavor and bouquet. The vine is best suited to cool areas of the coastal region.

Clusters are small, cylindrical, and well filled. Berries are small-medium, spherical, greenish-yellow, and speckled with brown. Berries are juicy and aromatic in flavor, and ripen in early midseason. Vines are vigorous and moderately productive with cane pruning. (7,194 acres; 2,913 hectares.)

Zinfandel. A variety of unknown origin, not grown extensively in other countries. The wine is of medium acidity and color and has a characteristic flavor. The vine is best suited to cooler districts for the production of dry wines, but is also grown extensively in the intermediate central valley region. Clusters are medium sized, winged cylindrical, and well filled to very compact. Berries are of medium size, spherical, reddish-black to black, juicy, and ripen in early midseason. Vines are moderately vigorous and highly productive. Head training is recommended. (29,616 acres; 11,990 hectares.)

Other Wine Grapes. Information on other wine grapes grown to a lesser extent in California is shown in Table 6-4.

Recent Evaluations of Wine Grapes and Wine in Climate Regions IV and V

Most California wine grape variety evaluations previously discussed were made in order to apply recommendations based on them to large areas of the state. These recommendations were sound, although the approach has disadvantages. For example, a variety recommended for Region IV might not be well suited to all parts of that region, which include delta areas of moderate climate as well as Sierra Nevada foothills having much greater climatic ranges. Kissler et al. (1973) have recently refined the evaluation technique by studying wine grape varieties grown in an area of similar climate and soils.

Region IV. For many years in the Lodi district, the principal viticultural and winemaking operations produced table grapes and dessert wines. In recent years, however, due to a decrease in demand for table grapes and dessert wines and an ever increasing demand for table wines, these studies were concentrated on table wine varieties.

Table 6-5 gives the variety and wine quality evaluation for a trial plot of grapes established at Lodi. These data and recommendations should be used in conjunction with those made earlier, to make the best decisions on planting and other problems.

Region V. Trials were also made in Madera, Fresno, Tulare, and Kern Counties (Ough et al., 1973). Although the area encompassed by the evaluations is quite large, the climates are similar, and the main differences in various areas of the region are caused by soil differences or by different irrigation or cultivation techniques. These differences were considered by the investigators, and their evaluations are based on experimental data and also on information gathered from the winemaking industry. A consensus of opinion was used in making the evaluations.

Viticultural characteristics of wine grape varieties are presented in Table 6-6, and an evaluation of various varieties for winemaking is given in Table 6-7.

Table 6-4 Other Wine Grapes Grown to a Lesser Extent in California

Variety	Berry Color	Time of Maturity	Region for Productivity	Region for Best Growth	Acreage* (1974)	Hectares (1974)
Calzin	Black		High		99	40
Carnelian	Red		High	IV	2,067	837
B-5 (Centurian)	Red		High	III, IV, V	126	51
Chasselas doré	White	Early	Medium	I, II	71	29
Early Burgundy	Black				905	366
Emerald Riesling	Dark green	Late	High	II, III	2,846	1,152
Feher Szagos	White	Medium	Very high	V	478	193
Flora	White	Early	Medium	I, II, III	422	171
Folle blanche	White	Early	Medium	II, III, IV	247	100
Fresia	Red	Early	Low	III	75	30
Gewürztraminer	Pink		Medium	I	2,175	881
Grand noir	Red	Medium	Medium		241	98
Gray Riesling	White	Medium	High	II, III	1,698	687
Green Hungarian	White	Late	High		360	146
Grignolino	Red	Early	Medium	II, III, IV	241	98
Malbec	Red	Medium	Medium	II, III	434	176
Malvasia bianca	White	Medium	Medium	IV, V	960	389
Mataro	Black	Medium	Medium	I, II	1,698	687
Merlot	Black	Medium	Medium	II	3,988	1,615
Muscadelle	White				62	25
Muscadelle du Bordelais	White				103	42

Muscat blanc (Muscat Canelli)	White		Medium	IV	620	251
Muscat Hamburg	Black	Medium			98	40
Nebbiolo	Red	Late	Medium	III, IV	383	155
Pedro Ximenes	White				562	228
Perelli 101					321	130
Peverella	White	Medium	Medium	III, IV	444	180
Pinot blanc	White		Medium	II, III	1,296	525
Pinot St. George	Red	Late	Medium	II, III	654	265
Red Veltliner	White	Late	Medium	II	89	36
Refosco (Mondeuse)	Red	Medium	Medium	II, III, IV	298	121
Royalty	Red	Medium	Medium	IV, V	2,978	1,206
Saint Émilion (Ugni blanc; Trebbiano)	White	Late	Medium	II, III	1,562	632
Saint Macaire	Red	Medium	Medium	IV, V	245	99
Sauvignon blanc	White	Medium	Medium	II	3,193	1,293
Sauvignon vert	Greenish-yellow	Medium	High	II, III	832	337
Souzão	Black	Late	High	IV, V	260	105
Sylvaner (Franken Riesling)	White	Early	Medium	II, III	1,668	675
Tinta Madeira	Red	Early	High	III, IV, V	1,290	522
Touriga	Red	Late	Medium	IV, V	89	36
Valdepeñas	Red	Early	High	IV	2,531	1,025

* Figures include bearing and nonbearing grapes.

Table 6-5 Variety Evaluation of Grapes Made at Lodi, California*

Variety	Vigor	Cluster Density	Rot Potential	Production† tons per acre	Production† tons per hectare	Viticulture	Wine Quality	Evaluation‡
				White Varieties				
Aligote	Medium	Well-filled	Low	6–8	15–20	No problems	Good; fruity; non-varietal aroma	Good
Burger	Medium	Compact	Low	10–14	25–36	Excessive yields; low sugar	Poor	Poor
Chenin blanc	High	Compact	High	8–10	20–25	Very fruitful; may require cluster thinning; susceptible to Botrytis rot	Fruity; above average quality	Good
Emerald Riesling	High	Well-filled	Low	8–10	20–25	Quick delivery after harvest necessary	Wines tend to brown; tart; fruity; high in acid	Qualified
French Colombard	High	Well-filled	Low	9–12	22–30	Susceptible to shoot breakage	Good; fruity; rich in flavor; high in acid	Good
Gewürz-traminer	Low to medium	Compact	High	3–5	7–12	Excessive rot; poor production; hard to pick	Flat and coarse	Poor
Gray Riesling	High	Well-filled	High	5–7	12–17	Susceptible to Botrytis rot and thrips	Good if cropped normally and picked at proper maturity	Qualified
Grillo	Medium	Well-filled	Low	8–10	20–25	No problem	Makes coarse table wine; very good dessert wine	Qualified

Helena	Medium to high	Well-filled	Moderate	7–9	17–22	No problem	Poor and unbalanced	Poor
Muscat blanc	Low to medium	Well-filled	Low	4–6	10–15	Low production; tendency to sunburn and shrivel	Used for semi-sweet muscat-flavored table wine	Good
Orange Muscat	Medium	Well-filled	Low	6–8	15–20	No problem	Used for semi-sweet muscat-flavored table wine	Good
Peverella	High	Compact	Moderate	7–11	17–27	Not a consistent bearer; subject to rot	Ordinary wine; tart; high in acid	Poor
Sauvignon blanc	High	Compact	Moderate	5–7	12–17	Hard to hand-harvest; small clusters	Fruity; distinct flavor	Good
Tokay	High	Well-filled	Moderate	9–12	22–30	No problem	For sherry; not for table wines	Good
White Riesling	Low to medium	Compact	Moderate	3–5	7–12	Susceptible to rot	Recognizable varietal characteristics, but thin-bodied wine	Poor
				Red Varieties				
Barbera	Medium	Well-filled	Low	7–9	17–22	Spindly growth; many second clusters	Good to excellent; fruity	Good
Cabernet Sauvignon	High	Loose	Low	4–6	10–15	Buds out late; no rot problem following rains	Good varietal character	Qualified
Carignane	High	Well-filled	Low	8–12	20–30	Susceptible to mildew	Standard; used for blending	Good
Grenache	High	Compact	Moderate	9–14	22–35	Erratic production; poor wood maturity	Below standard blending wine; poor color	Poor

Table 6-5 (Continued)

Variety	Vigor	Cluster Density	Rot Potential	Production† tons per acre	Production† tons per hectare	Viticulture	Wine Quality	Evaluation‡
Gros Manzenc	Medium	Loose to well-filled	Low	5–7	12–17	Susceptible to spider mites	High color; low tannins; poor character	Poor
Petite Sirah	Medium	Compact	Moderate	6–8	15–20	No problem	Good color; soft wine	Good
Pinot Saint George	Medium	Compact	High	3–5	7–12	Thin-skinned; berries tend to rot	Undesirable flavor; high nitrogen	Poor
Refosco	Medium	Well-filled	Low	5–8	12–20	No problem	Thin wines; high tannins	Poor
Ruby Cabernet	High	Loose to well-filled	Low	6–8	15–20	Hard to hand-harvest	Distinct varietal character	Good
Saint Macaire	Medium	Loose to well-filled	Low	6–9	15–22	No problem	Very good for blending; high color; high acid	Good
Zinfandel	Medium	Well-filled to very compact	High	5–9	12–22	Susceptible to rot and spider mites	Good quality if handled properly; if overcropped low color, low acid, poor quality	Good

* After Kissler et al., 1973.

†Production figures given are considered ideal crop levels, using good cultural practices.

‡Good: Varieties that proved to be viticulturally sound and produced wines of acceptable quality. Qualified: Variety may make quality wine, but for lack of sufficient information, or because of problems of growing or winemaking, it may be of questionable value. Poor: By reason of viticultural or enological problems, or of sensory quality of the product, the variety is considered inadequate.

Table 6-6 Viticulture Characteristics of Wine Grape Varieties in Madera, Fresno, Tulare, and Kern counties of California*

Variety	Production		Harvest Period	Vigor	Other Important Characteristics
	tons per acre	tons per hectare			
			White Table Wine Varieties		
Burger	9–12 (as high as 16)	22–30	9/15–10/7	Moderate	Susceptible to bunch rot and overcropping. Total soluble solids tend to be low even with light crops.
Chenin blanc	9–12	22–30	8/20–9/25	High	Tight clusters can contribute to bunch rot. Minimized by early irrigation cutoff; pre-bloom gibberellin application, and use of fungicides. Best in fine, sandy loams.
Emerald Riesling	9–13	22–32	8/20–9/30	Exceptional	Crushed grapes very susceptible to browning, requiring special care during harvest. Adaptable to all soil types.
French Colombard	9–14	22–36	9/5–9/25	Exceptional	Excess vigor can be a problem on deep, fertile soils. Dense foliage can interfere with sulfur dusting and harvest operations. Occasional bunch rot associated with powdery mildew problems or Botrytis infections. May need Zn correction.
Green Hungarian	12–15	30–37	9/15–9/25	High	Susceptible to bunch rot. Irrigation cut off prior to harvest reduces bunch rot and breakdown. Fails to ripen properly.
Helena	No recent experience				
Peverella	8–11	20–27	9/10–9/25	High	Compact clusters sometimes subject to rot. Dense foliage.
Saint Émilion	10–13	25–32	9/5–9/30	High	Fruit "holds" well on vine.

Table 6-6 (Continued)

Variety	Production: tons per acre	Production: tons per hectare	Harvest Period	Vigor	Other Important Characteristics
Sauvignon blanc	5–7	12–17	8/15–9/25	Exceptional	Excess, dense growth can interfere with sulfur dusting and harvest operations. Susceptible to Botrytis rot from fall rains.
Sauvignon vert	8–10	20–25	9/1–9/15	High	Fails to mature properly to desirable sugar. Acid is low.
Sémillon	7–11	17–27	8/15–9/1	Moderate	Rot is sometimes a problem if fruit is held on the vines too long.
Thompson Seedless	8–11	20–27	8/2–10/1	High	Early harvest is desirable.
White Malaga	9–14	22–36	9/20–10/15	High	Principal advantages are ease of growing and harvesting and excellent holding characteristics of fruit on vine for late harvest and extending the crushing season.
White Riesling	3–5	7–12	8/15–9/15	Weak	Fruit breaks down rapidly when held beyond 19°Brix. Susceptible to bunch rot.
			Red Table Wine Varieties		
Alicante Bouschet	8–10	20–25	9/20–10/10	Moderate	Compact clusters sometimes subject to bunch rot. Vines are sensitive to overcropping.
Barbera	6–10	15–25	9/10–10/5	Moderate	Leafroll virus is widespread in vineyards originating from noncertified planting stock. Vineyard establishment from cuttings often results in below normal initial stands.
Cabernet Sauvignon	5–7	12–17	8/15–9/10	Moderate	Few problems but very moderate producer. Tends to lack good fruit color.

Calzin	5–11	12–27	9/1–10/5	Moderate	Resistant to red spider. Loose clusters and little rot.
Carignane	8–13	30–32	9/1–10/7	High	Susceptible to powdery mildew. Summer bunch rot problems in some vineyards require pre-bloom gibberellin treatment. Yellow vein virus, "unfruitful" vines found in many commercial plantings.
Gamay (Napa)	5–8	12–20	8/20–9/15	Moderate	Larger, compact clusters subject to bunch rot.
Gamay Beaujolais	See Pinot noir				
Grenache	8–14	20–36	8/20–10/7	Very high	Bloomtime Botrytis infections are sometimes a problem. Compact clusters sometimes rot. Subject to delayed bud break in spring and occasional poor fruit-set. Good tolerance to salt and boron.
Grignolino	No recent experience				
Malbec	5–9	12–22	9/15–9/30	Moderate	Subject to poor fruit-set under high nitrogen conditions.
Niabell	3–5	7–12	8/15–9/15	Low to moderate	Berries subject to uneven ripening. Exposed fruit sunburns during hot spells. Fruit shrivels if harvest is delayed. Resistant to powdery mildew.
Pinot noir	4–5	10–12	8/10–8/25	Low to moderate	Tight clusters subject to bunch rot, especially with delayed harvest. Fruit raisins badly with delayed harvest.
Pinot Saint George	No experience				
Petite Sirah	6–10	15–25	9/5–10/1	Moderate	Tight clusters subject to rot. Fruit tends to sunburn and raisin from hot spells or delayed harvest. Leafroll and corky bark vines prevalent in older commercial plantings.

Table 6-6 (Continued)

Variety	Production		Harvest Period	Vigor	Other Important Characteristics
	tons per acre	tons per hectare			
Royalty	7–10	17–25	9/1–9/30	Moderate	Moderate tolerance to berry mildew. Other viticultural characteristics or problems similar to Rubired, except that Royalty vines tend to be weaker, especially with soil limitations such as sandy texture.
Rubired	8–12	20–30	9/1–9/25	High	Tolerant to berry mildew. Young vines subject to collar rot. Young vines easily overcropped. Many commercial vineyards are infected with leafroll virus. Crop estimate is difficult because of high percentage of stems per cluster.
Ruby Cabernet	6–10	15–25	9/1–10/1	Moderate	Performs very poorly in soils with limiting factors such as sandy texture, compaction, nematode problems, or shallow depth. Irregular berry-set occasionally. Many commercial vineyards are infected with leafroll virus.
Salvador	7–11	17–27	8/15–9/15	Moderate	Resistant to powdery mildew and somewhat resistant to hoppers, mites, and grape leaf folder. High thrips populations sometimes stunt shoot growth. Tight clusters subject to bunch rot. Harvest should not be delayed. Difficult to prune—bushy growth.
Saint Macaire	7–10	17–25	9/15–10/5	Moderate	Grapes tend to sunburn if vines are dried up early. Medium-to-large clusters.

Valdepeñas	10–12	25–30	8/20–9/20	High	Heavy, dense leaf area and powdery mildew susceptibility requires especially good sulfur dusting practices. Vines leaf out late.
Zinfandel	5–9	12–22	9/7–9/25	Low to moderate	Susceptible to spider mites. Bunch rot problem during ripening requires prebloom gibberellin spray. Exposed ripening fruit subject to raisining.
			White Dessert Wine Varieties		
Fernão Pirès	7–9	17–22	8/20–9/15	Moderate	Sound fruit, good sugar and acid.
Palomino	8–12	20–30	9/10–9/30	High	If overcropped will not mature, and will rot. Generally picked at lower °Brix for sherry and blending material.
Mission	8–14	20–36	9/25–10/25	Very high	Considered susceptible to crown gall, and tends to attract sharpshooter leafhoppers, making Pierce's disease losses more common. Fruit holds well on vine, achieving high sugar levels.
Verdelho	7–9	17–22	8/15–9/10	Moderate	Produces sound, clean fruit that begins to shrivel if harvest is delayed.
Thompson Seedless, White Malaga previously listed					
Malvasia bianca	5–8	12–20	8/15–9/10	Moderate	Somewhat open growth can result in considerable fruit sunburn during May or June hot spells. Erratic, low producer.
Muscat blanc	6–8	15–20	8/15–9/5	Moderate	Exposed, ripening fruit subject to sunburn. Fruit will raisin on vine if harvest is delayed. Compact clusters can bunch rot, especially with delayed harvest. Variety "suckers" badly.

Table 6-6 (Continued)

Variety	Production: tons per acre	Production: tons per hectare	Harvest Period	Vigor	Other Important Characteristics
Muscat of Alexandria	3–6	7–15	9/7–10/15	Moderate	Subject to zinc deficiency. Exposed fruit subject to burn in hot spells. Overcropping is common on cordon-trained vines, resulting in alternate-bearing tendencies and shortened vine life. Affected adversely by high salt and boron.
Orange Muscat	3–6	7–15	8/20–9/15	Weak	Very susceptible to salt or boron damage. Generally poor foliage; fruit sunburns and raisins.
Red Dessert Wine Varieties					
Souzão	6–8	15–20	9/15–9/30	Low to moderate	Fruit doesn't hold well on vine and raisins badly if harvest is delayed. High boron and salt adversely affect growth.
Tinto cão	5–7	12–17	9/15–10/5	High	Small clusters, but quite fruitful.
Tinta Madeira	6–9	15–22	8/20–9/10	Moderate	Compact clusters require prebloom gibberellin spray to reduce bunch rot. Fruit tends to raisin on vine beyond 22°Brix. Adversely affected by high boron or salt.
Trousseau	No recent experience			—	Clusters tend to be very tight and to rot.
Carignane, Grenache, Royalty, and Rubired previously listed.					

* After Ough et al., 1973.

Table 6-7 Evaluations of Various Wine Grape Varieties Grown in Madera, Fresno, Tulare, and Kern Counties of California for Winemaking*

- White Table Wine Varieties
  - Good
    - Chenin blanc
    - French Colombard
    - Thompson Seedless (mainly for blending)
  - Questionable
    - Emerald Riesling, Sémillon (winery problems)
    - Peverella, Helena, Sauvignon blanc (further testing needed)
  - Poor
    - Burger
    - Flora
    - Green Hungarian
    - Saint Emilion
    - Sauvignon vert
    - Sylvaner
    - White Malaga (good for grape concentrate or white dessert wine)
    - White Riesling
- Red Table Wine Varieties
  - Good
    - Barbera
    - Carignane (for standard blending material)
    - Grenache (for rosé, limited)
    - Petite Sirah (careful attention to pH necessary)
    - Rubired (for color, blending)
    - Ruby Cabernet
  - Questionable
    - Royalty, Salvador (inferior in crop and color to Rubired)
    - Saint Macaire, Gamay (Napa) (need further testing)
  - Poor
    - Alicante Bouschet
    - Black Malvoisie
    - Cabernet Sauvignon
    - Calzin
    - Grignolino
    - Malbec
    - Niabell
    - Pinot noir
    - Pinot St. George
    - Valdepeñas
    - Zinfandel
- White Dessert Wine Varieties
  - Good
    - Mission
    - Muscat of Alexandria
    - Muscat blanc (for sweet table wine)
    - Palomino (for sherry)
  - Questionable
    - Fernão Pirès
    - Verdelho
  - Poor
    - Malvasia bianca
    - Orange Muscat
- Red Dessert Wine Varieties
  - Good
    - Carignane (for standard port)
    - Rubired (blending)
    - Tinta Madeira (poor production)
  - Questionable
    - Royalty
    - Souzão
  - Poor
    - Grenache
    - Muscat Hamburg
    - Tinto cão
    - Trousseau

* After Ough et al., 1973.

RAISIN GRAPES (AFTER JACOB, 1950)

The three main raisin grapes are Black Corinth, Muscat of Alexandria, and Thompson Seedless (Sultanina). Worldwide production of raisins and currants is shown in Table 6-8.

Black Corinth (Zante Currant). This variety probably originated in Greece. Clusters are small to medium, winged, and uniformly cylindrical. Berries are very small, spherical to oblate, reddish-black, mostly seedless, very juicy, neutral in flavor, with very thin, tender skins. They ripen early and dry easily into very small raisins of soft texture and pleasing tart taste. They are well suited to the central and southern parts of the San Joaquin Valley.

Fiesta. A medium sized, white seedless grape that matures 12 to 14 days before Thompson Seedless. The berries are oval, but rounder than Thompson Seedless. Skin is tender and flavor is good. Clusters are medium to large and not usually compact. Raisins made from Fiesta

Table 6-8 Worldwide production of raisins and currants in 1974 (in thousands of metric tons)*

Country	Raisins	Zante Currant†
Australia		3.6
Lexias	2.8	
Sultanas	52.1	
Greece	95.0	85.5
Iran	50.0	—
Republic of South Africa	5.7	0.7
Spain	4.7	—
Turkey	87.0	—
California (U.S.A.)	217.7	—
Total‡	515.0	89.8

* Compiled from reports of the Foreign Agricultural Service, U.S. Department of Agriculture, Washington, D.C.

† Preliminary.

‡ Excludes California Zante currants, shown above under raisins.

Note: None of the above statistics include production by Argentina, Chile, Cyprus, or Afghanistan, whose production may be substantial but is not readily ascertainable.

are slightly larger plumper, and more meaty than Thompson Seedless. Grows best when cane-pruned (Weinberger and Loomis, 1974; V. E. Petrucci, private communications). Cap stems are slightly more difficult to remove from Fiesta than from Thompson Seedless (V. E. Petrucci, private communication).

Fiesta was developed by the U.S. Department of Agriculture, Fresno. It was first released in 1973.

Muscat of Alexandria (syn. *Hanepoot*, in South Africa). An old variety originating in North Africa from which the raisins of Spain are made. It is an important raisin variety in Australia and a good table grape for home gardens and local markets. It has a pronounced muscat flavor, juicy pulp, poor shipping quality, and lacks an attractive appearance; and therefore it is relatively unimportant in table grape shipments to eastern markets. Although Muscat of Alexandria is used extensively for muscatel, a dessert wine, dry wines made from it are only standard or mediocre. It adapts only to hot regions, but may sunburn in the hot desert. Clusters are medium sized, shouldered, conical, loose, and often straggly. Berries are large, obovoid, dull green, normally seeded, and pulpy. The moderately tough skins are covered with a gray bloom. It ripens in late midseason, and the grapes dry into large raisins of soft texture and excellent quality. Vines are medium in vigor, highly productive, and are usually head-trained.

Sultana (Round Seedless). This grape is somewhat similar to Thompson Seedless but differs in having smaller, oblate to round berries, a few of which contain partly hardened seeds. It is of little importance and is inferior to Thompson Seedless.

Sultanina rose has a rose or pink color, but is otherwise almost identical to Thompson Seedless. It is useful only for home gardens.

Thompson Seedless. Over half the world's raisins and about 90 percent of those in California are made from this variety which originated in Asia Minor. It is called Sultana in Australia and the Republic of South Africa, and Oval Kishmish in Asia Minor. Thompson Seedless is also a leading table grape, widely used in wine making. It is best suited to warm to hot regions, and does poorly in cool areas.

Clusters are large, cylindrical, and heavily shouldered. Berries are small, oval elongated, seedless, and white. Vines are vigorous and productive and cane pruning is necessary. Thompson Seedless ripens early, has fair shipping quality, and the grapes dry easily into raisins of excellent quality.

ROOTSTOCKS (AFTER KASIMATIS AND LIDER, 1972)

Phylloxera-resistant rootstocks are often required to prevent damage from the root-destroying aphid of the grape phylloxera. In very sandy soils, nematode infestations may make it impossible to grow grapes unless nematode-resistant rootstocks are used. Although most rootstock varieties are hybrids produced by crossing two or more grape species, a few are direct selections from wild grapes. *Vitis girdiana* and *V. californica* are native to California, but are not useful for viticulture. These and some important varieties are described below.

Phylloxera-Resistant Rootstock Varieties

Riparia Gloire (syn. *Gloire de Montpellier; Gloire*) is a seedling selection from *Vitis riparia,* a phylloxera-resistant species native to the eastern United States. This old stock, is not drought-tolerant, and is therefore suitable only for moist fertile soils. It is not recommended for use in California. Leaves are large, entire, heart shaped, with a tendency to form three lobes. Serrations are sharp, alternating in size, and distinctly enlarged at the apices of the lobe. Leaves are medium-green with a light tomentum on both surfaces, generally confined to the veins. The large petiole has uniform short hairs, and the petiolar sinus has a broad V-shape. The vine bears small clusters of male flowers.

St. George (syn. *Rupestris du Lot, Rupestris St. George*) is a variety of *Vitis rupestris,* a phylloxera-resistant species native to the eastern United States. This stock produces vigorous grafted vines and is drought-tolerant. It is recommended for the drier hillside locations of the nonirrigated coastal valleys. It is not resistant to nematodes or to oak root fungus. Cuttings root well, and the stock is easily budded or grafted. With low-bearing wine grapes, its high vigor has reduced yields. Leaves are small, entire, round, with a distinct, open or flat petiolar sinus. Uniform serrations are around the leaf edge. Leaves are light glossy-green, smooth, with a leathery texture and no tomentum. Vine grows upright, has compact, bushy appearance, and bears small clusters of male flowers only.

A X R #1 (syn. *Ganzin No. 1; Aramon X Rupestris Ganzin, No. 1; A X R G, No. 1; A X R*) arose as a hybrid between the species *Vitis vinifera,* var. Aramon, and the phylloxera-resistant species *Vitis rupestris,* var. Ganzin. This rootstock produces vigorous grafted vines that bear good yields of high-quality fruit. It is phylloxera-resistant but susceptible to nematodes. It performs well under irrigation and in the deeper, heavier soils of the floors of the coastal valleys, especially, with the lighter bearing varieties. It is recommended for raisin and table varieties in the heavier, phylloxerated soils of the San Joaquin Valley. Cuttings of this stock root quite readily, and it buds and grafts easily. Leaves are medium to small, entire, round to a short heart shape, with a distinct, curved, open V-shaped petiolar sinus. They are light glossy-green, smooth with a leathery texture and no tomentum. Shoots are semierect to trailing; internodes medium to long; dormant canes fairly large in diameter; bark tan to light-brown; some lengthwise cracking at the base of large canes. Vine is vigorous, semierect, spreading, with fairly heavy shoot growth. It bears male flowers profusely in small to medium sized clusters.

1202 (syn. *Couderc 1202; Mourvédre X Rupestris No. 1202*) is a variety arising as a hybrid between the fruiting type Mataro (Mourvédre) and *Vitis rupestris*. It is phylloxera-resistant but susceptible to nematodes. Although rarely used in California, it can produce excellent vines with table varieties in the heavier soils of the San Joaquin Valley. Similar to A X R #1, but less vigorous. Cuttings root readily, and it buds and grafts easily. Leaves are small to medium-small, entire, round to heart-shaped, with deep, narrow U-shaped petiolar sinus, and a tendency to close. Small, uniform serrations are around the leaf edge. Leaves medium to dark glossy-green, smooth, with a thin, leathery texture and no tomentum; slight tendency toward three-lobing. Vines are fairly vigorous, semierect, with dense appearance, and bear perfect flowers that develop clusters of small black berries.

99-R (syn. *Richter 99; Berlandieri X Rupestris, No. 99*) is a variety arising as a hybrid between the species *Vitis berlandieri* and *V. rupestris*. It is a low-vigor stock, drought-tolerant, and resistant to phylloxera but not to nematodes. Produces moderately small vines that bear heavily. It is suitable for use on hillside plantings that show drought conditions, or in more fertile, shallow valley-floor locations. Because of its low vigor, it is generally not recommended for commercial use. Although it has excellent tolerance to high-limestone soil conditions, this is of little or no benefit in California vineyards. This stock readily roots its cuttings, and is easy to bud and graft.

The leaves are small, round, entire, with a very distinct shallow, open U-shaped petiolar sinus. Serrations are small, sharp, and uniform. Upper surfaces are deep bluish-green, contrasting with light-green veins; lower surfaces have a metallic brownish-green cast, texture is tough or leathery, but smooth, without tomentum. Vine bears male (pollen-producing) flowers in small clusters.

3306 (syn. *Couderc 3306; Riparia X Rupestris, No. 3306*) is a variety arising as a hybrid between the species *Vitis riparia* and *V. rupestris*.

This phylloxera-resistant rootstock produces moderately vigorous, grafted vines that bear good crops. It is not recommended for California vineyards because other stocks are more suitable. Leaves are small, entire, round to a blunt heart shape, with a tendency toward three-lobing. The serrations are distinct, somewhat irregular, and enlarged at apex. Leaves are medium-green, smooth-textured, dense, short, and upright hairs uniformly cover the petioles and veins on the lower surfaces. There is only light tomentum on blades and veins on the upper surfaces; petiolar sinus is deep and open V-shape. Bears male flowers on small, inconspicuous clusters.

3309 (syn. *Couderc 3309; Riparia X Rupestris, No. 3309*) is a variety arising as a hybrid between the species *Vitis riparia* and *V. rupestris*. This phylloxera-resistant stock is used extensively in other countries, but is not recommended in California because other stocks perform better. Leaves are small, entire, round to a blunt heart shape, with a tendency toward three-lobing. Serrations are distinct, irregular, and somewhat larger than 3306; apex is prominent, medium-green, smooth textured. In contrast to 3306, the upper surfaces are glabrous; lower surfaces are also glabrous except for discrete tufts of long hairs at the junctions of large veins. Petioles are glabrous, petiolar sinus is deep and open V-shape. Bears male flowers in small, inconspicuous clusters.

5A (syn. *Teleki 5A; Berlandieri X Riparia 5A*) is a hybrid of *Vitis berlandieri* and *V. riparia* produced in Hungary by Teleki. It is phylloxera-resistant and has a high tolerance to lime soils. The latter quality is useful in parts of Europe but not in California. Its performance in California has been erratic in field trials: in the nonirrigated coastal valleys, it was surpassed by one or more of the phylloxera-resistant stocks. In trials in phylloxerated sites in the interior valleys, it has produced vigorous, heavy-bearing scions. Leaves are large, entire and heart-shaped, with a slight tendency toward three-lobing. Serrations are uniform but shallow. Leaves are medium-green, but darker on upper surfaces. Mature leaves are medium to dark-green with glabrous upper surfaces; lower surfaces have short upright tomentum, usually found only along the veins and petiole; petiolar sinus moderately open U-shape; petioles have a reddish tinge. Produces female flowers that may develop into small, loose clusters with round, black berries.

SO4 (syn. *Selection Oppenheim No. 4*) is a variety selected in Germany from the *Vitis berlandieri X V. riparia* hybrids of Teleki. It is a vigorous phylloxera-resistant stock, but more field testing must be done in California before it can be recommended. May be useful for heavier soils of irrigated interior valleys. Leaves are large, entire, heart-shaped, with a tendency toward three-lobing. Serrations are sharp, relatively uniform in size and shape, enlarged at the tips of the lobes, medium-green. Older leaves are slightly rough, short, with uniform tomentum on lower surfaces and tendency to tuft, especially along the veins. Upper surfaces and petioles have sparse, light tomentum with tufting tendency, slight curves, and a V-shaped petiolar sinus. Bears small clusters of male flowers.

Summary of Effectiveness of Phylloxera-Resistant Stocks. Table 6-9 sums up the evaluations of stocks for the northern coastal valleys of California (Lider, 1958). Note that five rootstocks showed high performance. *Aramon X Rupestris Ganzin #1* is generally the most vigorous and productive stock, and gives more beneficial results than *Rupestris St. George,* the most popular stock used commercially (Lider, 1958).

Nematode-Resistant Rootstock Varieties

Dogridge is a variety of *Vitis champini,* a nematode-resistant species native to north-central Texas. This stock is very vigorous, nematode-resistant, and moderately resistant to phylloxera. Because of its high vigor, scions frequently show symptoms of zinc deficiency. Dogridge is recommended for use only in the lighter, less fertile sandy soils of the irrigated interior valleys. It is most suitable for heavy-bearing wine varieties and where cultural practices have been adapted to use the vigorous growth. Cuttings root with difficulty, but the rootings bud and graft readily.

Leaves are medium sized, moderately to distinctly three- to five-lobed; upper surfaces are lightly tufted with long hairs; lower surfaces are moderately heavy with tomentum, particularly with heavy tufts along veins and petioles. Serrations very shallow, even, rounded; and medium-green; petiolar sinus is deep, open V-shaped. Vine is very vigorous, spreading, and prostrate. It produces female flowers that develop small, compact clusters of medium-sized black berries.

Freedom is a new introduction from the U.S. Horticulture Field Station at Fresno with characteristics similar to those of Harmony.

Harmony (syn. *US 16-154*) is a cross between a selected seedling of 1613 (#39) and a selected seedling of Dogridge (#5), made in 1955 at the U.S. Horticultural Field Station at Fresno. Vines grafted on Harmony are more vigorous than vines on 1613, but less strong than those on Dogridge and Salt Creek rootstocks. Although Harmony has shown greater resistant to root-knot nematode and phylloxera than 1613, it is not immune to either. Cuttings of this stock root readily, and it buds and grafts easily. In the San Jaoquin Valley, Harmony seems suited to

Table 6-9 Performance of Phylloxera Resistant Rootstocks in the Northern Coastal Valleys of California*

Rootstock	Degree of Phylloxera Resistance	Performance†	Remarks
Rupestris St. George (synonym, Rupestris du Lot)	High	Generally satisfactory	Still useful on shallow soils.
Riparia Gloire de Montpellier	High	Unsatisfactory	Not currently recommended.
Dogridge	Moderate	Unsatisfactory	Inconsistent growth habits; difficult to root.
Solonis × Othello 1613	Moderate	Unsatisfactory in coastal valleys	Vines on these stocks weak and unproductive.
Solonis × Riparia 1616	Moderate		
Riparia × Rupestris 3306	High	Low in comparison to other stocks	Not recommended for further planting in California.
Riparia × Rupestris 3309			
Berlandieri × Rupestris 99-R	High	Generally satisfactory	99-R best of the newer stocks; 110-R and 57-R less productive; 44-R needs further testing.
Berlandieri × Rupestris 110-R			
Berlandieri × Rupestris 57-R			
Berlandieri × Rupestris 44-R			
Berlandieri × Riparia 420-A	High	Unsatisfactory	Neither stock promising in coastal valleys.
Berlandieri × Riparia 5-A			
Chasselas × Berlandieri 41-B	Moderate	Unsatisfactory	Difficult to root cuttings; vines weak and slow growing.
Bourrisquou × Rupestris 93-5	Moderate	Satisfactory	Generally a vigorous stock.
Vinifera × Rupestris XX	Low(?)	Unproven	Not recommended for further use in California.
Mourvedre × Rupestris 1202	Moderate	Satisfactory	Both stocks recommended for planting in coastal valleys.
Aramon × Rupestris Ganzin No. 1		Outstanding	A × R#1 is superior to all other stocks tested.

* After Lider, 1958.

† Based on annual measurements of trunk circumference, yields, and fruit quality.

all but the very lightest soils, and is particularly adapted to the Thompson Seedless variety for raisin and wine production. Early tests indicate that Harmony is probably satisfactory for table grapes. Leaves are medium to medium-small, slightly three-lobed and round; serrations are shallow, distinct, sharp, irregular; petiolar sinus is deep, open, slightly U-shaped. Upper surfaces and petioles are lightly tufted with gray tomentum; lower surfaces are very lightly tufted, medium-green with bright cast; petiole is long, one-half to two-thirds the length of the blade (J. H. Weinberger). Vines are moderately vigorous, semierect, and have dense growth. It produces female flowers that develop small, compact clusters of small black berries.

Salt Creek (syn. *Ramsey, Vitis champini Salt Creek*) probably originated as a *V. champini* type and is closely related to Dogridge. It should not be confused with the true variety, Salt Creek, which was selected from *V. doaniana.* This stock imparts great vigor to its scions, and is highly resistant to nematodes and moderately resistant to phylloxera. It has performed well with wine and raisin varieties in light sandy soils of low fertility. Since it is less vigorous than Dogridge, it has a greater range of use. Cuttings root with difficulty, but the stock buds and grafts readily. Suckering is less of a problem than with Dogridge, but disbudding is recommended. Leaves are medium to medium-small, slightly three-lobed, roundish; serrations uniform, distinct, shallow, sharp or acute. Upper surfaces are lightly tufted with tomentum; lower surfaces and petioles are moderately tufted, medium-green with bright glossy cast; petiolar sinus is deep, with open U-shape. Vine is moderately vigorous, and has a dense, upright habit. It produces female flowers that develop small, compact clusters of medium-small black berries.

1613 (syn. *Courderc 1613; Solonis X Othello 1613; Solonis-Othello*) is a variety arising as a hybrid between the species *V. solonis* and the fruiting variety Othello. This stock produces moderately vigorous scions, is resistant to the more prevalent strains of rootknot nematodes and moderately resistant to phylloxera. Suitable for wine, raisin, and some table varieties in all but the lightest soils in the San Joaquin Valley and southern California. Cuttings root readily, and it buds and grafts easily. Leaves are large, entire, with a tendency to form lobes; broad, with nearly straight sides. Serrations are distinct and fairly uniform, but enlarged at tips of lobes; dull gray-green above, grayish with heavy tomentum below; Petioles and upper surfaces are tufted with tomentem; petiolar sinus is open with broad U-shape. Vine spreading is prostrate with vigorous growth, and produces female flowers that develop into small compact clusters of small black berries.

Vitis Species Native to California

Vitis californica (syn. *California wild grape, Pacific grape*) is a species native to northern California, ordinarily found along stream banks at lower elevations. It has little resistance to phylloxera, nematodes, or oak root fungus. It bears fruit in irregular, loose small-berried clusters of no commercial value. It crosses naturally with cultivated vines, and hybrids can occasionally be found in areas adjacent to vineyards. These hybrids are not useful for rootstocks. Leaves are medium to large, round, entire, with a slight tendency toward three-lobing; serrations are somewhat irregular and rounded. Upper surfaces are dark green with a gray-green cast; lower surfaces are gray-green with profuse tomentum. Mature leaves are heavily tufted with tomentum on the upper surfaces, and densely matted on the lower ones; young leaves are covered with tomentum on both surfaces; petiolar sinus is a deep, narrow U-shape. Vine is vigorous, climbing, attaches to trees and shrubs, frequently reaches 30 to 40 ft (9.14–12.20 m), and can form a dense canopy.

Vitis girdiana (syn. *Valley grape*) is closely related to though less vigorous than *V. californica* and occurs in southern California. No use has been made of this species. Leaves are similar to but smaller than *V. californica.* Shoots are moderately vigorous and densely covered with woolly-white tomentum at the tips.

Recommended Rootstocks for Nematodes

The rootstocks recommended for nematode resistant in California are Dogridge, Salt Creek, 1613, 1616, and 5A (Lider, 1959, 1960), as well as Harmony and Freedom.

NEW YORK VARIETIES (AFTER EINSET ET AL. 1973)

In New York State grapes are grown for use in juice, wine, fresh fruit, jelly, and jam. The Concord variety comprises about 75 percent of the New York grape crop delivered to processors. The leading wine varieties, Niagara, Catawba, and Delaware, total 16 percent, Elvira and Ives 3 percent, and the French hybrids 5 percent. The remainder, approximately 1 percent, is made up of miscellaneous other varieties. Descriptions of several varieties are given below.

Catawba is a late-ripening, red, American-type grape that requires the best vineyard sites in the most favorable locations to reach full maturity. Vines are vigorous, hardy, and productive, but the foliage is somewhat susceptible to fungus diseases that can be controlled by spraying. Catawba is an important ingredient in New York State champagnes and table wines.

Concord is a blue-black American-type grape with a tough skin that separates readily from the pulpy flesh (slipskin). Concord is a late midseason grape, and is also the standard by which both the vine and fruit of other varieties are judged. The use of this fruit for many purposes gives it a large market outlet. Concord is the only important variety for sweet juice, jelly, and preserves, and much is also used for wine production. Its pronounced fruity flavor makes it a good dessert grape. In a late season, and in less desirable locations, the variety may fail to attain full maturity on heavily loaded vines.

Delaware is one of the highest quality American grapes for table use and for white wine. It ripens two or more weeks before Concord, and has small compact clusters with small red berries. The tender skin is subject to cracking when fall rains occur near harvest time, and the variety is susceptible to fungus diseases. Delaware requires a deep, fertile well-drained soil for good vine growth, and vines may produce yields as high as those of Concord on such soils. On poorer soils and on old vineyard sites, a phylloxera-resistant rootstock should be used to ensure vigorous growth.

Niagara, the leading American-type white grape, is used fresh and for wine. It is less cold-resistant than Concord, and is moderately susceptible to the major grape diseases.

New York grapes of less commercial importance.

Clinton is a red wine grape with small berries and clusters that closely resemble the wild *V. riparia.* It is vigorous and has been used as a rootstock for less sturdy varieties.

Diamond resembles the fruit of Niagara and the vine of Concord. As a dry table wine it is one of the most distinctive and desirable of the American types.

Dutchess is a late-ripening white grape related to *V. vinifera.* It produces a white wine of high quality, is susceptible to disease, and may be injured by the low temperatures.

Elvira is a white wine grape with the American species, *V. riparia,* the River-bank or Frost Grape, in its ancestry. This variety is hardy, productive, and disease-resistant. Its thin skin and compact clusters may cause the berries to crack. Ripens at about the same time as Concord.

Fredonia is a black Concord-type grape that ripens about two weeks earlier than Concord. It lacks the typical Concord flavor desirable for making juice and jelly. Although it may exceed Concord in vigor and production, the fruit is suceptible to downy mildew.

Ives is a black grape of the *V. labrusca* or Fox Grape type used in red wines. It is vigorous and productive in good locations. A sturdy rootstock should be used in poor locations.

Isabella is an old, black *labrusca*-type grape used for wine. It is tender and subject to winter injury.

Missouri Riesling is similar in origin and appearance to Elvira. This wine grape matures after Elvira and Concord, but before Catawba, and is vigorous, productive, and hardy.

French Hybrids and Newer Wine Grapes (Einset et al., 1973)

French hybrids have been derived from crosses between *V. vinifera* and a number of wild American species. They were made by French hybridizers who were seeking phylloxera-resistant varieties that would produce wines less fruity and more neutral in flavor. These grapes are usually identified by the name of the originator and a number.

Nearly all American-type grapes or hybrids contain *V. labrusca* in their parentage, but the French have used mainly other American species with different fruit and plant characteristics. Thus the two groups of hybrids are quite distinct in appearance.

The shoots of most French hybrids grow more upright than American hybrids. The leaves usually have deeper sinuses, are more glossy, and lack the heavy tomentum common on the underside of American grape leaves and *V. labrusca*. The flavor of the fruit is generally more neutral, lacking the fruitiness of our native varieties.

Red French Hybrids

These are listed according to increasing color and season of ripening, from early to late. Data are taken from Einset et al., 1973.

Maréchal Foch (Kuhlmann 188-2) is a very early, small-clustered, small-berried black grape that produces an excellent Burgundy-type red wine. Vines are hardy and medium in vigor and production. It should be grafted on a resistant rootstock to ensure adequate vigor. Birds can be a problem.

Léon-Millot (Kuhlmann 194-2). The fruit, vine, and wine are similar to Foch. It is early ripening and has a high sugar content.

Cascade (Seibel 13053) is an early blue grape. Clusters are medium to large and loose. It is very productive and hardy, but the fruit attracts birds. The wine can be very good but may be light in color.

Baco Noir (Baco No. 1) is an extremely vigorous and disease-resistant variety used for red wine that grows well in arbors. Bud break occurs early in the spring and may be subject to late-spring frost injury.

De Chaunac (Seibel 9549) appears to be one of the best French hybrids for red wine. It is hardy, relatively free from disease, and is less susceptible to bird damage than some other varieties.

Chelois (Seibel 10878) is a vigorous, productive variety used for a claret-type wine. Vines are only moderately winter hardy, especially if overcropped.

Rougeon (Seibel 5898) is a commercial red wine grape that yields a highly colored wine used for blending. Vines are hardy but erratic producers.

Rosette (Seibel 1000) is an old French hybrid with hardy and moderately productive vines. The fruit lacks intense color and is used for rosé wine.

Chancellor (Seibel 7053) is widely grown in France. It is hardy and productive, and the wine produced is of the highest quality. The fruit is susceptible to early downy mildew.

Colobel (Seibel 8357) is a "teinturier," or a variety with highly colored juice, used for blending with paler wines. Vines are moderately hardy and productive.

White French Hybrids

These are listed according to increasing color and season of ripening from early to late.

Aurore (Seibel 5279) is widely planted, hardy, vigorous, and productive. The fruit ripens early, is of good dessert quality, and makes a very good wine.

Vignoles (Ravat 51) makes a fine Chablis-type white wine. The clusters are compact and the berries tend to crack in wet seasons. Vines are hardy but of medium vigor and production.

Verdelet (Seibel 9110) is a yellow-gold dessert and wine variety grown in the best locations. This grape tends to overbear and the crop must be thinned. It is subject to winter injury.

Seyval (Seyve-Villard 5-276) is a high-quality, midseason, white grape that produces a fine white wine. It tends to overbear and must be cluster-thinned for proper ripening and to prevent weakening of the vines.

Vidal 256 is a late, white, hardy, heavy producer. The wine is neutral in flavor and is rated good to very good.

Villard Blanc (Seyve-Villard 12-375) produces large, loose clusters. The wine is very good, and the fruit is useful as a dessert type.

Other Recent Wine Grape Introductions in New York (Einset et al., 1973)

Cayuga White, formerly identified as New York 33403, or G. W. 3, was named in 1972 by the Geneva Station. It produces a fruity, European-type white table wine of very good to excellent quality. The vine is vigorous, very productive, and moderately hardy.

Veeport, from the Horticultural Research Institute of Ontario at Vineland, Canada, makes a good dessert wine. It is productive and moderately vigorous.

Vincent, another Vineland introduction is a dark-blue grape with dark juice, and produces wine that is useful for blending. Its wine ratings have been high.

Vitis vinifera Grapes in New York (Einset et al., 1973)

White Riesling and Pinot Chardonnay are grown commercially on a limited scale in New York. Only the best sites with the least extreme winter temperatures should be used, and the *vinifera* should be grown on resistant rootstocks. Skilled management is required to offset winter injury to buds and trunks and to ensure commercially consistent cropping.

Other varieties recommended for trials are Cabernet Sauvignon, Gewürztraminer, Pinot noir, and White Riesling.

Other Varieties for the Home Vineyard and Roadside Market in New York (Einset et al., 1973)

Several varieties that mature over a season of from 8 to 10 weeks may be grown. Some can be stored under refrigeration for several months. Several varieties are listed below and in Table 6-10.

Bath is a productive black grape with a neutral flavor that ripens about a week before Concord.

Buffalo has the finest dessert quality of the early black grapes. Vines are moderately hardy and the fruit has good storage qualities.

Caco is an old, typically labrusca-type, red variety, reported to be a cross between Concord and Catawba.

Golden Muscat has golden fruit that produces large, juicy berries of high quality.

New York Muscat is a reddish-black grape with a rich muscat flavor.

Table 6-10 Dessert Varieties Recommended for the Home Vineyard and Roadside Market in New York*

Fruit Color	Very Early	Early	Midseason	Late Midseason	Late
White or green	Interlaken Seedless (E),† (S)‡	Ontario(A) Himrod (E), (S) Seneca (E)	Niagara (A) Lakemont (E), (S)		Golden Muscat (A)
Red			Delaware (A) Suffolk Red (E), (S)	Caco (A)	Catawba (A) Urbana (A) Vinered (A) Yates (A)
Reddish-black or black	Schuyler (E) VanBuren (A)§		Fredonia (A) Bath (A) Buffalo (A) N.Y. Muscat (A) Concord Seedless (A), (S)	Concord (A) Steuben (A)	Sheridan (A)

* After Einset et al., 1973.

Key: A§—American type, slipskin; E†—European type in fruit texture, nonslipskin;S‡—Seedless.

The fruit makes a good red muscatel wine, and vines are vigorous and hardy.

Schuyler is a very early black grape that ripens at least three weeks before Concord. Vines are very productive but tender to cold.

Seneca is an early white grape with *vinifera* type fruit of the highest dessert quality. Vines are moderately hardy and susceptible to mildew.

Sheridan is a late black grape that extends the season for those who like Concord-type fruit. The vines are vigorous, productive, and hardy.

Steuben is a bluish-black grape that ripens shortly after Concord. The flavor is sweet with a spicy tang. The vines are moderately vigorous, hardy, and productive. Steuben also makes good wine.

Urbana is a fine, late red variety with good storage quality.

Van Buren is the best early Concord-type grape. It is hardy, but slightly susceptible to downy mildew.

Vinered is a large-clustered, very handsome red grape introduced by Canadian workers. It is a late grape that ripens to maturity only in the most favorable seasons at Geneva.

Yates is a hardy, late red grape with juicy sweet flesh, tough skin, and good storage quality.

Seedless Varieties for New York (Einset et al., 1973)

Seedless grapes that are hardy enough to be recommended for trial in any but the most favored locations in New York State are of recent origin and have been produced by crosses between American-seeded and seedless *V. vinifera* varieties. Although most introductions from Geneva, New York are only moderately winter-hardy, they are much stronger than their seedless parents. Some promising selections are listed below.

Concord Seedless is probably a seedless mutation or sport of Concord. Clusters and berries are smaller than Concord, but the fruit matures earlier, has high flavor, and makes excellent preserves and pies.

Himrod is an early, white seedless variety that makes one of the most delicious dessert grapes. It is a cross between Ontario and Thompson Seedless. The clusters are large, loose, and irregular. This can be controlled by sprays of gibberellic acid, cane girdling, or thinning to increase berry-set and improve berry size.

Interlaken Seedless is also a cross between Ontario and Thompson Seedless. The clusters are medium sized and compact with small, white seedless berries that ripen a month before Concord. Grafted vines are preferred.

Lakemont (named in 1972) is another white grape that is a cross between Ontario and Thompson Seedless. The compact clusters ripen a week or two after Himrod.

Suffolk Red (named in 1972) is a large-berried, red seedless grape. It is moderately winter-hardy, has loose clusters, and responds well to sprays of gibberellic acid.

Grapes for home winemaking. Varieties recommended for the home winemarker in New York are listed in Table 6-11.

New York Rootstocks (Lider and Shaulis, 1974).

At least six species of *Vitis* are important among the American hybrids grown in New York, ranging from the highly phylloxera-resistant *V. riparia* to the nonresistant *V. vinifera* (Lider and Shaulis, 1974). This accounts for the varying resistance in some of the scion varieties.

Table 6-11 Varieties Recommended for the Home Winemaker in New York*

	Season at Geneva		
	Early	Midseason	Late
White wine	Aurore (FH) (Seibel 5279)	Delaware (A) Niagara (A) Cayuga White Seyval (FH) Vignoles (FH) (Ravat 51)	Dutchess (A) Catawba (A) Villard Blanc (FH) (S.V. 12-375) Vidal 256 (FH) Chardonnay (V) White Riesling (V)
Red wine	Maréchal Foch (FH) Léon Millott (FH) Cascade (FH) (Seibel 13053)	Fredonia (A) Buffalo (A) Bath (A) Baco Noir (FH) De Chaunac (FH) (Seibel 9549) Chelois (FH) (Seibel 10878) Rougeon (FH) (Seibel 5898)	Isabella (A) Steuben (A) Rosette (FH) (Seibel 1000) Chancellor (FH) (Seibel 7053) Vincent

* After Einset, et al., 1973.

Key: A—American type.
FH—French hybrid.
V—V. vinifera.

Baco noir, a *V. vinifera* crossed with a *V. riparia* hybrid, is used occasionally as a resistant rootstock in New York vineyards, and recently interest in its use has increased. The phylloxera resistance of Baco noir is high, but it is relatively low in cold hardiness when used either as a scion or a rootstock variety. Baco noir roots are also susceptible to infection by tomato ringspot virus. This variety is not recommended as a rootstock for New York vineyards.

Couderc 3309 is currently recommended for use in New York State. It has been an excellent stock on a wide range of soil types and varieties.

MUSCADINE GRAPES (ANONYMOUS, 1973)

These grapes are best suited for the southern states from the eastern third of Texas to the Atlantic seaboard. The fresh fruit is generally sold locally because the flavor and aroma deteriorate rapidly. It is also used for unfermented juice, pies, jellies, sauces, and wines. The latter have a distinctive flavor and are sold mainly to the specialty trade. The three species of muscadine grapes are *M. rotundifolia, M. munsoniana,* and *M. popenoei,* the important ones are *M. rotundifolia*. Muscadine grapes are insect- and disease-resistant.

Some of the leading varieties are the following:

BURGAW	Perfect flowers. Reddish-black fruit, fair to good in quality.
CREEK	Late maturing, reddish-black berries. Good variety for wine and culinary use.
DEARING	Perfect flowers. A fine-flavored, late green grape of medium size that develops high sugar content.
DULCET	Medium sized black fruit of good quality.
HIGGINS	An enormous white grape that outsells other varieties on the fresh fruit market.
HUNT	Best all-purpose black variety; a favorite for pies, sauces, jellies and jams.
MAGOON	Perfect flowers. A seedling of Thomas, which it resembles.
SCUPPERNONG	Best known and oldest cultivated variety of American grapes. Quality good; flavor is distinctive.

THOMAS Old standard grape with reddish-black berries that make flavorful unfermented juice.

TOPSAIL A large green grape with very high sugar content and pleasing flavor.

YUGA Late-maturing, reddish-bronze, high-quality, attractive fruit.

Chapter 7
PROPAGATION

REPRODUCTION IN GRAPEVINES

There are two types of reproduction in grapevines. One is sexual and involves the production of vines by seeds. The second type is asexual, in which vines are increased by vegetative means from parts of existing plants other than seeds. (Propagation refers to utilization of reproductive processes in the controlled perpetuation of grapes.) The basis for asexual reproduction is the ability of the grapevine to reproduce itself easily from cuttings. Sexual reproduction in grapevines is valuable in the development of new varieties by breeding (see p. 35).

In propagation by cuttings, a portion of the cane is cut from the parent plant, and this plant part is then placed under favorable environmental conditions and induced to form roots and shoots, which results in the production of a new plant that is usually identical to the parent plant (Fig. 7-1). Grapevines are propagated commercially by cuttings, buds, or grafts.

PROPAGATION OF VINES BY CUTTINGS AND ROOTINGS (JACOB, 1950)

Sections of canes are always used. These should be taken from healthy, moderately vigorous vines while the vines are dormant. The sections are cut directly from the vine or from brush that has been pruned from the vine. Brush should be used soon after pruning; otherwise the severed canes may dry enough so that cuttings made from them will not grow. Generally cuttings are ⅓ to ½ in. (8–13 mm) in diameter and 14 to 18 in. (36–46 cm) long (Fig. 7-2). The proper length varies with the type of soil. For example, in sandy soil longer cuttings are desirable so that their bases will be in moist soil. Cuttings of fruiting varieties should usually be at least ¼ in. (6.4 mm) in diameter at the small end. Generally the cut at the base of the cuttings is made perpendicular to

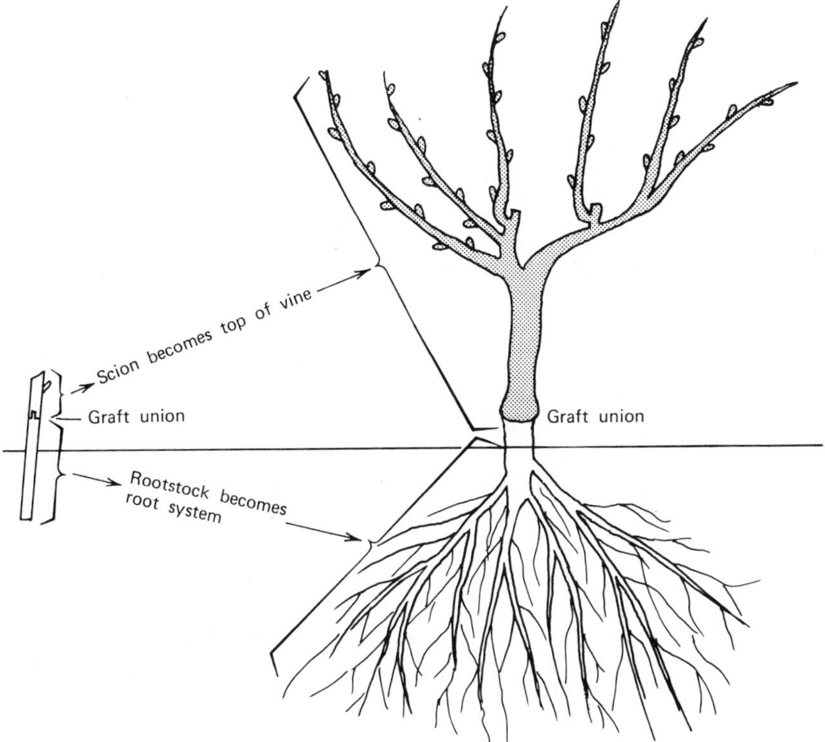

Fig. 7-1. In a grafted vine the shoot system arises from one or more buds from the scion, and the root system develops from the rootstock. The graft union remains where the scion and stock were joined for the entire life of the vine.

the cutting length just below the bud or node. The apical cut is made at an angle of about 45° at a distance of ¾ to 1 in. (19–25 mm) above the apical bud.

Cuttings should be planted as soon as possible after they are made. The nursery row is usually the best storage place. If they cannot be planted immediately, they should be stored in a cool place, preferably in moist sand or in a refrigerator. To facilitate handling and storage, cuttings are tied into bundles of 100 or 200 each (Fig. 7-2).

The nursery soil should be fertile, preferably a sandy loam, with irrigation available. Usually the cuttings are planted to the depth of the second bud from the apex of the cutting, and covered with loose soil (Fig. 7-3). Cuttings are planted in a straight row with the soil firmly settled around them. If they are placed in a trench, the soil must be

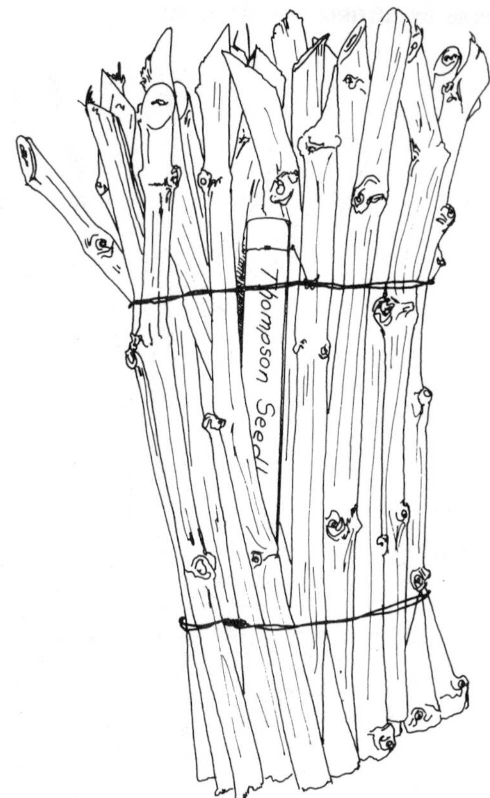

Fig. 7-2. Cuttings are usually tied into bundles of 100 or 200 each for storage.

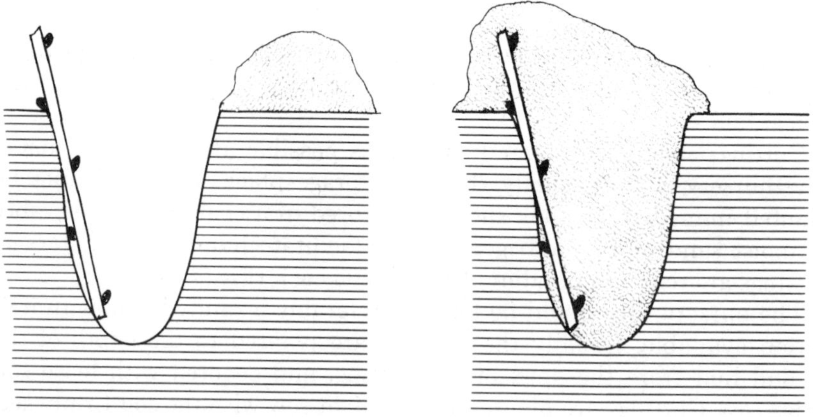

Fig. 7-3. Planting a cutting in the nursery row. View of only one cutting at end of furrow.

packed firmly around the base of the cuttings by tramping it down as the trench is filled or by irrigation. If the cuttings are inserted into the cut made by a chisel or subsoiler, the soil around them must be settled by irrigation immediately after planting. In fertile soil in a hot region the cuttings can be placed as close as 2 in. (5 cm) apart in the row, with the rows 4 ft (1.2 m) apart. In a less fertile soil, or in a cooler region, 3- or 4-in. (7.6–10.1 cm) spacing in the row will produce larger rootings.

To produce well-matured vines before autumn frost occurs, irrigation should be frequent in the spring and early summer, less often in late summer, and not at all in the last 6 or 8 weeks or growth. The ridges of soil over the apical ends of the cuttings should be left until the cuttings have rooted and made appreciable apical growth. Then the ridges should be removed to prevent development of surface roots.

Rootings (cuttings grown for one season in the nursery row) can be dug any time after the leaves fall (Fig. 7-4). They should be sorted according to size into at least two grades, and tied into bundles of 25 or 50 each for convenience in handling. Until used, they should be stored by heeling-in in moist sand or soil in a cool location. When possible, the rootings should be moved directly from the nursery to the vineyard for planting. Rootings with less than 6 in. (15.2 cm) of well-matured apical growth and at least one root ⅛ inch (3.3 mm) in diameter growing from the basal node of the cutting should not be used.

GREENHOUSE PROPAGATION BY GREEN CUTTINGS

The cheapest and most rapid technique to propagate vines is probably greenwood propagation in the greenhouse. The cuttings are grown under intermittent mist. This technique, along with the greenhouse techniques for growing bench grafts, has revolutionized the propagation industries in California and in Europe.

PROPAGATION USING ROOTSTOCKS

About 20 percent of the grape acreage in California is planted on rootstocks (Lider, 1963). This is necessary to combat root-infesting pests, principally the grape phylloxera (*Phylloxera vitifoliae*) and several strains and species of plant-parasitic nematodes of which the root-knot nematodes (*Meloidogyne* spp.) are the most serious. As old, low-producing vineyards are being replanted, the need increases for stocks resistant to soil pests common to the old plantings. This requires

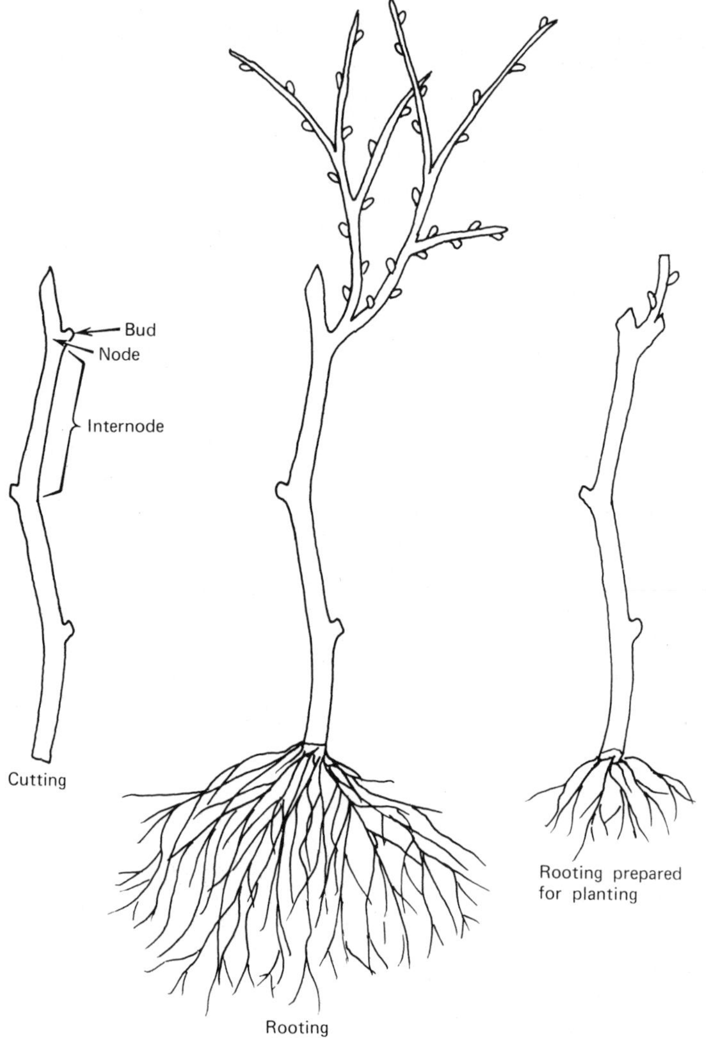

Fig. 7-4. Diagrams of a cutting, a cutting after one growing season in the nursery (rooting), and the rooting prepared for planting.

good techniques by which the scion varieties can be grafted onto resistant rootstocks. The main methods for establishing vineyards propagated on rootstocks are field budding (chip budding), bench grafting, and field grafting in the spring on rootstocks growing in the vineyard. For spring grafting, whip, cleft, notch and bark grafting can be used.

FIELD BUDDING (CHIP BUDDING) (LIDER, 1963; HARMON AND WEINBERGER, 1962)

Rootstocks

Strong, well-developed rootings of the proper rootstock variety for the soil type and root pests present are required. Rootstock cultivars may produce smaller canes than those of the fruiting cultivars. Cuttings of large to medium diameter should be used. Canes should be made into cuttings 16 to 18 in. (40.6 to 45.7 cm) long, so the rooted stock can be planted in the vineyard having 3 or 4 in. (7.6 or 10.1 cm) of the trunk aboveground. The inserted chip bud must be placed above ground level, and above the ridge when it is used in the vine row (Alley, 1975).

Rootstocks should be disbudded before planting. This is done most easily by disbudding before rooting in the nursery. All buds except the apical bud of the cutting should be removed. Such disbudding will prevent future growth of rootstock suckers, but does not reduce rooting capability of the rootstock.

Strong rootstock rootings should be used for vineyard planting. Direct planting of cuttings is not recommended, as rootstock cuttings root more poorly than most fruiting varieties and poor stands may be obtained. The stock rooting should be planted so that the apical 3 to 4 in. (7.6–10.1 cm) is above the permanent soil surface. This places the inserted bud slightly above ground level to discourage development of scion roots at or above the graft union. The entire plant should be covered to a depth of 3 or 4 in. (7.6 or 10.1 cm) with a mound of loose moist soil, which will prevent drying or sunburning of the trunk in the graft area in hot weather.

Stakes should be set in the vineyard before planting the rootstocks to permit proper placement of the stock by the stake and avoid possible damage if stakes are driven in later. Trellising, if desired, can be done at any time between budding and the following spring. Do not irrigate after budding; allow stocks to enter dormancy normally so their wood will be well matured.

Buds

Bud sticks (sections of canes about 20 to 24 in. (51–61 cm) long) are needed to supply mature buds for grafting. These are cut from the current season's growing shoots. The mature buds and wood are brown. The mother vine of the variety you wish to propagate should be free of virus diseases. Since grape mutations (bud sports) can occur, cut the sticks from vines bearing normal fruit and having foliage characteristic of the desired *vinifera* variety. Cut fresh bud sticks as needed, and protect them from drying by immersing them in water, wrapping them in damp cloth or burlap, or placing them in plastic bags. If bud sticks cannot be used immediately, they can be wrapped in damp cloth, placed in a polyethylene bag, and stored at 32°–34°F (0° or 1°C) several months until needed. The best method is to obtain virusfree stock and budwood.

When to Field Bud. Chip bud during late summer or early fall when the vines are actively growing. Budding in early summer is recommended to allow a long time for the bud to "callus in."

Budding in early spring at the start of the second growing season is as successful as early fall budding. Earlier chip-bud failures may be rebudded in the spring. Dormant buds may be collected in winter and kept at 32°–34°F (0°–1°C) until needed. Late spring budding is less desirable, since the delay in budding markedly reduces the top growth during the first season.

How to Field Bud. Select a smooth location on the stock 1 to 2 in. (2.5–51 mm) above soil level for insertion of the scion bud. Starting at this spot, make an oblique downward cut into the stock, then make a second cut ¾ to 1 in. (19–25 mm) above the first cut at an acute angle that meets the first cut at its lower end (Fig. 7-5). Remove the chip that is formed.

Repeat this procedure on the bud stick. Start the bottom cut just below a bud. Start the second cut above the bud and extend it downward behind the bud. The chip containing the bud and the cut in the rootstock should be as near the same size as possible.

Push the chip bearing the bud into the notch on the rootstock, matching the cambium layers (the line between bark and wood) as closely as possible. Hold the bud in place by wrapping it with a rubber budding strip. Start wrapping above the bud and proceed downward. This forces the bud chip firmly into the notch on the stock. Cover the budded part of the stock with 6 to 8 in. (15.2–20.3 cm) of moist soil, so that the bud is protected from drying while it is uniting with the stock.

Plastic tape may be used for field budding instead of using a rubber budding strip, according to C. J. Alley (1975). One-half inch (13 mm) white plastic budding tape is wrapped over the cut surfaces of the bud shield, but the bud itself should remain exposed so that it can grow (Fig. 7-6). Buds are placed about 3 in. (7.62 cm) above ground level facing the stake. After budding the vines are not mounded over with soil.

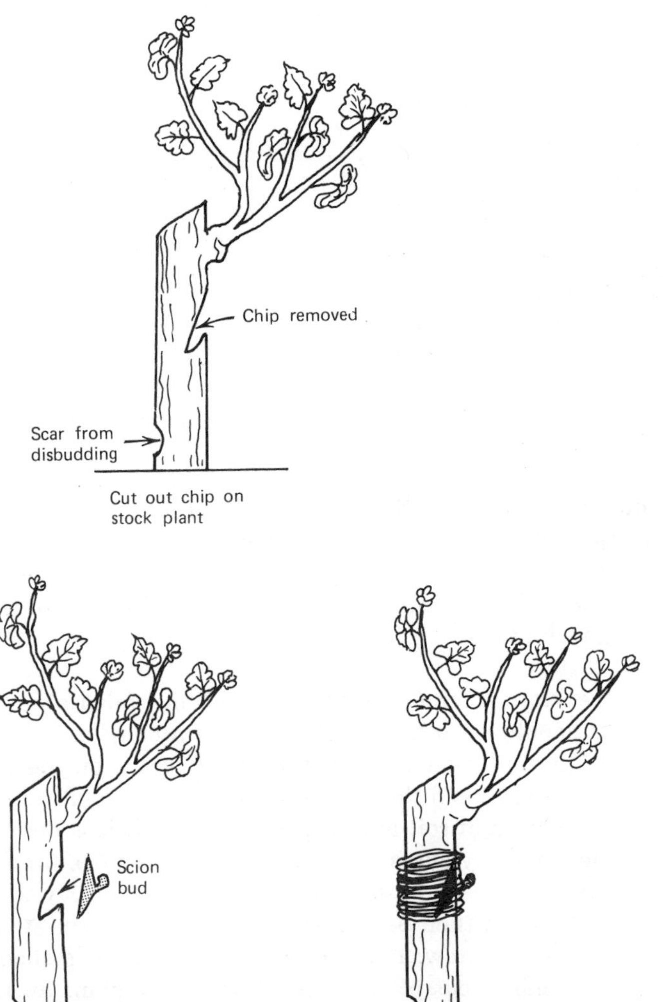

Fig. 7-5. Steps in the chip budding of grapevines.

The tape will not girdle the vine in the early part of the year since it tends to stretch with the growth of the stock.

There are several advantages in using tape for field budding. It is unnecessary to dig a basin around the rootstock or to cover the bud with soil after budding. There is no mound of soil hardened by winter rains

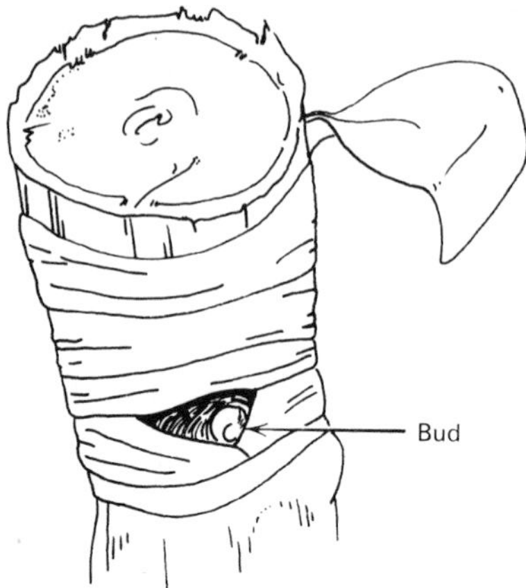

Fig. 7-6. White plastic tape can be used for tying chip buds (After Alley, 1975.)

that must be removed in the spring, the bud location is clearly marked by the tape when the top of the rootstock is cut off, and irrigation after budding is not detrimental to callusing of the bud. Also, bud growing is not delayed with the onset of warm weather as may occur with the mounding technique.

Care of the Vines after Budding. The soil mounded around fall-budded stocks should be left throughout the winter (Fig. 7-7). Chip buds inserted in the fall soon will unite with the stock, but do not begin to grow until spring. When the rootstock vine begins to grow in the spring, remove the mound of soil and inspect the bud. If it is dead, rebud the stock immediately by cutting a new notch on the stock for the new bud chip. The bud will usually start growing within 2 to 3 weeks. The soil around the bud can then be removed, and the top of the stock cut back to force the bud to grow. If the bud is alive and has a shoot ½ in. (13 mm) long or more, cut off the top of the stock plant an inch or two above the chip but at an angle on the opposite side from the bud taking care not to injure the bud or chip bud. Carefully trim off any scion roots growing from the scion bud and cut the budding strip below the bud. Place a collar or milk carton over the stock and bud for protection, and partially cover the outside with soil to hold it in place.

After cutting back the tops from fall- or spring-budded stocks, cover the stump again with 1 to 2 in. (25–51 mm) of loose soil. If the soil is heavy and forms a solid coat from spring rain, coarse sand can be used to cover the buds. This prevents the hard layer from forming over the buds and permits development of a straight shoot.

One to three shoots may develop from the *vinifera* bud. When they are about 8 to 10 in. (20.3–25.4 cm) long, remove all but one vigorous shoot. Care must be taken to prevent breaking the shoot you wish to save. The retained shoot is then trained to form the permanent trunk of the vine. Inspect the bud unions each year and remove any scion roots that develop. If they are not removed scion roots are likely to dominate the rootstock, and the vine will be susceptible to damage from grape phylloxera and neomatodes.

HOW TO GRAFT TO CHANGE VARIETIES

There are several methods in addition to field budding, to graft fruiting varieties onto rootstocks, and the type used usually depends on the size of the stock. Various grafting tools are shown in Figure 7-8.

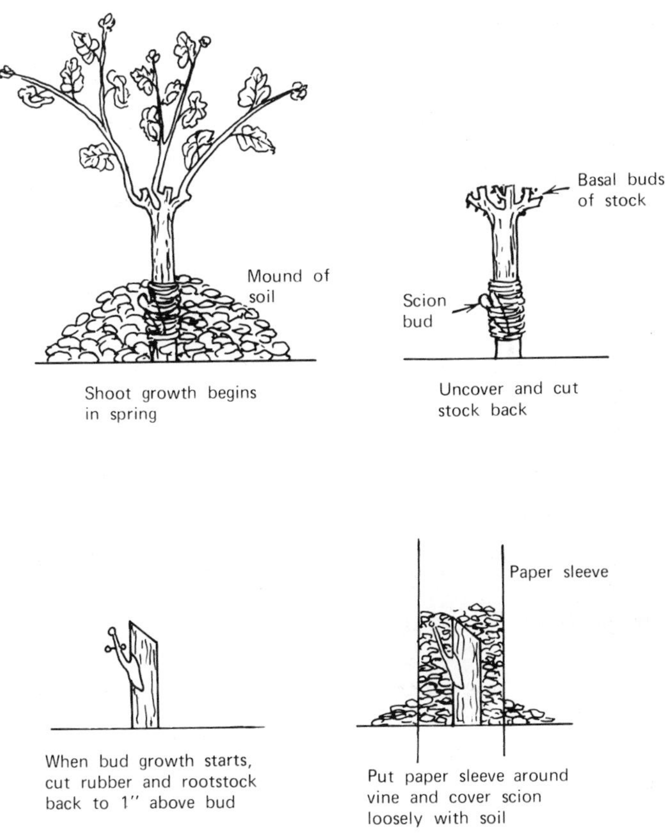

Fig. 7-7. Steps recommended for the care of stocks after field budding.

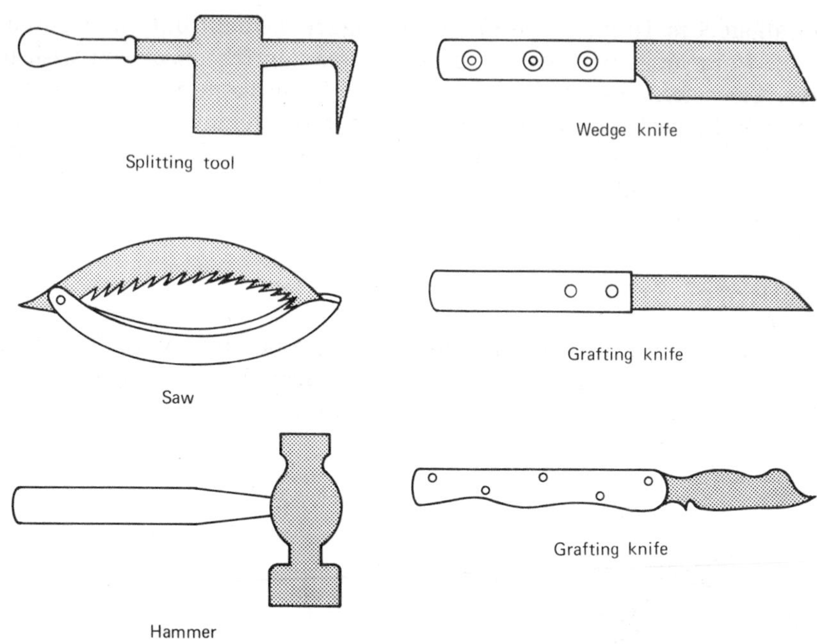

Fig. 7-8. Various tools used in grafting.

Whip Grafting

The whip graft gives best results when the stock is less than ⅝ in. in diameter (Jacob, 1950; Hartmann and Kester, 1975). Cuts are made according to Figure 7-9. Sloping cuts (15–25°) of the same length are made on the stock and scion, after which another, shorter incision is made beginning about one-third of the distance from the point to form the tongue. Opening out the tongues by bending them over with the knife as it is withdrawn aids in fitting the scion on the stock, so that the line between the bark and the wood (the cambium or region of growth) of the scion coincides with the corresponding line on the stock as completely as possible. The graft is then tied firmly with budding rubber, raffia, or string. If rubber is used, it must be cut and removed after the graft has grown together firmly before it girdles the graft union.

Cleft Grafting

Vines with trunks over 1 in. (2.54 cm) in diameter are suitable for cleft grafting (Jacob, 1950). The vine is sawed off so that about 2 in. (5.08

cm) of smooth, straight grain is left at the top of the stump (Fig. 7-9). If one saws at or near a place where the grain of the wood is crooked, obtaining a good fit is difficult.

Vines of fruiting varieties grafted to change the variety may be sawed off anywhere around ground level. When resistant stocks are grafted to a fruiting variety the graft must be above ground level. If it is placed below the ground, scion roots will form and the resistant stock may die.

After the trunk is cut off, the stump is split to a depth of 1 or 1½ in. (2.54 or 3.81 cm) with a grafting tool. After splitting, the small end of the grafting tool is placed in the cleft to hold it apart so that the scion can be inserted. The wedge-shaped scion usually with a long taper, is

Fig. 7-9. Diagrams of whip, cleft, notch, and bark grafts.

inserted so that the cambium of the scion coincides as much as possible with the cambium of the stock. For this to occur, the outer surface of the scion must be set in slightly from the outer surface of the stock.

The scion should be cut with a sharp knife and inserted in the stock before it dries even on the surface. Generally scions of two buds are used. One scion is used if the trunk is an inch or less in diameter. When two scions are used, the weaker developing shoot is removed as soon as the other is tied to the stake. Raffia or cotton string is wrapped around the top of the stock to hold the scions in place. Trunks over 1½ in. (3.81 cm) in diameter usually require tying. If a rubber strip is used for tying, it should be removed after the scion and stock have grown together.

Side Whip Graft

Recently the side whip graft has been introduced as an alternative method to the split graft for use on stocks ¾ to 1½ in. (19–38 mm) in diameter (Alley, 1975). Stocks of this size are too large to chip bud and whip graft completely. The side whip graft has a greater cambium-to-cambium contact of the cut surfaces than does split grafting (Fig. 7-10).

Notch Grafting

With notch grafting, scions are shaped to fit into a V-shaped notch on the side of the stock extending from the top of the stump downward for 1 to 1½ in. (2.54–3.81). The width and depth at the top of the notch depends on the diameter of the scion. First, make a saw cut as long and as deep as the desired notch. Then widen the notch at the top and taper it to a point at the lower end (Fig. 7-9). The cut surface should be smooth and straight to obtain a good fit with the scion. The angle formed at the bottom of the notch by the cut surfaces should be from 70° to 90°.

The scion must make a good fit in the notch so that the cambiums of the stock and scion coincide as much as possible. The angle that the cuts of the scion make should be more obtuse than the angle of the notch, to ensure good contact when the scion is placed on the stock. Small flat-headed nails can be used to hold the scion firmly in place.

Wedge Graft

The wedge graft is an improvement over the notch graft as it is easier and more rapid to make (Alley, 1975). It requires less skill than the

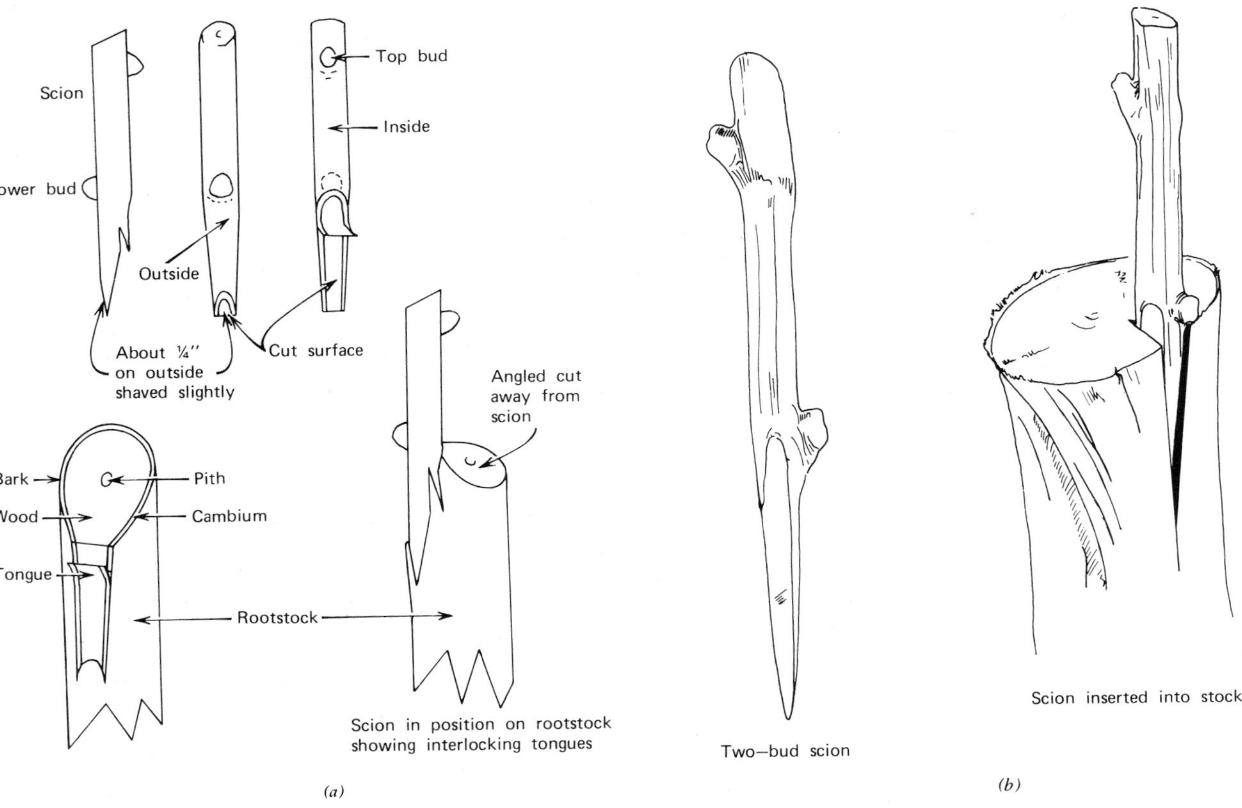

Fig. 7-10. Diagram of side whip graft (*a*) and wedge graft (*b*) (After Alley, 1975.)

notch graft and can be performed at a high level on the trunk. A diagram of the technique for wedge grafting is shown in Figure 7-10. The wedge graft can be made earlier in the spring than the bark graft because the latter type cannot be made until the bark "slips," around the end of April.

Bark Grafting

Bark grafting is performed in the spring when the bark on the trunks of older vines slips (is easy to peel off) (Jensen et al., 1970). This doesn't occur until late April in the San Joaquin Valley. One must wait until the bark slips in all locations on the trunk before grafting. The optimum date for grafting varies from year to year and with the variety of grape.

After the top of the vine is cut off, the dead, loose outer bark is removed from the grafting site (Fig. 7-9). Prepare the scion by making a straight, slanting cut about 2 in. (5.08 cm) long, usually on the side across from the bud, and a cut on the opposite side about one-third as long. Hold the long cut of the scion against the trunk and slit the bark on both sides as wide as the diameter of the scion and two-thirds as far down the trunk as the scion length. This bark flap is then peeled away from the trunk at the top, but remains attached at the bottom. Thus approximately the top third of the flap is cut off, usually while it is still on the trunk before peeling back. Insert the scion against the trunk and underneath the flap. Use two flat-headed wire brads or a staple gun to secure the scion, one through the bark flap and one through the scion near the top of the cutoff trunk. The graft is then covered with white water-soluble latex paint.

This method has not been as successful as cleft or notch grafting. The healing is somewhat inferior, since the growth of the scion is on the outside of the trunk. Also, the cutoff area of the trunk does not readily cover or heal.

Protecting the Graft

A stake should be placed by a grafted vine soon after planting. The graft should then be covered immediately with a mound of moist, well-pulverized soil. Care should be taken not to disturb the scions. In hot dry weather, scions should be covered to a depth of 2 or 3 in. (5.08 to 7.62 cm) to prevent drying. A wide mound of soil should be used since narrow mounds may not remain moist enough for good growth of the shoots. The mounds must not be disturbed by cultivation until the unions are formed. If rains form a hard crust over the mound, the crust should be carefully broken.

Suckering Rootstocks

Sucker shoots may develop from improperly disbudded stocks. When the grafts have started to grow and scion shoots can be tied to the stake, suckering should be started. At first the tender shoots may be pulled by hand without removing any soil, provided that they are not entangled with the scion. If grafts are slow in starting and if the suckers are vigorous, one must sucker before the scion has grown much, and extreme care must be taken.

Training of scion shoots

The growth of shoots on large vines is generally rapid and vigorous, and management should be the same as in training vigorous, ungrafted vines.

HIGH-LEVEL GRAFTING

High-level notch or bark grafts can be made on *Vitis vinifera* grapevines 2 to 3 ft (0.6 to 0.91 m) above ground (Alley, 1964). This makes it unnecessary to grow a new trunk from near the soil surface. The best time for notch grafting in Region V is about a month after bud burst prior to bark slippage (Jensen, 1971), although earlier grafting is usually more successful. The scions should be painted to reduce moisture loss. Black asphalt-water emulsion grafting compound followed by an application of white, interior water-soluble latex paints are effective. Two scions on opposite sides of the trunk should be used; if only one scion is used, only one side of the trunk (below the scion) will remain alive. Bark grafting is sometimes less successful than notch or cleft grafting.

GREEN GRAFTING (CARLSON, 1972; HARMON, 1954)

In green grafting new, succulent green current-season growth is used for scions that are grafted onto the new growth of the stock (Fig. 7-11). Green grafting may be used during late spring and early summer to establish *vinifera* grapes on resistant roots. It can be done early as shoots are large enough and firm enough to handle. Select shoots that have a white pith, preferably ⅜ in. (9.65 mm) in diameter or larger. The shoots should be the same size as the stock shoot. Early grafting will result in more growth than later grafting.

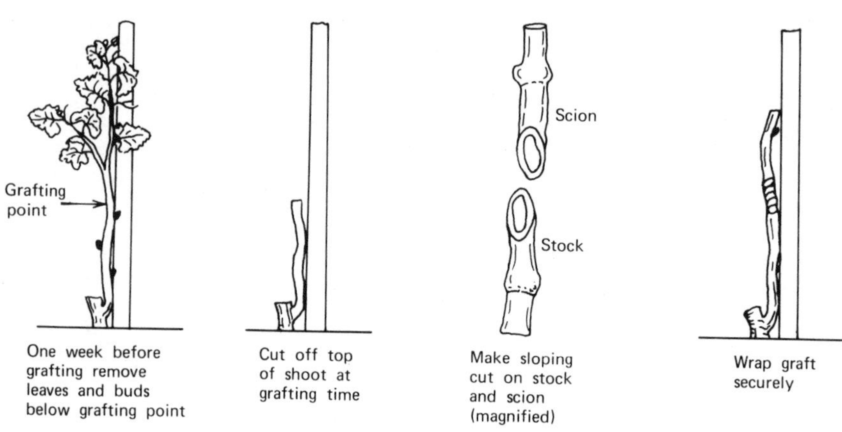

Fig. 7-11. Techniques used for green grafting of vines.

One-bud scions are cut from new shoots. The leaves should be removed from scions, taking care not to injure the bud in the leaf axil. Scions must be kept moist in damp packing material.

One week before grafting, cut off all leaves and buds on the rootstock below the place to be grafted. The cuts will bleed but will heal before grafting time. Irrigate the soil before grafting. At grafting time cut off the tops of the stocks, then make sloping cuts 1 to 1½ in. (3.81 cm) long on both stock and scions of the same size. Cuts are matched to get the best possible contact between cambium layers of scion and stock. Wrap the union with polyethylene tape. After the first wrap, check to make sure the union is still aligned, then complete the wrapping. The graft must be securely wrapped and completely covered for a successful union. Often a drop of water will appear at top of the scion within about 30 minutes. Buds usually grow within 2 weeks, and shoot growth must be tied to a stake so that the graft will not break. After grafting, remove suckers that develop.

Another technique is to collect bud sticks of the scion in the dormant season and graft one-bud scions on green shoots in the spring. This method has been used to some extent in the Republic of South Africa.

BENCH GRAFTING (Jacob, 1950)

This grafting is done indoors and was, for many years, the most common method of producing grafted vines in which the desired fruiting variety is placed on a rootstock resistant to phylloxera or nematodes, or both.

Collection and Storage of Stocks and Scions (Jacob, 1936)

The wood for scion and stock cuttings should be selected in the same manner as for cuttings to be planted in the nursery (see p. 104). The canes for scions and for stock cuttings are usually made up in lengths of 30 in. (0.76 m) or more. At grafting time, the long pieces are cut into lengths of about 11 in. (2.79 cm) for the stocks and single-bud pieces for the scions.

Rootstock and scion material for bench grafting may be stored in sand, sawdust, or plastic bags as are cuttings for nursery planting. (If stored in sand, they must be washed free before use to insure against damage to the grafting knives.) Cuttings can be dipped into fungicides such as Benlate (benomyl) or chinosol to prevent development of molds and rots.

Preparation of Stocks and Scions. Just before grafting, the stock and scion materials are removed from storage and washed. If they are dry they should be soaked in water for 24 hr. Cut the stocks into segments about 11 in. (2.79 cm) long, making the basal cut just below the lowest node, and slanting the upper cut about 1 in. (2.54 cm) or more above a node. Remove buds on the stocks with a knife or small pruning shears to prevent the growth of stock suckers in the nursery and the vineyard.

The scions are cut to a single bud with not over ½ in. (1.27 cm) of internode above and from 1½ to 2 in. (3.81–5.08 cm) below the bud. Stocks and scions should be about the same diameter for best results.

Whip Grafting. Grafting can be done by hand or by machine. For hand grafting, short whip graft (Fig. 7-9) is the preferred type. There are several types of machines available: one has saw blades separated by spacer discs that control the depth and width of the cuts. The stocks and scions are pushed endwise (radially) against the blades (Fig. 7-12). The table or guide is so constructed and adjusted that when the stock is cut against the right-hand guide and the scion agaist the left-hand guide, or vice versa, a good fit between the two is obtained, provided they are of the same size and the end of each has been cut square across before the slots are cut on the machine.

With the machine, three men can make between 700 and 1000 grafts an hour. Good, straight hardwood, carefully graded according to size, is necessary for the most rapid work.

Bench-Grafting Rootings. If 1-year-old rootings are to be bench-grafted, they must first be washed, the length of the trunk shortened to

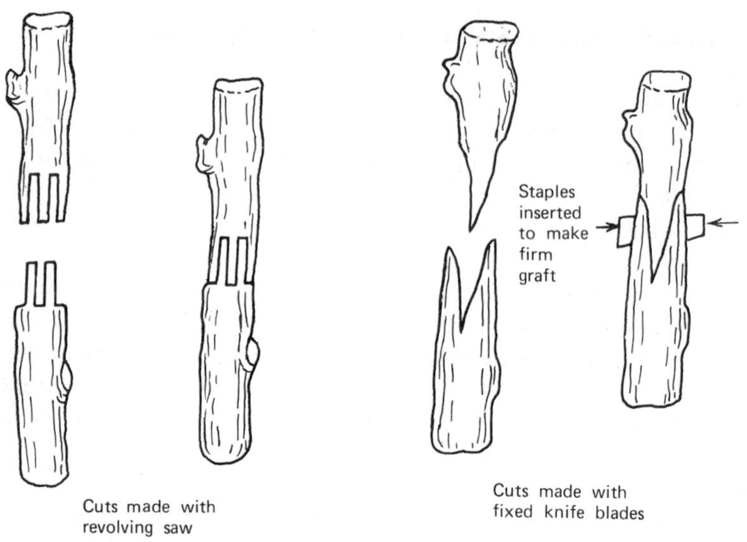

Fig. 7-12. Two types of machine grafting used for bench grafts.

about 12 in. (30.4 cm), and completely disbudded. The roots are usually shortened to a length of 3 to 4 in. (7.62–10.2 cm) for convenience in handling and planting. One-bud scions can be grafted on them by the long whip or other suitable method. They are usually callused in sand and replanted in the nursery for 1 year before being placed in the vineyard.

Callusing. When bench grafts are made, the stocks and scions are completely dormant. If the grafts are planted out-of-doors directly, many will dry out or be injured by cold or excess moisture before they can begin active growth. Therefore, in countries with severe climates, such as France and Germany, newly made grafts should always be placed where moisture, temperature, and aeration can be controlled until the unions are made.

Sand Callusing. The long whip or machine grafts are usually callused in moist sand or sawdust. First, they usually are tied in bundles of 10 for conveniece in handling. The sand must be sufficiently moist to support plant growth, yet not too wet to work easily between the bundles of grafts.

The callusing bed may be in the open shed or out-of-doors on the south side of a building. A callusing bed in the open must have a glass, canvas, board, or other type of cover so that it can be protected during rainy or cold weather and opened during dry warm weather. The bundles of grafts can be placed in the sand in rows in a vertical position, scions up, and sand worked in between the bundles. A layer of sand 2 or 3 in. (5.08 or 7.62 cm) deep is first deposited on the bottom of the bed, then sand is added until the tops of the grafts are covered with 3 or 4 in. (7.62 or 10.2 cm). All graft unions should be at the same level. The sand must be slightly moistened frequently to wet the portion that has dried out. The bed should be covered during rainy weather and on cold nights.

Another method of sand callusing, is to deposit a single layer of bundles of grafts horizontally on sand. The grafts are covered with another layer of sand, and then another layer or grafts is added. This is followed by alternate layers of sand and grafts until the stratified pile reaches a depth of 3 or 4 ft (0.61–1.2 m). However, the vertical placing of the grafts is easier than the horizontal method.

Hot-Room Callusing. In hot-room callusing grafts are packed into boxes, and peat moss, coarse sawdust, or wood shavings can be used as a packing medium. After the grafts are packed in the box, they are stored at 75°–85°F (24°–29°C). At this temperature the grafts rapidly callus and join at the union. Hot-room callusing is not used commercially in California except for the greenhouse grafting technique (p. 125).

Planting Grafts

The planting of callused bench grafts requires more care than nursery cuttings. The soil should be warm, moist, and pulverized, and weather should be mild.

All suckers from the stocks and all roots from the scions must be removed. Scion shoots and stock roots should be cut back to 1 in. in length; if they are longer it will be difficult to save them. The grafts must be kept covered with wet sacks or other suitable material between the time they are removed from the sand or sawdust and the time they are planted.

Grafts are usually planted in trenches so that the unions are just above the soil level, and the scions are immediately covered to a depth of 2 or 3 in. (5.08 or 7.62 cm) with a wide ridge of soil. A planting distance of 3 in. (7.62 cm) in the row is suitable in a fertile sandy loam or loam soil that can be well irrigated. In a poorer soil, or where irrigation water may be scanty or lacking late in the summer, a spacing of 4 or 5 in. (10.2 or 12.7 cm) in the row will produce larger vines. Although the rows may be spaced as close as 3 ft (7.62 cm) apart, 4 ft (10.2 cm) makes planting, cultivation, irrigation, and digging easier.

A planting board is necessary for planting in a trench. A 1 in. × 4 in. (2.54 × 10.2 cm) board, 16 ft (4.88 m) long, marked with a saw cut or shallow notch every 3 or 4 in. (7.62–10.2 cm) according to the spacing desired in the row may be used. An 18-in. (0.46 m) crosspiece through which a small iron bar or a heavy spike can be inserted holds the board in place. The board is laid on the edge of a straight trench made to a depth of at least the length from the union to the base of the stock. Each graft is placed opposite a saw cut (or notch) with the union just above the guide. The trench is half or two-thirds filled with soil packed around the stocks by tramping. It is then filled with loose soil, followed by covering the scions with 2 or 3 in. (5.08 or 7.62 cm) of fine soil. The ridge of soil should be broad and regular as a narrow ridge dries out too quickly. Irrigation should follow within a day or two but should not wet the tops of the ridges.

The ridges may dry out enough to injure the scions, and this may require one or more additional irrigations before the shoots come through the soil.

Removing Scion Roots and Stock Suckers. As soon as shoots begin to develop, roots that have started from the scions should not be removed. Since these roots keep the scion alive and allow the union to be completed, they should not be removed too soon. The scion roots, however, use carbohydrates made by the scion leaves, and this will weaken the stock roots if the scion roots are removed too late. July is often a good time for their removal. When removing the scion roots, the rubber tying material should be cut and removed on all grafts so that growth of the union will not be impeded.

Stocks properly disbudded before grafting will produce very few suckers, but those that do develop must be extracted as soon as they appear above ground. When the scion roots are removed, any suckers present should also be carefully destroyed. For the latter operation it is necessary to remove the soil from around the grafts to below the unions. After these procedures the soil should be replaced immediately to the original level, and an irrigation made. About one month later any new scion roots or stock suckers should be removed. Usually after the middle of September the soil does not need to be replaced around the scions.

RAPID METHOD FOR GRAFTING (GREENHOUSE-FORCED BENCH GRAFTS)

In recent years large plantings have resulted in a heavy demand for bench-grafted vines. Workers in Germany, France, the United States, and elsewhere have been developing rapid methods for producing bench grafts. The objective is to grow bench grafts in the greenhouse that have shoots and a root system and can be planted directly into the vineyard under favorable climatic conditions.

The steps involved in a method developed in California are discussed below (Weinberger and Loomis, 1972).

Grafting. Graft a scion on a rootstock using a long whip or other type of graft. Carry out the procedure discussed under bench grafting. One technique is to use a machine that makes a V-shaped cut with two knives (Fig. 7-12). The scion is cut to a V point by inserting it on one side of the machine, and the corresponding V notch is cut into the rootstock by inserting it onto the opposite side. The two pieces fit perfectly when their diameters are equal. Stapling the union is more rapid than tying. Other types of grafts that are widely used are the omega cut and the short whip graft.

Callusing the Graft. This should be done at 75–85°F (24°–29°C) and near 100 percent humidity. Callus is the white healing tissue that is necessary to form a living contact between scion and stock. Grafts can be packed for callusing in moist peat moss, vermiculite, perlite, or pine wood chips, materials that soak up water rapidly and drain well. Callusing can also be done without packing material if the humidity is almost 100 percent. Two to three weeks suffice for the union to be well callused. Ample oxygen is needed to keep grafts alive and for callusing, so the boxes must not be airtight. Since packing material must drain well, shallow boxes with good ventilation given excellent results. Callused grafts can be stored for callusing in moist peat moss or pine sawdust (not redwood sawdust), polyethylene bags, or boxes lined with polyethylene to prevent moisture loss.

Waxing. To prevent moisture loss while the union knits, the scion and union are dipped in a low melting-point paraffin at a temperature just above the melting point. After cooling, the dip can be repeated if necessary. If the wax is too hot it may injure the bud.

Planting in Tubes. After callusing, the buds are ready to grow and require a growing environment. Almost any good potting mixture or sandy loam soil is adequate for root development. The planting tube can be made out of 15-lb roofing paper cut in rectangles 6 × 7 in. (15.2 × 17.8 cm). The paper is curled into a tube 6 in. (15.2 cm) high and 2 in. (5.08 cm) in diameter, and stapled. Various other kinds of tubes made of milk carton paper or plastic impregnated papers are available commercially (Fig. 7-13). Although a 6-in. (15.2 cm) high tube is deep enough for most grafts, a 7-in (17.8 cm) tube may be preferred for very long grafts. In cross section a 2 in. × 2-in. (5.08 × 5.08 cm) tube is

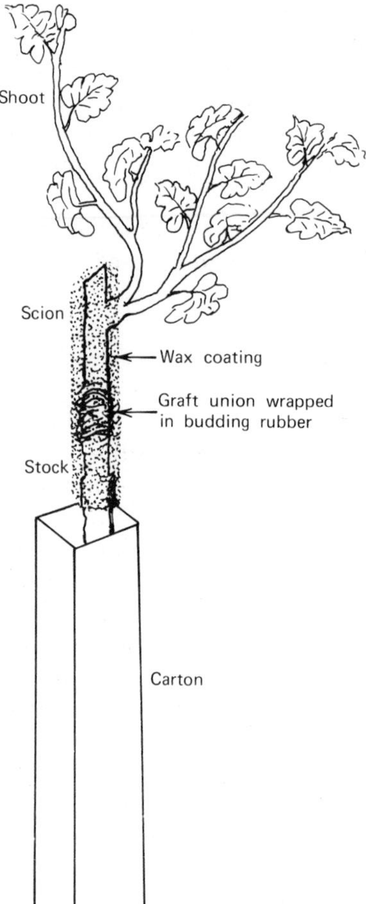

Fig. 7-13. Carton containing a rooted *vinifera* grape ready for planting in the vineyard.

large enough for root development. Square tubes of heavy paper are available commercially in several different widths and lengths.

The grafts should be planted with the basal ends 2 or 3 in. (5.08 or 7.62 cm) from the bottom of the tube to allow room for root development. Roots already formed on the grafts should be cut back to about 1 inch in length for convenience in planting in the tube. Roots of easy-to-root varieties can be completely removed at planting time.

Growing the Grafts. The tender plants must be protected from the sun and have proper humidity for 2 or 3 weeks. About 30 percent of normal sunlight will result in good growth. A greenhouse or plastic house with a 70 percent shade cloth makes an excellent growing environment. The temperature may range from 70–100°F, (21°–38°C) but humidity must be 70 percent or more to keep the plants from drying out. Evaporative cooling or mist nozzles may be utilized.

With easily rooted rootstocks, bud and root growth begin almost simultaneously, but with slow-rooting rootstocks bud growth may begin weeks ahead of root growth. After root development begins, shoot growth accelerates rapidly. Humidity control is essential to keep the graft alive until active roots develop.

When roots begin to develop, the plants must be hardened off before planting in the vineyard. Light exposure should be increased, humidity decreased, and heating stopped. Before planting in the vineyard, they should be exposed to full sunlight and wind. Grafts with foliage growing to top and strong roots emerging from the bottom of the tube are then ready for transplanting to the vineyard (Fig. 7-13).

Transplanting to the Vineyard. The entire tube is buried in the soil and water is applied over the top of the tube. Roots emerge through the bottom of the tube. Full contact with the soil can be obtained by sliding the tube up 2 or 3 in. (5.08 or 7.62 cm) after the graft is set in the ground. Another technique is to use perforated tubes by which roots can grow through the holes. Irrigations should be frequent enough to avoid plant water stress.

Planting in the vineyard can be done any time after danger of frost is past. Early planting is very desirable since transplanting is less of a shock in cooler weather; it also provides a longer growing season and results in larger vines the first year. Transplanting in hot summer weather usually results in poor growth. If graft should fail after transplanting, it should be replaced by another.

Timing and Advantages of Greenhouse Grafting. The time from grafting to planting usually is about 2 to 2½ months. If planting is planned for May 1, grafts should be started to callus in late February, callused for 2 to 3 weeks, grown under controlled environmental conditions for 4 to 6 weeks, and then hardened off for at least 1 week.

Advantages of this method are stapling of the unions instead of tying, and waxing of the graft unions to permit planting with the union exposed aboveground to prevent the production of scion roots. The use of planting tubes also ensures a high percentage of successful grafts under favorable controlled conditions, and transplanting is thus facilitated. The controlled conditions under which the grafts are grown after callusing are the principal reasons for better survival of grafts compared with the usual nursery practices.

LAYERING

This technique is used mainly to replace missing vines in the vineyard on their own roots and to propagate varieties that can be rooted only with great difficulty (Jacob, 1936).

Replacing Missing Vines in a Vineyard

Rootings cannot be used because they cannot compete with the surrounding vines and make poor growth. The best technique for replacing a missing vine is to use a layer. A long, vigorous cane from a vine adjacent to a missing vine in the row is bent down into a hole or trench about 10 in. (25.4 cm) deep (Fig. 7-14). The apical end of the cane projecting after the hole is filled is in the correct location to replace the missing vine. The apex is cut back so that one or two buds are above ground. A wire is often placed snuggly around the cane between the parent vine and the buried portion so that the cane is partially girdled as the new vine grows. This enhances the growth of the new vine.

All growth should be removed on the layered cane, except for a single shoot to form the new trunk. New shoot growth on the portion of the cane between the ground and the parent vine should be removed. After three years, the new vine can develop by itself and can then be cut from the parent vine.

Layering is performed during the dormant season.

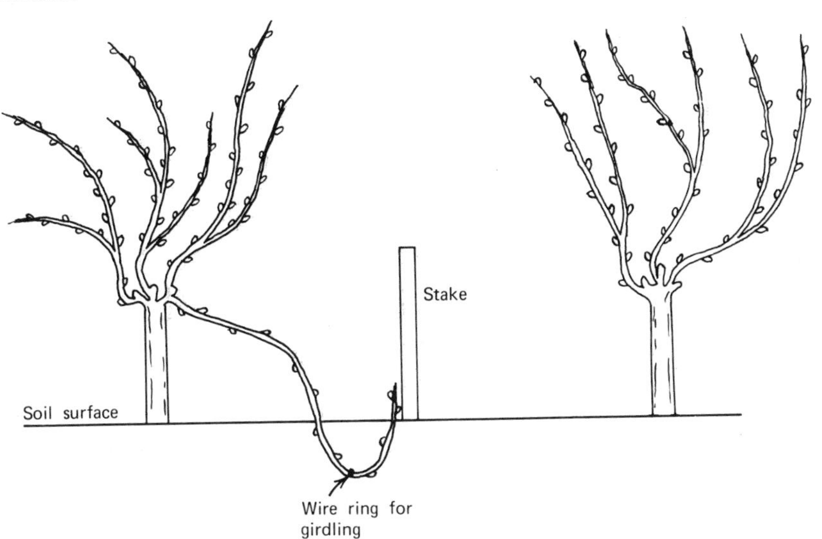

Fig. 7-14. Diagram showing technique of replacing a missing vine by layering.

Propagation of Difficult-To-Root Varieties

Some varieties, such as the rotundifolia group, will not root or will root only poorly from cuttings. Simple layers as described above, trench layers, or mound layers can be used. For trench layering, ditches 8 to 10 in. (20.3–25.4 cm) deep that radiate from the parent vine are dug in early spring. A cane is placed in each trench, held in place by pegs, and covered with 2 or 3 in. (5.08 or 7.62 cm) of soil. As shoots arise from the buried cane, soil is added until the ditch is full. The cane roots at each node, and the following winter rooted plants are obtained by digging up the cane and cutting it into sections each containing a shoot.

Mound layers are made by covering the heads of low-headed vines with soil during the growing season, leaving the apices of the shoots exposed. Each shoot roots near its base and can be severed from the parent vine the following winter as a rooted vine.

The easiest way to root *rotundifolia* is to grow green cuttings under intermittent mist.

Chapter 8 SOILS

Soil is that part of the earth's surface in which plant roots grow. In sustaining plant life, its major functions are to supply mineral nutrients and water and serve as a medium for root anchorage. It is difficult to grow grapes without soil (hydroponics) either with nutrient solutions only or with sand or gravel.

The four major components of soil—mineral materials, organic matter, water, and air—are intimately mixed, but are not independent in nature. If one system is changed, there is a change in all. An average silt loam soil may be composed of about 50 percent pore space (air and water); the solid space is made up of about 45 percent mineral matter and 5 percent organic matter.

SOIL PARENT MATERIAL

Parent materials of soils consist of rock and unconsolidated materials from which the mineral components of the soil have arisen. The physical and chemical properties of soil are determined largely by climate and parent material. While some materials are residual (not moved from the place they were formed), others have been transported by water, wind, gravity, or ice. *Alluvial* soils, for example, have been deposited in the bottomlands along rivers. Sometimes landslides carry unconsolidated materials down slopes that later become *colluvial* soil parent materials.

MINERAL CONSTITUENTS OF SOIL

The mineral or inorganic part of the soil is comprised of small rock fragments and minerals of different sizes, and may be *original* (e.g., sand derived from quartz) and relatively unchanged from the parent materials, or *secondary* (e.g., clay) and formed by weathering of less resistant minerals.

Soil texture refers to the size of the individual mineral particles (Buckman and Brady, 1960). Textural designations generally used to describe soils are sand, silt, or clay, and the textural triangle shows the relative percentages of these components (Fig. 8-1). For example, a typical loam soil might have 40 percent sand, 40 percent silt, and 20 percent clay. A "heavy" soil refers to one that is high in clay and other fine particles, and a "light" soil is one composed of much sand and coarse particles but little clay. Silts are intermediate in size and have properties midway between sand and clay, the finest of the mineral fractions.

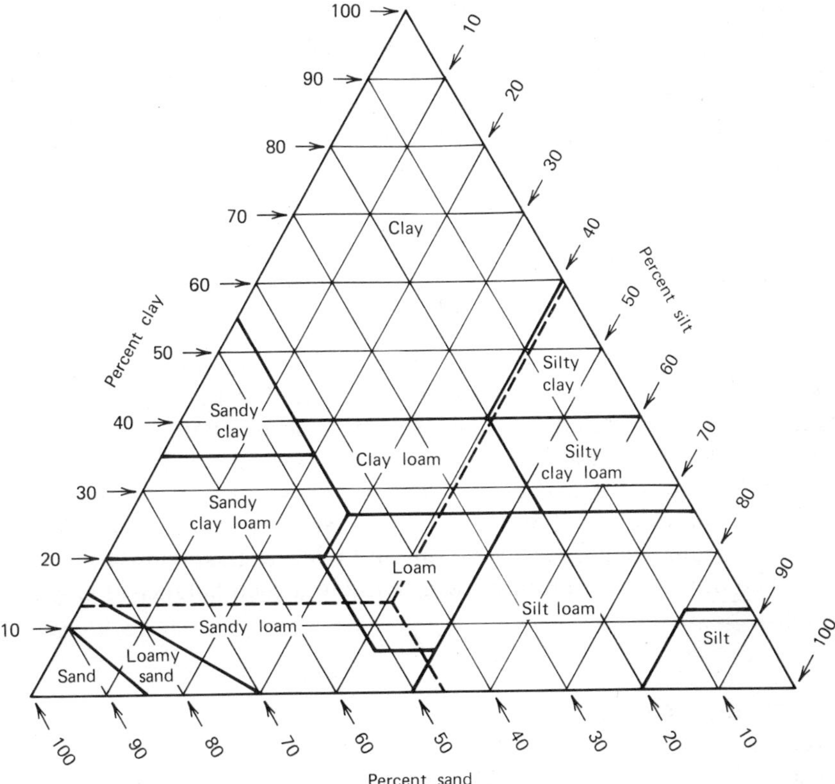

Fig. 8-1. Soil textural triangle. Percentages of sand, silt, and clay in soil can be plotted on this diagram to ascertain the textural class of a particular soil. (After Wildman and Gowans, 1974.)

ORGANIC SOIL MATTER

Organic soil matter consists of partially decayed and partially resynthesized plant and animal residues. While it may comprise only about 1 to 5 percent of a mineral topsoil, it has a great influence on soil properties. Many of the valley grape soils such as Hanford, Delhi, Tujunga, and Hesperia contain 1 percent or less organic matter, which is continuously being attacked and decayed by soil microorganisms and is thus a transitory soil constituent that must be replenished by the addition of plant or animal residues. Humus is an organic, colloidal fraction consisting of resistant products that result from decay of original tissues such as leaves and plant tissues.

Organic matter has a high water holding capacity that acts like a sponge in the soil. It is also a source of minerals made available during its decomposition. Small particles of organic matter cover mineral particles and keep them from sticking together, thus helping to maintain a good soil structure. A muck soil contains 10 to 40 percent organic matter and a peat soil about 40 to 100 percent. When such soils are properly managed, they can be extremely productive.

COLLOIDAL SOIL PARTICLES

Clay and organic matter are the finer parts of the soil and exist in the colloidal state. Such particles are characterized by a large surface area per unit of weight, and have surface charges that attract ions and water. Chemical reactions and nutrient exchanges occur at the surfaces. The capacity of a soil to retain and exchange cations such as H^+, Ca^{++}, and K^+ is termed base exchange. The attraction of ions to negatively charged colloidal surfaces prevents essential, positively charged nutrients from leaching from the soil and releases them slowly for plant use. The best soils for grape growing contain a good balance of clay and humus.

WATER AND AIR IN SOIL

Water in the soil and some dissolved salts comprise the soil solution that is important in supplying plant nutrients. Air is the other component of the pore spaces in the soil. As one increases the other decreases, and there must be an optimum balance between the two for optimum plant growth. Soil air may often consist of isolated, unconnected pore space. The carbon dioxide content of the soil is greater than that in the atmosphere above it because of the decomposition of organic matter in the soil. Oxygen in the soil is less than in the air above it because it is used in respiration by roots and microorganisms.

If too much water is added to the soil, the air is forced out of it and plant roots may be deprived of oxygen. Flooding of grapes during the dormant season does not appear to be harmful, although prolonged flooding in the growing season can be injurious.

LIVING ORGANISMS IN SOIL

The soil contains many living organisms ranging from bacteria to worms, insects, rodents, and roots of plants. Excluding roots of higher plants, the weight of soil organisms can be as high as 6000 lb per acre (6725 kg per hectare) (Janick et al., 1974). Insects and earthworms can physically disintegrate plant residues, and smaller organisms such as bacteria and fungi can then completely decompose these residues. Nutrient elements such as phosphorus and nitrogen are released in the decomposition process, although microorganisms also require such elements for their growth and can convert them into organic compounds not available to higher plants.

SOILS FOR GRAPES

Grapevines are well adapted for many types of soils. In various parts of the world grapes are grown commercially in almost all types of soils from shallow to deep, gravelly sands to clay loams, and in soils high to low in fertility. However, heavy clays, very shallow, poorly drained soils, and soils containing relatively high concentrations of alkali salts, boron, or other toxic materials should be avoided.

The deeper and more fertile soils usually produce the heaviest crops and are therefore preferred for raisins, common wine grapes, and table grapes such as Thompson Seedless and Tokay. Soils of limited depth are preferable for varieties such as Emperor and Malaga. Wine grapes often produce high quality fruit on infertile or rocky soil; thus good grape crops can often be grown commercially where other crops are unsuccessful.

DRAINAGE

The grapevine cannot grow well in a wet soil; the vine does not like "wet feet." The water table should not be too close to the surface of the soil; otherwise the soil must be tile-drained.

CHARACTERISTICS OF GOOD SOIL

To a grape grower soil is a reservoir to supply the needs of his vines. Water, air, nutrients, and growing space are soil requirements for grapevines, and sufficient growing space is needed for roots to explore the soil for these ingredients.

The best soils for grapes are loams to loamy in texture, well structured, moderately deep to deep and well-drained, uniform, and free from harmful salt accumulations, damaging soil pests, and soil pathogens. Relatively flat land is preferred.

A loam soil is a mixture of sand, silt, and clay particles containing about 35–45 percent sand, 35–40 percent silt and 10–25 percent clay. Loams and fine sandy loams are often the best soil textures for agricultural soils. Sands cannot retain much water or supply many nutrients, but good aeration may allow roots to penetrate deeply. Clays have good water holding capacity and nutrient supplying ability, but root penetration may be limited by poor aeration. Silts are similar to clays in many respects.

SOIL TEXTURE

A person looking for land to grow grapes should obtain a soil map that shows soil texture in the various areas from the local soil conservation service or agricultural extension office. Another method of determining soil texture is to feel the moist soil: sand particles feel gritty, silt particles are smooth and slick, and clay particles are sticky and plastic. Commercial laboratories can accurately analyze the percentage of sand, silt, and clay, and a textural triangle can then be used to determine the textural type (Fig. 8-1). Thus a soil containing 3 percent clay, 42 percent silt, and 50 percent sand, found at point A on the textural triangle, would be considered a loam soil.

Loams and fine sandy loams are considered excellent texture, while sand, silt, and clay are often poor (Neja and Wildman, 1973). Coarse-textured sandy soils cannot hold as much water or supply it as rapidly to the vine as can the finer textured loams. As soil depth increases, however, the preference of loams over fine sands decreases. For example, a 6-ft (1.83 m) deep, fine sand can be equal or even superior to a 6-ft (1.83 m) deep loam soil.

SOIL STRUCTURE

Soil structure refers to the arrangement of soil particles into aggregates. Most crop soils have a compound structure in which the aggregates stick together. Good agricultural soil has a crumbly nature dependent upon the soil texture and presence of humus. Very wet or dry soils must not be cultivated, lest the structure be damaged. When wet soils are worked, dry clods form that are difficult to work into the soil. When soils are worked too dry, a powder is formed that causes crusting and sealing against water intake. Granular structure is poorer with increasing proportions of sand and decreasing organic matter.

Fine-textured silts and clayey soils can sometimes limit the depth of rooting of irrigated grapes. Since shallow soils, say 30 inches (76.2 cm) deep, require frequent irrigation, a good irrigation setup is required.

Soil texture is only one of several aspects of the soil environment, and soil structure, depth, uniformity, and compaction must also be considered.

SOIL DEPTH

Two major factors that usually limit or prevent deep rooting are lack of pore space for roots to grow into and lack of air caused by extremely wet soil. A tight layer of soil can also block root penetration. For example, if a subsoil layer has an increase in clay content, as in a claypan soil, water may accumulate above this layer and roots may be injured because of poor aeration. This condition is known as *waterlogging*. A man-made compact layer occurring in the upper two feet (60.9 cm) of soil can sharply decrease the rate of water penetration (Figs. 8-2, 8-3).

Another type of barrier to root growth is an abrupt change from a fine or moderately fine topsoil to a coarse-textured subsoil. Water will not move from the topsoil to the subsoil until several inches of soil above the interface are saturated. This lingering saturated zone remains

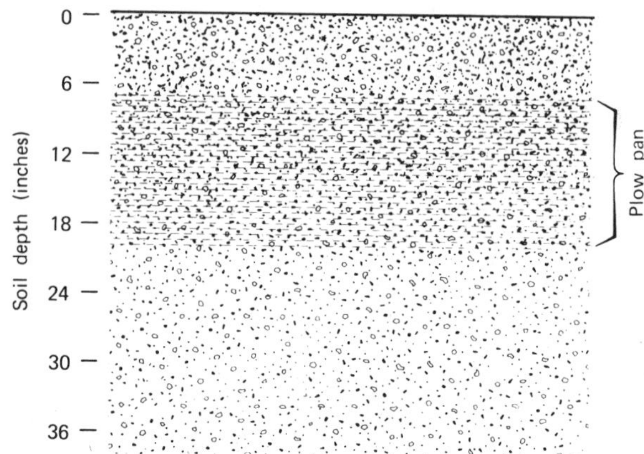

Fig. 8-2. Soil profile showing man-made soil compaction. (Redrawn from Neja and Wildman, 1974.)

because particle-to-particle flow of water is slow from the upper to the lower layer. Good drainage is essential to obtain root depth, and perched water tables block root growth.

Sometimes very dense, unfractured rocklike layers called *hardpan* occur. Such cemented hardpans are impervious to both water and roots. During the winter, rainfall can accumulate above the hardpan but can-

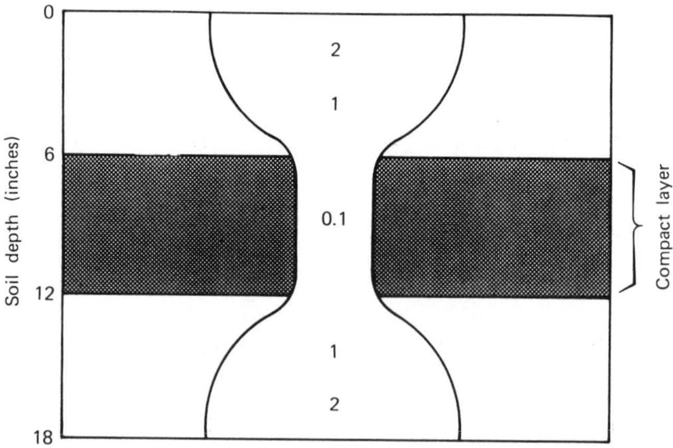

Fig. 8-3. Water movement through a compacted soil. Note rate of movement in compacted soil is 20-fold slower than that near the soil surface or at depth of about 16 inches (40.6 cm). (Redrawn from Wildman and Gowans, 1974.)

not pass through it. Unless the hardpan is mechanically shattered so that drainage is improved, vines may grow poorly (Wildman and Gowans, 1974).

Methods of detecting topsoil depth limitations include a direct study of the soil profile by appearance and feel, a laboratory analysis of texture and bulk density (weight per unit volume of a core of oven-dry soil), and a study of the rate of water infiltration at different soil depths (Fig. 8-4).

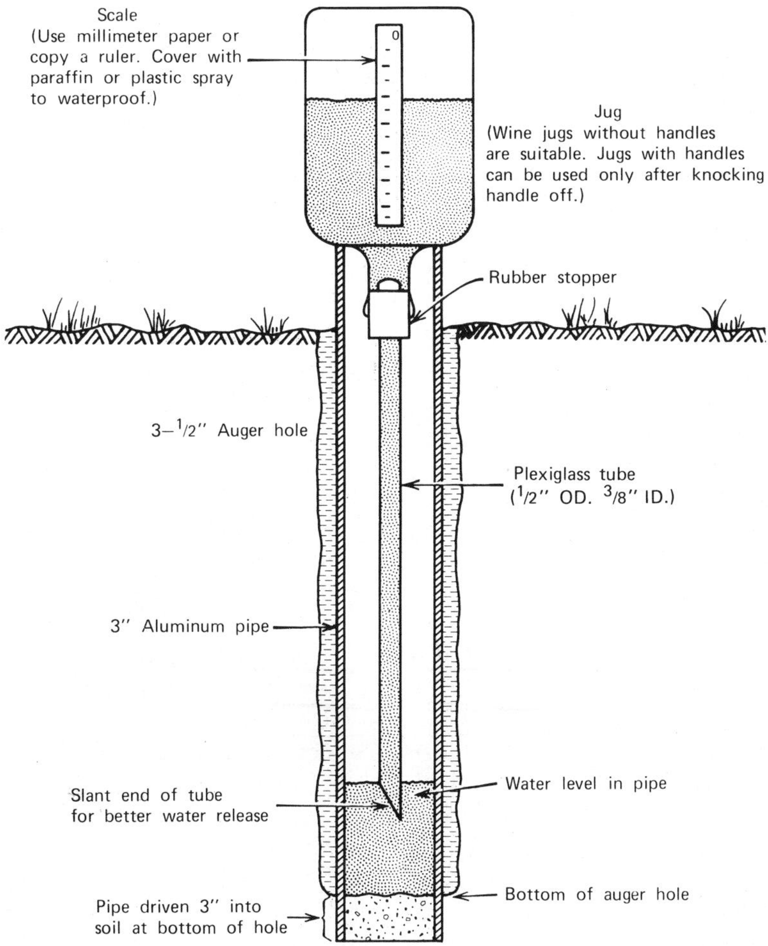

Fig. 8-4 Diagram of a mini-infiltrometer, used to measure rate of intake of water at various levels in the soil. (After Neja and Wildman, 1974.)

SOIL UNIFORMITY

The uniformity of the soil reservoir is based on evenness of expected root depth and of topsoil across the field. Depth of rooting can be estimated by mapping the differences in soil structure and texture, with depth, over the field. Generally depth uniformity is considered good if deepest soils are no more than 1½ times the shallowest depth, and is considered unacceptable if deepest soils are greater than twice the depth of the shallowest soils.

A grower makes the best of a nonuniform soil by using several techniques. Vineyard blocks should be laid out so that there is even texture and depth within each, leaving nonuniformity among blocks. Irrigation systems should be set up to allow irrigation of individual blocks. The uniformity of soil depth can be increased by plowing and ripping. One should use grape varieties and/or rootstocks that are adapted to grow well on the particular type of soil.

SOIL COMPACTION

Compacted soil results when improper cultivation practices break down the natural soil structure. Sometimes irrigation water may remain on the soil surface for 2 days or a week after normal irrigation because the large soil aggregates have been broken down, destroying the large pore spaces essential for rapid water percolation and good soil aeration. When deep cuts and fills are made in land grading, old soil surfaces and compact layers may be buried deeply and act as hardpans.

The best means of avoiding compaction is to cultivate as little as possible and only when the soil water content is intermediate, not too wet or too dry. Keep all traffic off the freshly loosened soil. The size of water-conducting pores can be increased by planting permanent cover crops; deep rooted grasses that do not require tillage are best and can increase water intake rates 5 to 10 times over a 5 to 20-year period. In some cases, however, a sod cover in vineyards may have deleterious effects on vine development and fruit production.

Soil compaction results in a loss of good structure, and often shows up as a slight layering or increasing density just below the usual depth of cultivation (Fig. 8-3). Man-made compaction often extends to a depth of 2 ft (60.9 cm) (Neja and Wildman, 1974).

ROOT GROWTH

Root growth can be restricted by lack of pore space (tight soils) for roots to grow into or by lack of air in soil that is too wet. Water movement through compacted soil can be very slow. In coastal soils, perched water tables that can cause root pruning often determine root depth (Neja and Wildman, 1973).

Abrupt changes in structure and/or texture can cause a barrier to root growth. For example, if a fine or moderately fine topsoil is underlaid with a coarse sand subsoil, roots will not grow across the abrupt boundary. Irrigation water will not pass from the upper layer into the sand until several inches of water above the interface is saturated because of the higher capillarity of upper finer soil compared to that of lower sandy soil.

Temporary saturation above the interface can limit aeration in the area just above the juncture of the two textures and thus restrict root growth. Abrupt boundaries in stratified soils break up the normal downward percolation of water, resulting in poorly aerated zones above the boundaries. A uniformly mixed soil is a much better root medium than a stratified soil.

IMPROVING PHYSICAL SOIL FACTORS BEFORE PLANTING

Frequently the soil reservoir available to grape roots can be increased, but success depends upon the texture and structure of the soil, the location and extent of the soil restriction and the type of deep tillage equipment available for use. It is generally harder to eliminate or alleviate soil restrictions that have a high clay content.

Deep tillage refers to loosening, breaking up, or mixing restricting subsoil layers located below the depth of normal or ordinary cultivation. Its purpose is to break up man-made compact layers in the top 2 feet of soil, break through natural subsoil claypans, hardpans and dense layers, mix stratified soils, and eliminate abrupt boundaries between unlike soil textures (Wildman et al., 1974). Types of equipment usually used in deep tillage include rippers, chisels, subsoilers, slip plows, disc plows and moldboard plows. The rippers and subsoilers, names which are often used interchangeably, operate at a depth of 2 to 7 ft (0.61–2.13 m) or more, where they break up hard layers by cracking and shattering; little mixing or dislocation of layers occurs. They are most effective in moderately dry, brittle soils and least efficient in moist sands and clays. Very dry soils require tremendous power to rip and may produce large clods that are difficult to manage.

Slip plowing is usually done to a depth of 3 to 6 ft. (0.91–1.83 m). A slip plow consists of a verticle ripping shank with a 12–15-in. (30.5–38.1 cm) wide inclined beam extending to the rear from the ripping point at an angle up to the soil surface. Chunks of subsoil are torn loose by the point, slide up the beam, and are permanently moved from their original position. Surface soils fall into the channel produced, resulting in some permanent mixing and thus improving water and root penetration (Wildman et al., 1974).

Disc plows are valuable for relieving soil compaction and for mixing shallow clays in the upper 2 feet (60.9 cm) of soil. Moldboard plows developed in the 1950s could plow 4 to 6 ft (1.2–1.8 m) deep and effectively loosen and mix soil. Today, only a few 3- and 4-ft (0.91 and 1.21 m) plows are still in operation.

The success of the ripping operation can best be checked by backhoe appraisal. Often slip plowing or ripping along a future vine row may be beneficial, unless you plan to disc plow to 18 to 24 in. (45.7–60.9 cm) after deep tillage. It is important that young vine roots penetrate quickly into loosened and well-aerated soil.

Backhoes are probably the most universally available and easily transportable type of equipment listed for soil modification, and are adaptable mainly to small acreages.

Wheel trenchers are usually used to dig trenches for water and drain lines, and have been used to a limited extent to loosen and mix soil layers before planting a vineyard.

Chapter 9 WATER

Water most closely approximates a universal solvent: it dissolves minerals present in the soil and serves as the pathway for solutes to enter the vine and move through the tissues. Water is a raw material for photosynthesis and is essential for maintenance of cell turgidity, without which cells cannot function properly. It has a very high specific heat which tends to stabilize protoplasm temperature that is subjected to heat or cold. Water in the soil is continuous with that in the plant, and the entire mass of water is in constant upward movement since the shoot constantly loses water to the atmosphere. Nearly all of the water (over 99 percent) moving upward is lost by transpiration, and only 0.1 to 0.3 percent is converted to chemical compounds. It serves as the cooling system for the plant, keeping the leaves nearer to air temperature through transpiration. This protects the leaves from heat injury and keeps them nearer to optimum temperature for photosynthesis.

PROPERTIES OF WATER

Water has a high *latent heat of evaporation,* which refers to the energy required to change water from a liquid to a gaseous state. When water reaches the boiling point, the temperature will not rise further until it changes from the liquid to the gaseous state. Before this change occurs, the water will absorb 539.3 cal per gram. The surface temperatures of the earth tend to be stabilized because of the high latent heat of evaporation of water.

The *latent heat of fusion* refers to the energy required to melt ice. When the temperature of ice has risen to 32°F (0°C), it will absorb 79.9 cal per gram before melting; however, 79.7 cal of heat are released when 1 gram of water changes to ice.

Water contracts as it is cooled, but it expands as the temperatures are lowered from 39°–32°F (4°C to 0°C). It also has a very high surface tension. Water is a fairly inert medium in which chemical reactions can occur without involving the water itself.

ATMOSPHERIC MOISTURE

The amount of invisible water vapor in the air is usually expressed as relative humidity, defined as the percentage of the maximum quantity that the air can hold at the prevailing temperature. Warm air can hold more water than cold air. *Dew* can form at night when atmospheric moisture condenses on cold surfaces, if the sky is clear enough so the soil and plants can lose enough heat to cause their temperature to fall below the dewpoint. Dew is usually an unimportant source of water for grape growing.

Clouds and fog are visible vapors that consist of water droplets or small ice crystals which result from cooling of the air to a temperature below its dewpoint. When cooling is due to upward movement of air from the land surface into colder levels of the atmosphere, clouds are formed; when air cooling occurs at or near the land surface, fog is formed. Prolonged presence of water vapors can reduce solar radiation enough to reduce food manufacture by the leaves.

SOIL MOISTURE

Soil water can be divided into three classifications. *Hygroscopic* water is tightly bound to the surface of the soil grains and cannot move either by gravity or capillary action. *Capillary* water is that part, in excess of the hygroscopic water, which exists in the soil's pore spaces and is retained against the force of gravity in a soil that can undergo unobstructed drainage. *Gravitational* water is that portion, in addition to the hygroscopic and capillary water, which will drain from the soil if favorable drainage is provided (Fig. 9-1).

Water can be classified as unavailable, available, and gravitational when based on availability to the grapevine. Gravitational water usually drains rapidly from the root zone, and unavailable water is held so tightly by capillary forces that it is not available to plant roots. Available water consists of the difference between unavailable and gravitational water.

Water drains from soil because of constant gravitational pull. A much longer time is required for clay soils to drain than for sandy soils, so that a sandy soil might lose its gravitational or free water in 1 day,

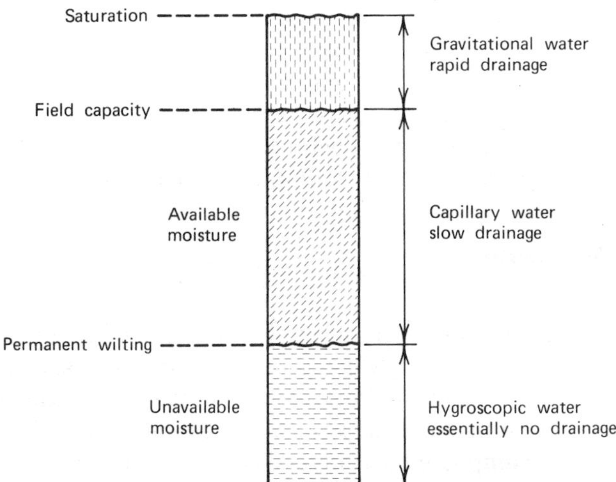

Fig. 9-1. Classes of soil-water availability to plants, and drainage characteristics. (After Israelson and Hansen, 1962.)

while a clay might require 4 or more days for drainage. The drainage rate is fastest immediately after irrigation.

After all the gravitational water has drained from the soil, the soil moisture content is said to be at *field capacity*. This point is usually reached from 1 to 3 days after the soil has been thoroughly wetted by irrigation or rain. A simple way to determine field capacity is to pour water on the soil surface, allow it to drain for 1 to 3 days depending on soil type, then measure the soil water content.

The permanent *wilting point* refers to the soil moisture content when plants wilt permanently, and is at the lower end of the available moisture range. Plants wilt when they are unable to extract enough water from the soil to supply their needs. Plants may wilt temporarily on a hot windy day and then recover in the cooler part of the day. Field estimates of the wilting point can be obtained by measuring the soil water content in soils with permanently wilted plants, although it may be difficult to find plants in a condition of permanent wilt. A rough estimate of the wilting percentage can be obtained by dividing the field capacity by a factor varying from 2.0 to 2.4. Since the factor depends on the amount of silt, 2.4 should be used for soils with a high silt content.

Available moisture refers to the difference in moisture content between the field capacity and the permanent wilting point. This is the moisture available for plant use. Generally, finer textured soils have a wider range of water between permanent wilting point and field capacity than do coarse-textured soils. In any soil, an increased rooting depth in the soil profile as a whole can compensate for a narrow range of available water in other horizons. Grapes often have very deep roots. Within the range of available water, the degree of availability tends to decrease as soil water content declines. Frequently the range of water for survival is greater than that available for good growth.

Measuring Soil Moisture

Farmers in irrigated areas can determine how deep the soil moisture is by the resistance of the soil to penetration. One can insert a shovel or push a half-inch (12.7 mm) pointed steel rod into the ground. The depth that the rod can be pushed into the soil will indicate the depth of wetting, if seriously compacted soil layers exist to resist the rod's penetration.

Samples of soil can be obtained throughout the root zone of the soil with a soil auger or a soil tube. Soil moisture content can be estimated by examining and feeling the soil. There are also several more accurate methods. In the gravimetric method, soil samples are usually placed in cans with tight lids, weighed and oven-dried to a constant weight, and reweighed to determine the loss of water. The percent of water on a dry or wet weight basis can then be calculated.

Another method to express soil moisture is to determine the tension with which it is bound to the soil. Tension is expressed in atmospheres of pressure (bars), and a tensiometer measures the tension by which soil water is held in the soil. It consists of a porous ceramic cup filled with water that can be buried to a desired depth in the soil in the vicinity of roots. The cup is connected by a water-filled tube to a manometer or vacuum gauge. As the soil dries, it sucks water out through the porous wall of the cup, creating a partial vacuum inside the tensiometer that can be read on a manometer or a vacuum gauge (Fig. 9-2). This power of the soil (soil suction) to withdraw water from the tensiometer increases as the soil dries and the gauge reading rises. Tensiometers operate satisfactorily only up to tensions of −0.6 bar (−60 centibars), and these are most widely used in medium (loam) to light (sandy) textured soils.

Tensiometers also indicate extremely wet or saturated soil conditions. A reading of 0 to −5 centibars indicates a wet soil in which vine roots will suffer for lack of oxygen; readings above −10 centibars assure the grower that the soil is drained of excess moisture.

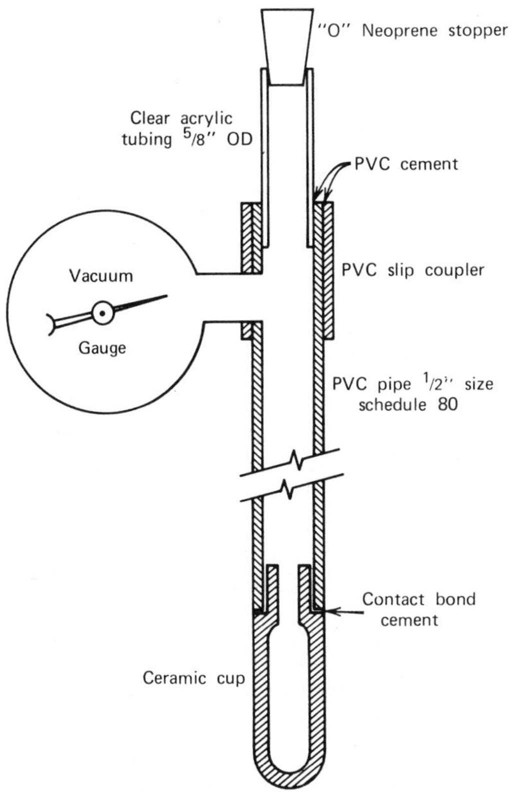

Fig. 9-2. Schematic drawing of a tensiometer (After Flocker, 1973.)

Irrigation

Water may be applied to soil in several ways. In *surface irrigation* water is distributed over the surface of the soil in furrows or basins, or by flooding. Application of water under pressure as simulated rain is termed *sprinkler irrigation*. There are both portable and permanent sprinkler systems. In *subirrigation* water is distributed to the soil below the surface and reaches the vine by upward capillary movement. *Drip irrigation* refers to the addition of water to soil by droplets emitted from small holes in plastic or other types of tubes.

In California water is often distributed by *furrow irrigation*. Usually there are two or three to a middle between rows usually spaced 10 to 12 ft (3.05–3.67 m) apart. Sometimes crosschecking is used to slow the flow of water and improve penetration. The wetted perimeter is increased, resulting in a more even distribution of water by using two wide-bottom furrows. In table grape vineyards semipermanent furrows are often made, and are renewed each year at the beginning of the irrigation season. In summer the soil is not cultivated and weed growth is mowed frequently. Water penetration is improved by reducing compaction from equipment, and the vineyard is less dusty.

Furrow irrigation on steep terrains is difficult and expensive; contouring is necessary, and care must be taken to prevent run off and destruction of furrows.

Flood irrigation can be used where large heads of water are available and there is a salinity problem. Flood irrigating in a wide flattened area between the raised berms along the vine rows is increasing in popularity. It spreads the water more widely for better penetration, eliminates the need to establish furrows, and provides a convenient surface for mowing under grass culture. *Cross-basin irrigation* is seldomly practiced currently due to high costs.

Sprinkler irrigation. For the past 10 to 15 years there has been much expansion in the use of permanent sprinkler systems. This is due in part to increased labor costs and difficulty in finding experienced irrigators. Growers can also design sprinkler systems to be used for frost protection, heat suppression, and to apply fertilizers, with the irrigation water (Meyer and Marsh, 1970). Cultivation of furrowing and ridging required for surface type irrigation systems are unnecessary. Under sprinkler irrigation, weed control can be accomplished using preemergence herbicides. Only enough land smoothing to provide drainage of excess surface water is required.

Sprinkler irrigation is especially useful on hilly soils, on shallow soils that cannot be leveled, and where land was not leveled for irrigation before planting. It is also used where a sprinkling system has been installed for frost protection.

Water conservation can be enhanced by sprinkler systems. A more uniform distribution of water and elimination of runoff losses result from a well-designed sprinkler system. A sprinkler system is 85–95 percent efficient, a border check system 60–90 percent, and a furrow-irrigation system 50–90 percent efficient. Irrigation efficiency is a measure of the uniformity of water application over the entire field and the prevention of water losses. Sprinklers achieve a more uniform distribution of water than does furrow irrigation, since the latter often leaves dry areas within the vine row.

Good control of the amount of water applied to a block of grapes can be obtained with sprinklers, making it possible to avoid both drought and a too wet condition. Irrigation time periods (12, 18, 24, or 36 hr) can be regulated according to the depth of soil wetting desired. Sprinklers with proper design and controls can also be operated at any time and location when the vines require water.

Sprinklers located above the vines are called overhead sprinklers and are the preferred type. This system requires tall riser pipes to raise the sprinklers above the vines. In windy locations, care must be taken to select the proper type of sprinkler and proper spacing. The water rate applied to the soil (precipitation rate) must be less than the rate that the soil can absorb the water (infiltration rate), otherwise water will run off the surface and result in nonuniform distribution, water logging of low areas, and perhaps soil erosion. Precipitation rates can vary from 0.1 in. (2.54 mm) per hr or less to 0.25 in. (6.35 mm) per hr, depending on soil texture. Many growers use a rate of 0.10–0.15 in. (2.54–3.81 mm) per hr, as this rate produces no runoff on coarse-textured soils and is also satisfactory for many fine-textured soils, which tend to compact, surface-seal, or crust, and require low application rates to prevent runoff.

Although the risers are usually about 18 in. (45.7 cm) above the vines so that foliage does not stop the spray stream, they must be low enough so that mechanical harvesting equipment can clear them (Meyer and Marsh, 1973). Risers should not be over 70 to 72 in. (1.78–1.83 m) high, and are usually secured to the grape stakes or to separate stakes between vines. A sprinkler spacing of 32 × 48 ft (9.75 × 14.63 m) is recommended for good uniformity of water application with vines spaced at 8 × 12 feet (2.44 × 3.66 m). Vines planted at 7 × 12 ft (2.13 × 3.66 m) require a 35 × 48 ft (10.7 × 14.6 m) sprinkler spacing.

Portable sprinklers are difficult to move over trellises and undulating terrain. When sprinkling is done with saline water, it should be carried out at night to minimize salt damage.

Drip Irrigation. This refers to the frequent slow application of water to soil through mechanical emitter devices located at selected points along water delivery lines. The flow of water across the soil is limited, and runoff does not occur. Most of the movement to wet the soil between emitters occurs by capillarity beneath the soil surface. The volume of soil wetted is usually far less than that accomplished by other methods of irrigation.

Most emitters have a fixed rate of output with a specified water pressure, and some have adjustable rates. A rate ranging from ½ to 2 gal (1.89–7.57 liters) per hr is most common. Most emitters lie on the soil surface, but some are buried at a shallow depth; the tubes of others are strung above the soil surface.

The emitters are part of, or connected to, a small (⅜–¾ in. (9.65 × 19.05 mm) plastic lateral line. These connect to a plastic main line that receives water from a head, which is the control system where water is measured, filtered, treated with fertilizer in solution, and regulated for pressure and timing of application.

A grower can often irrigate his vines with less than half the usual amount of water and also sometimes obtain greater yields. Labor costs are also lower with the drip irrigation than the other systems, and it is possible to inject fertilizers into the irrigation water and thus avoid the cost of ground application. Frequent irrigations maintain a soil moisture content that does not vary between wet and dry extremes, and most of the soil is kept well aerated.

Adequate filtering with screens or sand is essential to keep the water clean. Irrigations must be frequent, light, and not too long, so that local excessive wetness in soils is avoided. This also minimizes algae growth in laterals. Lateral lines must be designed to carry the maximum expected flow rate with little or no loss of pressure. One should not apply phosphate fertilizers through a drip irrigation system, as phosphate reacts with calcium to form a precipitate that can clog emitters. Iron has also been a problem in drip irrigation. A high concentration of calcium carbonate in the water also tends to concentrate in the emitters. Algae are also a nuisance when drippers are used and their removal by screens and chemical treatments is recommended. Silt and clay should not be introduced to the drippers.

In the San Joaquin Valley of California there are large areas of soil that exhibit slow penetration of water. A few of these soils have an infiltration rate of only .03 in. (0.76 mm) per hr. These soils, whether clay or sand, tend to develop a thin crust of silt or clay on the surface that impedes the rate of water infiltration. Such practices as cultivating the soil before irrigation to break up the crust, sod culture, deep ripping, and use of gypsum which are used to improve water penetration, all have disadvantages (Aljibury, 1973). In the San Joaquin Valley the potential of drip irrigation to increase water penetration appears most important among the various advantages of the system.

In the San Joaquin Valley many soils have a poor and unstable structure that deteriorates upon contact with water. Water may cause deterioration of soil aggregates by hydration, which disperses the soil structure through swelling, and by physical breakdown of aggregates due to the erosive action of the moving water (Aljibury and Christensen, 1972). The dispersed particles are carried into the upper soil pores and precipitate on the surface as a crust. Rate of water penetration can be increased by using high electrolyte (dissolved salts) irrigation water and addition of gypsum to the soil. The rate of water penetration is in proportion to the degree of soil crust prevention: a cover crop and organic matter added to the soils improves the rate of water penetration as long as a crust does not dominate. Penetration can also be improved by disturbing the soil crust by mechanical means.

VINE SYMPTOMS RESULTING FROM A WATER DEFICIT

Shoots grow rapidly in the spring and early summer, and the rate of growth during this period is a sensitive indicator of soil water availability (Vaadia and Kasimatis, 1961). As soil water approaches the wilting point, the length of shoot growth decreases and internodes near the shoot tip become shorter. The normal yellowish-green color changes to a dark green. Often as the stress increases, tendrils become flaccid and droop. Although leaves do not usually wilt, the older ones turn yellow, leaf margins become desiccated and curled, and the leaves gradually die and absciss. Under severe or prolonged stress in the hotter areas, the growing points of the shoots may dry. When irrigation is resumed during the growth season, however, the vine will begin new growth.

A sudden reduction of water from the vine can cause wilting of leaves and succulent shoots followed by abscission of basal leaves. This may occur on vines growing in shallow soil because of a sudden rise in temperature. Overall vine growth is reduced under drought conditions.

Young clusters at prebloom or bloom stage dry out easily, but more mature berries persist longer under drought conditions. Under stress, berries become wilted or shriveled. Berries that have colored contain much sugar and are much more resistant to drying than less mature ones, since the fruit can probably obtain water from leaves and other portions of the plant.

Stomatal closure occurs at a leaf water potential of −13 bars, but rate of shoot growth is inhibited at lower negative tensions (Smart, 1974). A reduction in trunk diameter also results when leaf water potential reaches about −7 bars. With accelerated water stress, the angle made by the petiole and lamina midrib increases from about 50 to 80°. The drooping of the leaves results from a change in turgor at the junction of the petiole and the lamina, and is a good indicator of water stress (Smart, 1974). Recently matured leaves should be used for measurement of leaf angles (Fig. 9-3).

EFFECTS OF SEASONAL DROUGHT

During the dormant period, vines under low winter rainfall conditions can be irrigated to make up the water deficit. The soil serves as a storage place for the water. This practice is often performed in California, where rainfall averages less than 12 in (30.5 cm). Although flooding of vines for several weeks during the dormant season is apparently not harmful, prolonged flooding in the active growing season can kill all or part of the root system and vines will show water stress, especially in hot weather. Present evidence indicates that sprinkler irrigation during bloom does not adversely affect set.

During the period following fruit-set, when berry growth is rapid, withholding water results in smaller berries. This procedure is sometimes used to reduce berry size and cluster compactness and thus to decrease the prevalence of rot resulting from berry splitting at matura-

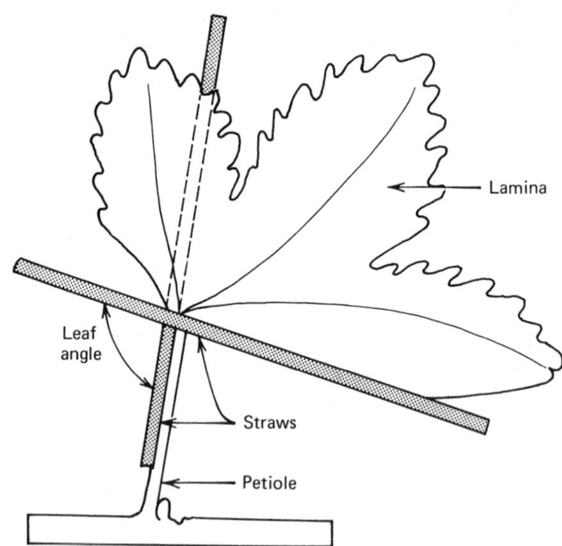

Fig. 9-3. Measuring leaf angle with two lengths of attached straw. View of underside of leaf drooping downwards to the right, as moisture tension increases. (Redrawn after Smart, 1974.)

tion. Berry size reduction as well as reduced shoot growth occurred in Thompson Seedless vineyards in the San Joaquin Valley when preharvest irrigation was cut off in early July (Christensen, 1975). Yields and/or soluble solids of fruit are usually reduced in grapes growing under drought conditions.

When vines are girdled at fruit-set to increase the size of berries of seedless varieties, such as Thompson Seedless, an irrigation should be made prior to or immediately after the operation to protect the vine in case of water stress brought on by sudden hot weather. Table grapes sometimes crack during maturation. This can often be avoided by irrigating very lightly or using only short periods of sprinkling during maturation. Such berry splitting may be aggravated by restricted transpiration that occurs in humid regions.

After harvest, vines grown under desert conditions require irrigation to maintain the established leaf surface. It is best that new shoot growth does not occur late in the season. Often vines of vigorous varieties will continue to grow or start new growth after harvest and the wood will fail to ripen. Such shoots may be subject to low temperature injury in the winter. Also, late growth utilizes food materials (carbohydrates) that should be saved for the active growth of the subsequent season. Irrigations after midsummer can be reduced or omitted to harden the growth of inherently vigorous varieties. In this case a late irrigation after temperatures have become too low to permit initiation of new growth is needed to furnish the vine's water requirement, even though it is low at this period.

EFFECT OF IRRIGATION ON GRAPE QUALITY

The keeping quality of Thompson Seedless and Emperor in cold storage was unaffected by high soil-water conditions (Hendrickson and Veihmeyer, 1951). With Tokay, however, a deeper color of red fruit was obtained when soil was near the wilting point for a relatively long period. Winkler, et al., (1974) state it is brightness, not density of color, that make an attractive table grape. An ample but not excessive supply of water at ripening seems to favor brilliant color; with insufficient water. more color develops, but it is dull and less attractive.

Special Irrigation Problems

Vines growing in light soils infested with parasitic nematodes that attack the roots should be irrigated more frequently than soils without nematodes, since the additional water helps compensate for the injured root system that is unable to absorb sufficient water. Vineyard growing areas with a high water table must be furnished with tile drains to remove the water. Where the soil is high in salinity that can result from use of irrigation water with a high salt content, leaching must be used to drain the salt. The land must be leveled to a flat grade and irrigated frequently by flooding to cover the soil surface. Irrigations should also be made in the autumn to reduce salt concentrations, even though there may be no need to supply soil water. Grapevines damaged by salinity slowly recover after the salts are leached.

Insect buildup and outbreaks of powdery mildew (*Uncinula necator*) are usually not affected by surface applications or by sprinkling. However, sulfur should be reapplied as soon as possible after sprinkler irrigation to reestablish normal protection against powdery mildew infection.

Amount of Water Required for Vineyards

The time of irrigation and amount of water to apply is determined by the water needs of the vine, availability of water with which to irrigate, and the capacity of the rootzone soil to store water. The approximate consumptive use of water (amount of water used by transpiration plus the amount lost from the ground around the plant by evaporation) must be replaced by the annual rainfall plus the amount of irrigation.

Water use reported in Table 9-1 includes rainfall stored in the soil of the rootzone and the added irrigation water, but it does not account for surface runoff or deep percolation. Water used by growers is greater than the needs shown by the consumptive data. Doubtless there is a desire to be safe and use plenty of water if it is available to avoid deleterious effects of sudden hot spells.

The water-holding capacity of various soils varies greatly (Table 9-2). Soils with a high water-holding capacity need less frequent irrigation.

SALINE SOILS

A saline soil is caused by the presence of too much soluble salt. Salts are originally formed by weathering and breaking down of rocks, and are often transported by moving surface water to low-lying or irrigated areas where they accumulate. Salts may accumulate because there is too little rainfall for sufficient leaching of the soil or because soil drainage is inadequate. Runoff water may accumulate in low areas,

Table 9-1. Comparison of annual rainfall, consumptive use, and approximate amounts of water applied in the commercial production of grapes in various areas of California and Arizona*

Area	Climatic Region†	Range of Annual Rainfall‡ inches (cm)	Amounts of Irrigation Water, acre-inches/acre (hectare-cm/hectare) Wine	Raisin	Table	Approximate Consumptive Use, acre-inches/acre (hectare-inches/hectare)
Hot desert		2–8 (5.1–20.3)	—	—§	48–72 (301–451)	43.6 (274)
Hot	V	6–10 (15.2–25.4)	36–42 (226–264)	30–36 (188–226)	42–48 (264–301	—
Warm	IV	12–18 (30.5–45.7)	12–18 (75.3–112)	—	18–24 (113–151)	18.8–20.9 (117–131)
Moderately warm	III	12–32 (30.5–81.3)	6–12 (37.7–75.3)	—	—	20.2 (127)
Moderately cool	II	12–32 (30.5–81.3)	6–12 (37.7–75.3)	—	—	—
Cool	I	12–32 (30.5–81.3)	4–6 (25.1–37.7)	—	—	—

* After Kasimatis, 1967.

† Climatic regions based on the summation of degree-days above 50°F (10°C), April 1 through October 31.

‡ Total rainfall, not corrected for evaporation. Rainy season, September through April.

§ Grapes not grown for this purpose in this area.

Table 9-2. Range of variability in the water-holding characteristics of some California soils on which vineyards are grown*

Soil Type	County	Depth of Sampling (ft) (in m in parentheses)	Field Capacity (%)	Permanent Wilting (%)	Inches of Available Water Feet of Depth mm per foot in parentheses
Hanford sandy loam	San Joaquin	0–12 (0 to 3.66 m)	6.3–8.4	2.7–3.4	0.53–0.91 (13.5–23.1 mm)
Hesperia loamy fine sand	Kern	0–8 (0–2.44 m)	9.0–14.7	4.4–6.6	0.76–1.30 (19.3–33.1 mm)
Delhi sand	Stanislaus	0–5 (0–1.52 m)	3.5–4.2	1.3–1.4	0.43–0.64 (10.9–16.3 mm)
Yolo loam	Yolo	0–12 (0–3.66 m)	15.7–34.0	8.6–11.9	1.10–3.14 (27.9–79.8 mm)

* After Kasimatis, 1971.

evaporate, or be lost by transpiration by plants, and leave their salts in or on the soil. This continuous addition of salts to the soil will eventually result in a salt problem unless adequate drainage is done to leach the salt in the rootzone. Leaching in this case is best accomplished during the dormant season, when surface evaporation and vine use of water is low and there is ample time for leaching. Some areas require installation of underdrains.

Factors that Accentuate Salt Accumulation. Several factors that can accentuate salt accumulation in soils and thus reduce yields (Neja, et al., 1974) include the following:

1. Irrigation water of poor quality (high salinity).
2. Poor water management that allows too much salt to accumulate in the root area.
3. Original salt content of soil was high.
4. Poor soil drainage characteristics. Shallow soils, claypans or compact layers that restrict downward water movement and root penetration. Salty perched or high water tables.
5. Semiarid climate with low rainfall and high temperatures, resulting in a high demand for water. Low rainfall causes an insufficient winter leaching, and large amounts of water applied may then add more salts.
6. Heavy applications of fertilizers.
7. Infrequent, heavy water applications during the growing season to soils with shallow claypans. This may cause water to remain above the claypan and can cause roots to rot, thus decreasing water available to the vine.
8. If soil is dried out excessively after infrequent irrigations, salts can concentrate in the soil solution.
9. Uneven topography and variable depths of soils over impervious soil layers can collect drainage water and form water tables.

Effect of Saline Soils on Vines. Vines growing on saline or salty soils may be unable to absorb water rapidly enough for their requirements. This can result in decreased vine growth, yield of fruit, and quality. Previous crop history is of limited use in predicting salinity damage, or sodic or toxic soil conditions because various crops respond differently to such soils. The best way to appraise the salinity status of vineyard soils is to obtain a soil and water analysis.

Sodic Soil Condition (Sodium, Alkali, or Black Alkali Soils). When there is a high amount of sodium in proportion to the amount of calcium and magnesium, a sodic soil condition usually results. Such soils have poor soil structure and water permeability that reduce water percolation, root development, and may impair drainage. When there is an excessive uptake of sodium, leaf burn may occur. When calcium and magnesium are high and sodium is low, there usually is good soil structure and permeability.

Boron and Chloride Toxicity. Excessive absorption of chloride, boron, or sodium can cause leaf burn. A marginal leaf burning proceeds inward, and excessively high levels can result in defoliation and even death of the vine. Although boron is essential for plant growth, it is toxic in higher than normal concentrations; more than 2 to 3 ppm makes it dangerous for irrigation purposes.

Sprinkle irrigation over the vine with water containing an excess of these salts can result in damaging salt absorption. Windblown sprays or mists are especially hazardous because salts concentrate in the spray, fall on the leaves, and are absorbed and not washed off by runoff. Intermittent wetting and drying of the leaves by a slowly rotating sprinkler head can also cause a problem. The adverse effects can usually be reduced by sprinkling only at night, when humidity is relatively high and the drying rate is reduced.

Laboratory Soil Tests. Several laboratory tests useful for determining whether a soil is "salt affected" (Neja et al., 1974) are the following:

1. Saturation percentage (SP), a measure of texture.
2. Electrical conductivity (EC_e), for evaluation of salinity.
3. Exchangeable sodium percentage (ESP) (estimate), to evaluate sodium hazard.

   3a. Calcium plus magnesium (Ca + Mg), used in ESP test.

   3b. Sodium (Na), used in ESP est.
4. Boron (B), to check on boron toxicity.
5. Gypsum Requirement (GR), to confirm that an indicated sodium problem—ESP estimate—is real, and to estimate amount of gypsum needed.
6. pH, a measure of acidity or alkalinity.

Separate soil tests for chlorides are not usually required for grapes. Generally, critical appraisal of chloride levels are considered to be satisfactorily included in the electrical conductivity appraisal. For grapes, EC_e values above 2.7 mmho/cm (about 1700 ppm) indicate possible problems associated with excessive salts and also with toxic levels of chlorides or sodium.

Electrical conductivity is measured in millimhos per centimeter, which is, for example, about 640 ppm of table salt (sodium chloride). Water with EC_e values less than 1.8 mmho/cm (1150 ppm) should not be injurious to grapes.

Soil Sampling. Soil sampling must be based on a particular problem or soil condition. Separate samples should be taken at locations representing a specific set of conditions, either poor or good plant growth areas and/or from shallow or deep soil areas. Samples may be taken at 12-in. (30.5 cm) depth increments for soils of uniform texture. For layered soils, a given sample should be confined to a single textural layer rather than mixing the differing textures into one sample. At least 1 lb (0.454 kg) of soil should be taken per sample, which should be taken to the laboratory within 24–48 hr, or air dried before the trip.

QUALITY OF WATER FOR IRRIGATION

Salinity. Irrigation waters contain varied quantities of soluble salts such as the bicarbonate, chloride, and sulfate salts of calcium, magnesium, and sodium, and traces of other materials. The total *salinity* (salt content) of most irrigation water is usually the most important single factor in evaluating its quality, and generally ranges from 70 to 3500 ppm.

Saline soils contain sufficient soluble salts to interfere with crop productivity. Usually they are only slightly alkaline in reaction (pH 7.0–8.5), contain little exchangeable sodium, and have good physical properties. The excess soluble salts can usually be leached from these soils with large amounts of ordinary irrigation water.

Hard water contains appreciable quantities of calcium and magnesium. Such water is called "hard" because it is difficult to make work up a lather with soap. Calcium and magnesium react with soap to form a curdlike material that leaves greasy rings in washbasins. After water is softened by removal of calcium and magnesium salts, sodium salts remain.

WATER QUALITY AND SOIL PERMEABILITY

Water that contains a high proportion of sodium salts (alkali hazard) has an adverse effect on water intake rates and physical properties of soil. Sodium concentration is second only to total concentration of salts in evaluating water quality. The sodium hazard of irrigation water can best be evaluated by the sodium adsorption ratio (SAR) defined below, where concentrations of the three soluble cations are given in milliequivalents per liter.

$$\mathrm{SAR} = \frac{\mathrm{Na}^{+}}{\sqrt{\dfrac{\mathrm{Ca}^{2+} + \mathrm{Mg}^{2+}}{2}}}$$

The SAR of irrigation water indicates the ESP equilibrium of a soil with the water.

The SAR and salt content of the water have the main impact on permeability of soils. A high soil ESP with a high SAR ratio causes poor physical soil properties such as stickiness. Irrigation waters with a SAR of 6 or less are usually not hazardous for grapes on deep, uniform soils managed by good irrigation practices. An SAR value of 6 to 9 may be hazardous to vines, and a value above 9 should be avoided. High carbonate (CO_3^-) and bicarbonate (HCO_3^-) can increase permeability problems.

Analysis of irrigation waters is generally a satisfactory guide to their suitability for agricultural use. Of greater importance, however, is a thorough knowledge of drainage, leaching, salt tolerance of grapes, irrigation methods, and soil management practices. Plants do not grow in irrigation water but in a complex plant–soil–water system.

LEACHING OF SOILS

One can irrigate in the fall to fill the soil, and facilitate leaching by winter rainfall. Winter irrigation can be done for leaching in dry areas. Leaching to remove excessive accumulations of salt in the rootzone is usually successful if drainage is adequate. Where drainage is poor, tile drains can be used to facilitate leaching.

VINEYARD WATER MANAGEMENT

Management varies according to location, soil, and type of grape grown, but there are certain general practices for most vineyards (Kasimatis, 1967). In the dormant season the soil should be filled with rain or supplementary irrigation. During most of the growing season readily available water should be maintained for full production. Usually vines are damaged less by salinity if irrigations are more frequent and soils are not allowed to dry out nearly to the wilting point. When table fruit is girdled, irrigation should be utilized to protect the vines. With table grapes and raisins, maturation can be enhanced by limiting the soil area to which water is applied. With wine varieties, irrigations should be continued when needed until harvest. If bunch rot is a problem, water may be withheld during the period of rapid berry enlargement (Stage 1) and restricted during maturation. After midsummer water may be withheld to enhance ripening of wood on varieties that continue growth in the fall. Leaching must be performed on soils where salinity is a problem.

Chapter

10

MINERAL NUTRITION

The bulk of the vine is water. The remaining dry matter consists mainly of carbon, hydrogen, and oxygen. However, a small amount of the dry matter consists of other elements that are indispensable to growth. In addition to carbon, hydrogen, and oxygen, 13 other elements are essential to the growth of higher plants including grapes.

The *major elements* required in relatively large amounts include nitrogen, phosphorus, potassium, calcium, magnesium, and sulfur. *Minor elements* are required in extremely small quantities and include boron, iron, manganese, zinc, molybdenum, copper, and chlorine.

Many elements are not essential for plant growth often occur in plants; they include cobalt, sodium, vanadium, fluorine, iodine, silicon, and aluminum.

Carbon (C), *hydrogen* (H), *and oxygen* (O_2). Large quantities of carbon dioxide and oxygen from the air and water absorbed from the soil furnishes carbon, oxygen, and hydrogen, which form the major portion of the numerous organic compounds in the vine.

ROLE OF ESSENTIAL MACROELEMENTS

Nitrogen (N) is a component of proteins essential for protoplasm and also occurs as stored food in plant cells. It is a constituent of chlorophyll, amino acids, alkaloids, and many plant hormones. The growth rate of plants is determined mainly by available nitrogen.

Phosphorus (P) is a component of some plant proteins. Most phosphorus is found in rapidly growing plant parts such as meristematic regions, and in developing seeds and fruit.

Calcium (Ca) is a constituent of the middle lamella of cell walls. It affects cell membrane permeability and colloid hydration. Calcium combines with some toxic organic acids and renders them harmless by making them insoluble, favors translocation of amino acids and carbohydrates, and stimulates root development.

Magnesium (Mg) is a component of chlorophyll. A deficiency can cause chlorosis (yellowing) of foliage.

Potassium (K) is required for functions such as synthesis and translocation of carbohydrates, cell division, development of chlorophyll, opening and closing of stomata, protein synthesis in meristematic cells, and reduction of nitrates. Potassium migrates from older to younger tissues, and can move from leaves into maturing fruit.

Sulfur (S) Many plant proteins contain sulfur, which must be oxidized to sulfates to be utilized by plants. It is also usually absorbed in this form.

ROLE OF ESSENTIAL MICROELEMENTS

Boron (B). The quantity of boron in the soil must be small or injury may result. When the compound is deficient, darkening of tissues and various growth disturbances and abnormalities can occur.

Iron (Fe) is essential for the synthesis of chlorophyll although it is not a component of it. Iron deficiency can result in chlorosis.

Manganese (Mn) deficiency can cause chlorosis, which is expressed in the leaf by a mottled appearance. Manganese is known to activate many enzymes in plants, and may play a structural role in the chloroplast membrane system. It is also involved in fatty acid synthesis and in photosynthesis.

Zinc (Zn) is a component of certain enzymes and plays a role in synthesizing the growth hormone, indoleacetic acid. "Little leaf" of grapes and deciduous trees and mottle leaf of citrus are well-known diseases caused by zinc deficiency.

Molybdenum (Mo) is part of an enzyme system that reduces nitrate to ammonia, and is importance in synthesis of protein. It is required for root nodule bacteria of legumes and also perhaps for vitamin synthesis.

Copper (Cu) may be important as a component of enzyme systems that utilize proteins and carbohydrates. It is involved in nitrogen metabolism, and is found mainly in young root tips.

Chlorine (Cl) is important in photosynthetic reactions and may act as an enzyme activator for one or more reactions.

DEFICIENCY SYMPTOMS FOUND IN VINEYARDS (SMITH ET AL., 1964)

Vitis labrusca, V. rotundifolia, and *V. vinifera* are usually quite similar with regard to visual symptoms of nutrient deficiencies that may develop. Other species probably show similarities too.

Nitrogen deficiency results in light-green foliage and reduced growth of shoots. Symptoms may be intensified by light pruning which results in more shoots per vine. Visual symptoms of nitrogen deficiency in *vinifera* may not be apparent until the deficiency becomes severe. At harvest time a reduction of 25 percent or more of the normal amount of foliage produced may result from nitrogen deficiency.

Magnesium deficiency symptoms first appear as red or yellow (depending on variety) leaf coloration between the main veins of the leaf. The basal leaves are the first to develop chlorosis.

Potassium deficiency symptoms are similar to those of magnesium deficiency. Color loss in leaves caused by low potassium or magnesium begins at the leaf edges, but with potassium the color change is first to a pale green or bronze instead of to the yellow or white seen in magnesium deficiency. Finally the tissue dies, and this symptom is referred to as "leaf scorch." Also, with potassium symptoms, leaves near the middle of the shoots first show chlorosis, which occurs on the apical leaves of lateral shoots. Severely deficient vines have compact clusters with small berries, and maturation is uneven or retarded.

Another symptom resulting from potassium deficiency is "black leaf," in which initially the interveinal upper leaf surfaces exposed to the sun, and later and lower surfaces, blacken. Since overcropping of low potash vines seems to be related to the occurrence of black leaf, potash applications can alleviate but not eliminate the symptoms.

Manganese and zinc. Manganese deficiency occurs only rarely in California vineyards; it is most commonly found on poorly drained soils and those with a high pH. Yields are rarely reduced, but it is important to distinguish manganese deficiency from zinc deficiency. Manganese deficiency causes chlorosis between the primary and secondary veins of basal leaves. These veins are bordered by an irregular band of green tissue that includes the tertiary veins, giving them a herring bone outline. In contrast, zinc deficiency symptoms appear on the apical leaves of the main shoot and on the leaves of lateral shoots that develop in summer. Since lateral shoot leaves are especially small, "little leaf" is the name given to zinc deficiency. The angles of the basal leaf lobes with the petiole of zinc-deficient vines are much more obtuse or widened than the normal ones. Leaf chlorosis appears as a fading of green color between the many small veins throughout the leaf. Low zinc also causes poor fruit-set, straggly clusters, and the presence of small, green, immature "shot" berries. Affected vines usually appear on sandy or high pH (alkaline) soils. Land leveling cuts and old barnyard and poultry yard areas are also more susceptible to zinc deficiency.

Boron deficiency can result in necrosis of many flower clusters with a light fruit-set in others, which may occur with a high percentage of shot berries, some of which may shatter off in midsummer. Leaf chlorosis first appears as blotchy yellow areas on apical leaves between the primary and secondary veins. The loss of green color may progress to leaf margins and also between the major veins. There may be a terminal die back of shoots in early summer that stimulates lateral shoots, which may be stunted but show no other symptoms. Boron deficiency frequently occurs on coarse-textured soils irrigated with canal or well water very low in boron. It may also appear in high rainfall areas low in native soil boron.

Iron deficiency symptoms occur on apical leaves earlier in the season and with a different pattern from that of zinc deficiency. The chlorotic condition of low iron is more severe and, except for fine green lines around the veins, the leaf becomes yellow or creamy white. Iron deficiency is seldom a problem in California, but it is widespread in the northeastern and northwestern United States where the more susceptible *V. labrusca* and American and European hybrid varieties are widely grown.

FERTILIZERS

Detection of N Deficiency in the Field (see p. 169 for discussion of laboratory testing). This element is the nutrient most likely to be deficient, but crop yields may be reduced before deficiency symptoms appear. For Thompson Seedless and perhaps for other varieties such as Tokay, Emperor, Carignane, Zinfandel, Muscat of Alexandria, and Petite Sirah, the nitrate color test is a good method to determine the nitrogen status of the vine (Fig. 10-1). Petioles are collected at bloom, when about two-thirds of the calyptras have fallen from the flowers (Cook and Kasimatis, 1959). At this time the greatest differences in nitrogen content in petioles between nitrogen-deficient vines and well-fertilized vines occurs. The leaf blades are detached from the petioles and discarded. Then a ½ to 1 in. (1.27–2.54 cm) lengthwise cut is made through the bulbous basal ends of the petioles, exposing about 1 in. (2.54 cm) of cut surface. One or two drops of indicator solution (1 g of pure, colorless diphenylamine in 100 ml of concentrated sulfuric acid) is then added to the cut surface. (Sulfuric acid is extremely caustic and must be used with care.) If 5 out ot 20 petioles show a blue color after 5 or 10 sec, the vineyards have sufficient nitrogen; if 4 or less turn blue, a nitrogen defi-

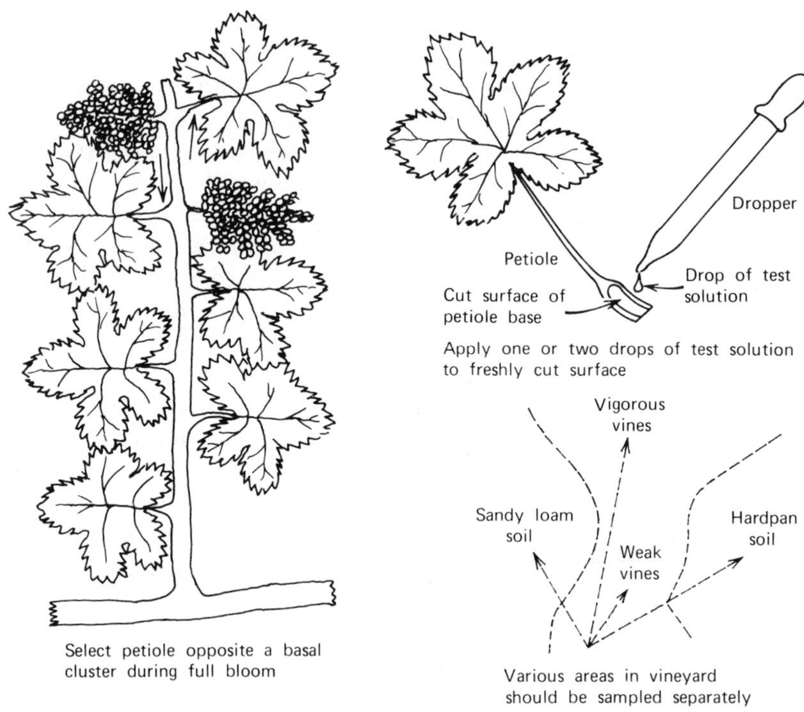

Fig. 10-1. Technique used in petiole test to detect nitrogen status of the vine.

ciency is present. If 15 or more petioles out of 20 show an immediate blue-black color, there may be an excess of nitrogen that could reduce crop yield. In vineyards with both weak and strong vine-growing areas, it is important to test each area separately. Even if vines are uniform a sampling should be confined to 4 or 5 acres.

Response to Nitrogen. The need is greatest and the response most rapid on sandy and gravelly soils. On heavier soils nitrogen availability and response will occur more slowly. On light sandy soils, 20–70 lb of nitrogen per acre (22.4–78.2 kg per hectare) may be required, but 20–40 lb per acre (22.4–44.80 kg per hectare) is enough for colored table varieties such as Emperor and Tokay where highly colored fruit is desired. The lower amounts should also be used on young nonbearing vines of low vigor (Cook, 1960).

If the nitrogen in the soil is excessive after fruit-set stage, too much shoot growth will occur. Dense foliage will make mildew and insect control more difficult, and will also reduce the development of flower cluster formation, which would result in a decreased crop the subsequent year. If excessive shoot growth persists during fruit ripening, it will decrease the amount of sugar accumulation in the berries. High nitrogen-induced shoot growth in the autumn may prevent the shoots from maturing properly, and a die-back of canes can then occur in the winter.

Generally, the least expensive fertilizer should be used (Table 10-1). However, in heavy rainfall areas such as the coastal regions of California, ammonium nitrate or urea may be most suitable because they resist too rapid leaching.

Since the vine's greatest demand for nitrogen is during rapid growth in spring through blooming, nitrogen should be applied in the rootzone when spring growth begins. The optimum time for application of fertilizer depends on rainfall or the irrigation schedule, soil texture, form of nitrogen applied, and ability to get equipment into the vineyard—often just before the rainy season, so that the fertilizer can be leached into the rootzone. In the irrigated San Joaquin Valley all forms of nitrogen may be applied in January and February, using the cheapest kind available. With applications made after bud break, however, ammonia forms or ammonium sulfate should not be used because these types must undergo a conversion of soil microorganisms before the fertilizer can be leached downward.

Steer manure contains about 30 lb (18.6 kg) of nitrogen per ton (about 2 cu yd or 1.53 cu m) on a dry weight basis. It is recommended when it is cheaper to apply than an equal amount of commercial fertilizer. Heavy use of poultry manure on sandy soils may cause zinc deficiency. Grape pomace (the skins and seeds left over after pressing) is equal to manure as a fertilizer. In California, pomace has the advantage of decomposing over a longer period than manure. It can be applied in winter at a rate of 6 to 8 yd per acre (11.3 to 15.1 cu m per hectare).

Potassium. Deficiencies have been associated with vineyard areas that have been scraped in land leveling (Fig. 10-2), and with very sandy soils (Cook, 1960). About 1500 lb of potassium sulfate per acre (1681 kg per hectare) will correct the deficiency by the second year after treatment, but 2000–3000 lb per acre (2241–3359 kg per hectare) are required to correct the deficiencies in one year.

Potassium sulfate is the most widely recommended fertilizer. Potassium chloride can be used in well-drained irrigated soils at 1300–2150 lb per acre (1456–2410 kg per hectare), but should be avoided in saline, shallow, or poorly drained soils because chloride toxicity can occur. Potassium fertilizer must be placed near the rootzone. A furrow

Table 10-1. Characteristics of Some Common Commercial Nitrogen Sources*

Nitrogen Carrier	Nitrogen (%)	Pounds per acre of Materials Required to supply rate of of Nitrogen needed (kg per hectare shown in parentheses)			Advantages and Disadvantages
		20 (9.07)	40 (18.1)	60 (27.2)	
Anhydrous ammonia	82	24 (26.9)	49 (54.9)	73 (81.8)	(a) In irrigation water: Fertilizer distribution depends on good water distribution and penetration. (b) Dry injection: Some loss if ground is trashy, cloddy, very dry, or sandy. Special equipment required for application.
Aqua ammonia	20	100 (112)	200 (224)	300 (336)	Same as for anhydrous.
Ammonium sulfate	21	95 (107)	190 (204)	285 (320)	Acid residue suitable for alkaline soils, but undesirable in very acid soils.
Ammonium nitrate	33	61 (68.4)	121 (136)	182 (204)	High nitrogen percentage. Half is immediately available; half is delayed.
Calcium nitrate	15.5	129 (145)	258 (289)	386 (433)	Immediately available.
Urea	46	43 (48.2)	87 (97.6)	130 (146)	(a) High nitrogen content. (b) Requires incorporation into soil to avoid surface loss into air.

* After Christensen, 1973.

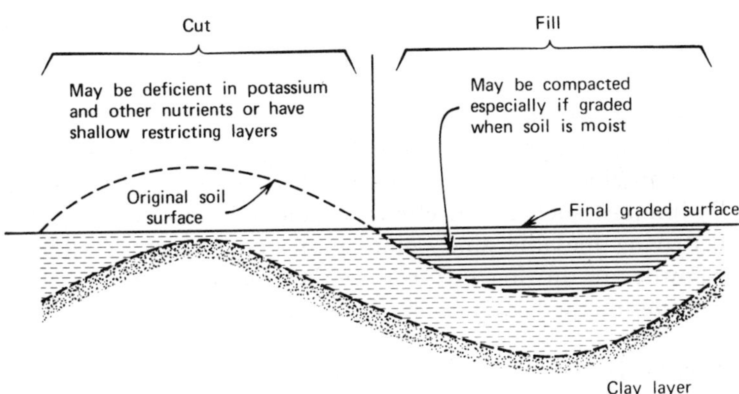

Fig. 10-2. Grading of land can result in temporary potassium and other nutrient deficiencies where the original soil surface is scraped away. (Redrawn from Wildman, 1974.)

can be made near vines, but not so close that too many roots are injured (Christensen, 1975). The fertilizer should be placed in the furrow and irrigated or left open so rains can leach it into the soil. Late fall to early winter are optimum times to fertilize. Measurable response to potassium fertilizer has only been obtained where visual deficiency symptoms have appeared.

Zinc. Zinc is the second most widely required fertilizer element in vineyards after nitrogen. Vines need only a trace of this element. One-half pound of actual zinc per acre is plenty, but it is difficult to get zinc into the vine because most soils fix or inactivate large quantities. On very sandy soils, vines can be fertilized with 2–3 lb (0.91–1.36 kg) of zinc sulfate per vine applied by the techniques used for potash. Spur-pruned vines can be successfully treated by daubing the pruning cuts with a solution of 1 lb of zinc sulfate (36 percent zinc) per gallon of water. Daubing is ineffective on cane-pruned varieties because the zinc moves too short a distance in the cane. It is best to daub within 3 or 4 hr after pruning. Daubing is ineffective if the pruning cuts bleed from root activity in warm weather (Christensen, 1968).

Foliage sprays of zinc sulfate varying from 4 to 10 lb per 100 gallons (1.81–4.54 kg per 378.54 liters) of water applied at least 3 weeks before bloom are moderately effective. Both upper and lower surfaces of leaves should be wetted. One spray may suffice to cure poor berry-set, but when vine growth is also stunted a second spray is necessary several weeks after bloom. In all sprays, half as much spray lime as zinc sulfate is required to prevent burning of foliage. Wetting agent or sticker should also be added to the spray mixture. Many growers prefer to use a neutralized or basic zinc sulfate commercial product instead of the zinc sulfate-lime spray tank mixture.

Chelated zinc compounds used as foliage sprays are usually no more effective than zinc sulfate when used at the same per acre application cost. Techniques such as injections of zinc sulfate solutions into vine trunks, driving metallic zinc points into vine trunks, or adding zinc sulfate powder in dusting sulfur are either ineffective or less effective than the other methods described (Christensen, 1968).

Phosphorus. In California, the use of phosphate fertilizer is of questionable value except when applied primarily to stimulate cover crops (Cook, 1960).

Magnesium. Transient leaf symptoms have been noted mainly on young vines planted in sandy soil, especially where cuts and fills have been made during land leveling (Fig. 10-2). These symptoms usually disappear after a few years, when the roots penetrate into soils richer in magnesium than the surface soil. Occasionally, deficiency symptoms are also noted in high lime and/or sodic soils.

Calcium. This nutrient has not been observed to be deficient in vines in California, although it is often used in soil amendment treatments such as gypsum (calcium sulfate).

Iron. The quantity of iron required to cure chlorosis is very small and can be applied by spraying plants with iron salt solutions, injecting iron compounds into the vine trunk, or by application of iron salts to the soil.

Manganese. Spraying with 4 lb manganese sulfate per 100 gallons (1.81 kg per 378.5 liters) of water or with a manganese chelate solution will reduce deficiency symptoms.

Boron. Boron deficiency can be overcome easily, but care must be taken not to produce toxic effects by overdoses. Boron materials contain about 35–65 percent boron trioxide (B_2O_3). One-half oz (14.2 g) of the stronger material or one oz (28.4 g) of the less concentrated forms may be used, dissolved in water and sprayed on the soil during the dormant period. For small areas a 1 oz (28.4 g) measure may be used to surface broadcast by hand.

Boron excess can occur in isolated spots in coastal vineyards and in areas on the western side of the San Joaquin Valley. Grapes are sensitive to an excess of boron. To be safe, the amount of boron in irrigation water should be below 0.5 ppm. Boron excess causes cupping, puckering, or wrinkling of leaves because of growth inhibition of young leaf edges while the rest of the leaf continues to grow. When mature leaves are affected, black or brown necrotic specks develop near the leaf margin that may spread and become continuous. The remedy for boron excess is to leach the soil with low-boron irrigation water.

LABORATORY PETIOLE ANALYSIS

Laboratory petiole analysis is presently the most widely used technique for establishing fertilizer needs or diagnosing nutritional problems. Petioles should be collected during bloom, as previously described (p. 163). Remove the blade and discard it, leaving only the petiole for analysis. Representative samples can usually be obtained by not sampling more than 10 acres (4.05 hectares) of vineyard. Sample areas with different soil types of weaker or stronger portions of the vineyard separately (Christensen, 1972). Use 75–100 petioles collected from separate vines distributed over each area. The petioles should be kept in a paper (not plastic) bag in a warm, dry location until delivered for analysis.

Uniform samples can be obtained at bloom, but it is usually midseason or early harvest before deficiency and/or toxicity symptoms are displayed by the vine. After bloom one should collect petioles of the most recently matured leaves (the first fully expanded leaves, about 5–7 leaves below the shoot apex). With toxicity symptoms, both petiole and blade should be retained because large, toxic amounts of boron and certain other elements accumulate in the blade. Collect both leaves showing symptoms and healthy tissues from an unaffected area. Laboratory analysis should include bloom time samples for nitrate nitrogen, total phosphorus, total potassium, and boron. Since zinc deficiency can be identified by visual vine symptoms, it need not be determined.

Arginine levels in grape berries may also prove useful for detecting nitrogen deficiencies in grapes. In Thompson Seedless, arginine levels in dormant canes and mature fruits correlate well with the amount of nitrate in petioles. Arginine in mature fruits also correlates with pruning weights (Kliewer and Cook, 1974). Critical deficiency ranges are 4–6 mg/g dry weight for arginine in dormant canes, and 400–500 μg/ml juice for arginine in fruits at harvest.

Table 10-2. Levels of Various Nutrients Found in Healthy Vines and in Vines with Deficiency or Excess Symptoms*

Nitrogen	Nitrate N (ppm)
Deficient	Less than 350
Adequate	500–1000
More than necessary†	Over 1200
Excess	Over 2000
Phosphorus	Total phosphorus (%)
Possibly deficient‡	Less than 0.10
Questionable	0.10–0.20
Normal	More than 0.20
Potassium	Potassium (% total or soluble)
Deficient	Less than 0.8
Questionable	0.8–1.5‡
Adequate	More than 1.5
Magnesium	Total magnesium (%)
Probably deficient	Less than 0.3
Normal	0.5–0.8
Zinc	Total zinc (ppm)
Deficient	Less than 15
Questionable	15–25
Normal	More than 26
Boron	Total boron (ppm)
Deficient	Less than 25
Questionable	26–30
Adequate	40–70
Toxicity	More than 150
Salt injury (chloride)	
Normal	0.01–0.30%
Toxicity	0.8 and higher

* After Christensen and Kasimatis, private communication.

† May be justified where heavy foliage is desired.

‡ Sample again about 2 months after bloom with petioles from most recently matured leaves. Levels below 0.5% K would be deficient, and a safe level would be above 0.8%.

Interpretation of Tissue Laboratory Analysis

Table 10-2 shows critical values for vine nutritional levels (Christensen and Kasimatis, private communication). The data for nitrate nitrogen applies only to Thompson Seedless and effects on total yield, effects on table fruit quality are not considered. For instance, excess levels for Emperor might be lower than the value given in Table 10-2 because of the adverse affect of high nitrogen on development of berry color. The other data are based on results with Thompson Seedless and/or several wine varieties with petioles taken at bloom time.

COVER CROPS

The main benefit of cover crops is that they prevent erosion of the soil by water and wind. It is difficult to maintain soil organic matter at a higher level than the native one present in the soil before plowing. An interplanted winter legume cover crop usually will not supply more than 20–25 lb per acre (22.5–28.1 kg per hectare) (Cook, 1960). Grass crops do not produce nitrogen. A cover crop can use up soil water that may be harmful in nonirrigated areas. If a green manure crop is plowed under in early spring, nitrogen in the soil may be lowered when the vines require it most, and may be released at a later time when low soil nitrogen is desirable. Organic matter decomposes so rapidly in a semiarid growing climate that cover crops have no effect on the amount of water the soil can hold. Cover crops can be used to reduce leaching in very sandy soils and to prolong the availability of nitrogen being released by decaying organic matter.

The rate of water penetration may be aided by cover crops, by improving physical condition of the surface soil, and by penetration of plow soles or plow pans by roots that later decay and leave channels for water penetration. Winter cover crops are planted after harvest but before the winter rains begin. Purple vetch (*Vicia atropurpurea*), barley, oats, rye, subterranean clover (*Trifolium subterraneum*), bur clover (*Medicago hispida*), sweet clover (*Melilotus indica*), field peas, and bell beans (small-seeded horse beans) are sometimes used.

Cereals are usually preferred on compacted soils because their fine root systems can sometimes penetrate compacted soil or hardpans. Legumes have tap roots, but roots of vetch and other legumes usually branch out laterally and grow above a compacted soil layer (Christensen and Fischer, 1968).

Part 3
VINE MANAGEMENT

Chapter 11 PRUNING

OBJECTIVES OF PRUNING*

Pruning refers to the removal of canes, shoots, leaves, and other vegetative parts of the vine. The three main objectives of pruning are as follows:

1. To establish amd maintain vines in a desired shape that will enhance productivity and facilitate various cultural operations such as thinning, harvesting, irrigation, cultivation, and disease and pest control.
2. To distribute the proper amount of wood over the vine and between vines according to vine capacity so that large crops of high-quality fruit can be maintained over the years.
3. To regulate the amount of crop to lessen the need for or without thinning. Pruning is the cheap way to reduce the number of clusters per vine and thereby increase the foliage-to-fruit ratio. Thinning is a more efficient but costly way to regulate crop. Flower-cluster thinning is an important means of reducing crop and is not expensive.

DEFINITIONS

One must understand the meaning of the following definitions to understand the explanations of pruning.

ARMS are branches older than 1 year.
FRUIT CANE is the basal part of a mature cane 8 to 15 buds long that produces fruit during the next growing season, and is removed at pruning time the following year.

* Much of this chapter was adapted from Jacob, 1950, and Winkler, 1959.

HALF-LONG CANES OR RODS are canes cut back to 5 to 10 buds. They are used in some foreign countries, and in California they are used for double pruning. Vine shape is more difficult to maintain with this system.
SPUR is the basal portion of the cane, from 1 to 4 buds long.
RENEWAL SPURS are intended to produce canes to be used for fruit canes the following season. They usually bear fruit.
REPLACEMENT SPURS are used to shorten or replace arms or branches, and usually bear fruit.
SUCKERS are water sprouts that develop from below the soil surface. Water sprouts from the trunk and main branches are also frequently called suckers.
TRUNK is the undivided main stem of the vine.

LENGTH OF BEARING UNITS

This is determined by the fruiting habit of the variety—by the fruitfulness of the buds on the cane and the size of the clusters produced. Usually spur pruning is used on varieties having large clusters and fruitful buds to the base of the cane, and cane pruning is used on varieties with small clusters or those in which buds on the basal portion are less fruitful.

GENERAL EFFECTS OF PRUNING ON VINE BEHAVIOR

One must be acquainted with the following pruning relationships to understand the recommendations for pruning:

1. *Depression of vine growth.* Pruning decreases the productive capability of the vine because a heavily pruned (with much wood removed) vine produces fewer leaves than a lightly pruned (with less wood removed) one, and also produce its maximum number of leaves and area of foliage surface later in the season. Thus the total annual photosynthate production by the leaves is less on a heavily pruned vine than in a lightly pruned one. Therefore smaller quantities of assimilates such as sugar and starch are formed, and the amount available for the nourishment of roots, stem, shoots, flowers, and fruit will be less in a heavily pruned vine.
2. *Pruning and vine capacity.* Within limits, capacity of a vine is directly proportional to total growth. (This may not necessarily be true

for very large, dense vines.) A large cane has a greater capacity than a small one, but it produces fewer fruitful buds because of its vigorous growth. Therefore a large cane should be pruned longer and carry more buds so that the shoot growth will be restrained.

3. *Pruning to produce a normal crop.* A vine is only capable of producing a certain quantity of fruit. The date of fruit ripening is determined mainly by heat units. The maximum amount the vine will bear without delaying maturation is termed a "normal crop." As the crop increases beyond this point, the first effect is delayed maturity. Further successive increases in crop result in low sugar and low acid content, and with some varieties in water-berries and/or drying of the tips of the clusters. Other symptoms are reduced growth, immature wood, and poor fruit-bud formation. The latter will limit the crop of the following year.

4. *Vine capacity is reduced by an overcrop.* Vines with a very heavy crop grow less vigorously than vines with a light crop, and vines which are overcropped are likely to have a lighter crop the following year. Years of excessive overcropping are often followed by lower yields.

5. *Relation of leaf area to vine capacity.* Total leaf area and number of shoots on a vine is directly related to vine capacity. A vine with few shoots that elongate rapidly appears vigorous, but another vine which is less active because of numerous shoots of slower growth produces a larger total leaf area capable of greater production.

6. *Relation of number of shoots to crop size.* The fewer the number of shoots, the more vigorously (rapidly) each shoot will grow. When a single shoot is trained up the stake, it grows far more vigorously than when other shoots are present. With fewer shoots the fruitfulness and crop production is less. Also, the fewer the number of arms, clusters, or berries, the larger are the retained respective parts.

7. *Pruning in relation to growth and fruiting.* Vines tend to grow mostly at the top due to apical dominance, and in pruning one must attempt to overcome the effect of position on growth. Growth on spurs or canes usually begins at the apical ends. In head-trained vines, spurs are formed and maintained near a common level so that there is no great vertical distance among clusters. On cordon-trained vines, spurs should be trained so that they are at the same level. Vertical cordons are not satisfactory because of apical dominance, and shading of the basal portions of the vines also weakens the lower arms. In cane pruning, canes should be bent down and tied in a horizontal position on the trellis.

The best canes are those that arise from 1-year-wood that has made regular growth since the beginning of the season, as shown by canes which have internodes of normal length for the variety.

TYPES OF TRAINING–PRUNING SYSTEMS

The different pruning-training systems worldwide are innumerable (Fig. 12-1). In California the main systems are head-trained-spur pruned, cordon-trained-spur pruned, and head-trained-cane pruned (Fig. 11-1).

Head Training–Spur Pruning

The vine has the shape of a small upright shrub, with a vertical trunk 1 to 3 ft (0.30–0.91 m) high that supports arms spaced around its head. At winter pruning, spurs are left to produce the shoots that will bear the next crop and furnish canes for the following year's spurs. Head train-

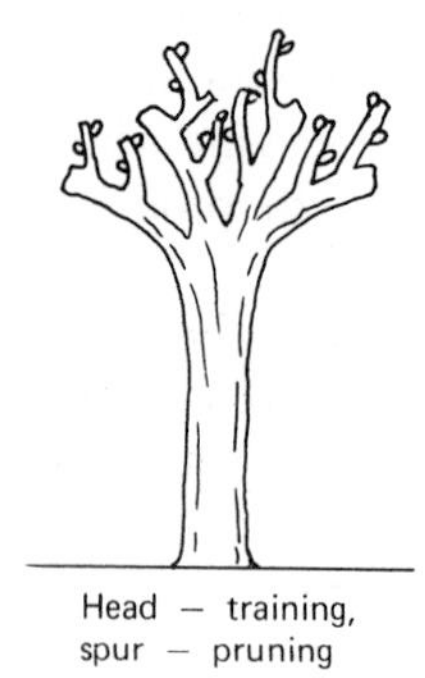

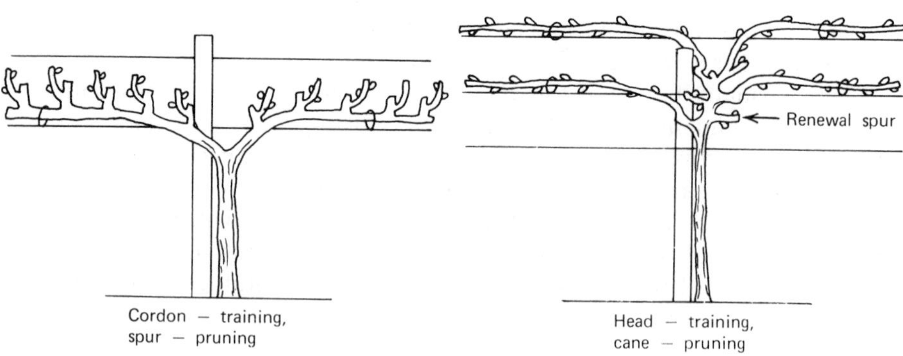

Fig. 11-1. Diagrammatic sketches of the three main training-pruning systems.

ing is used on many wine grapes. Its' advantages are simplicity of shape, ease of training, inexpensive to establish, and it does not require a wire trellis. Short stakes are required for about 10 years, or until trunks are rigid enough to be self-supporting. Cross-cultivation is possible to help in controlling noxious weeds.

Disadvantages of the head system are that it requires severe pruning which depresses the growth of the vine, and vines are slow to come into full production. The fruit often masses within a small area, causing bunch rot and poorer color on some varieties.

Cordon Training–Spur Pruning

The trunk of the bilateral, horizontal cordon rises vertically to about 8–12 in. (20.3–30.5 cm) below the lower wire of the trellis, and then divides into two branches that extend in opposite directions along the lower wire to within about 10 in. (25.4 cm) of the adjacent vines. The bends should be smooth and regular, and the branches should be straight. Shoots or spurs should not be retained at the bend of mature vines because they will become very vigorous and shade out the other spurs. Bearing units are spurs on small arms located at regular intervals on the horizontal branches (cordons). Spurs should be located on the upper surface of the cordons.

Advantages of the cordon system are that fruit is well distributed, hangs at about the same level from the ground, and usually colors well. This system is well adapted to mechanical harvest. Disadvantages are that it is laborious and costly to establish, and permanent support is required. It is well adapted for table grapes and large-clustered wine grapes.

Head Training–Cane Pruning

The shape is similar to that of head-trained vines, except that the head may be fanshaped in the plane of the trellis. Only two or three arms on each side of the head are usually needed. At winter pruning, fruit canes with 8 to 15 buds [2–4 ft (0.61–1.22 m) long] are retained for fruit production and old fruit canes are removed. Canes for use the following year are produced from the 3 or 4 renewal spurs (usually 2 buds long) retained in the head of the vine. Occasionally growers leave no renewal spurs because ample shoots develop from basal buds of the canes in some vineyards.

Advantages of cane pruning are that a full crop is obtained for varieties that have few or no fruitful buds near the base of the cane, such as in Thompson Seedless. A larger crop can be obtained for very small-clustered wine varieties such as Cabernet Sauvignon, White Riesling, and Sauvignon blanc, and the fruit is well distributed.

Disadvantages of cane pruning include the tendency of most varieties to overcrop fruit and the high cost of pruning and tying canes. A trellis is required.

TECHNIQUE OF PRUNING MATURE VINES

Head Training–Spur Pruning

Vines should be pruned according to their capacity (Fig. 11-2). In head-trained spur-pruned vines, the number and length of spurs left the previous year, the size of the canes, and the number of clusters produced during the current season may be used as a guide in determining the number and length of spurs to retain. The number of clusters produced may be determined by counting the stubs remaining where the clusters were cut off. A vine with canes of normal size that produced a good crop should be pruned to about the same number of spurs of similar length, as measured by the number of buds, as the previous year. If canes are larger than normal, they should be pruned less severely, leaving more or longer spurs or both, so that the vine capacity can be used fully for fruit production. Weaker than normal vines should be pruned more severely, leaving fewer buds, by retaining fewer or shorter spurs or both.

Spurs from vigorous canes should have more buds than those from weak canes. Basal buds, ¼ in. (6.35 mm) or closer to the base of the cane, should not be counted. Pencil-size canes should be cut back to 1 bud, and thumb-sized ones to 3 or 4 buds. The best canes are often the 2 or 3-bud spurs of intermediate size.

A more exact method of determining capacity is to prune the vine and weigh the pruning brush. The number of buds to be retained is proportional to the weight of brush. For example, in Thompson Seedless growing at Davis a formula of 33 nodes for the first 2⅕ lb (1 kg) of prunings with an additional 11 nodes for each additional 2⅕ lb (1 kg) might be appropriate (Lider et al., 1975). One must prune lightly enough before weighing so that there are plenty of buds left to make the final pruning adjustment. Early work on this system called balanced pruning, was done in New York State (Kimball and Shaulis, 1958).

Cordon Training–Spur Pruning

Cordon-trained vines are pruned to spurs, and the same techniques are used as for head training-spur pruning.

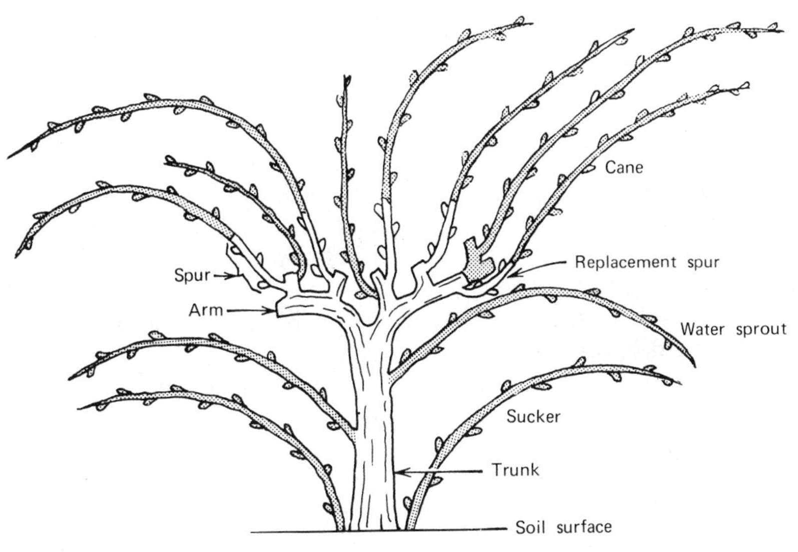

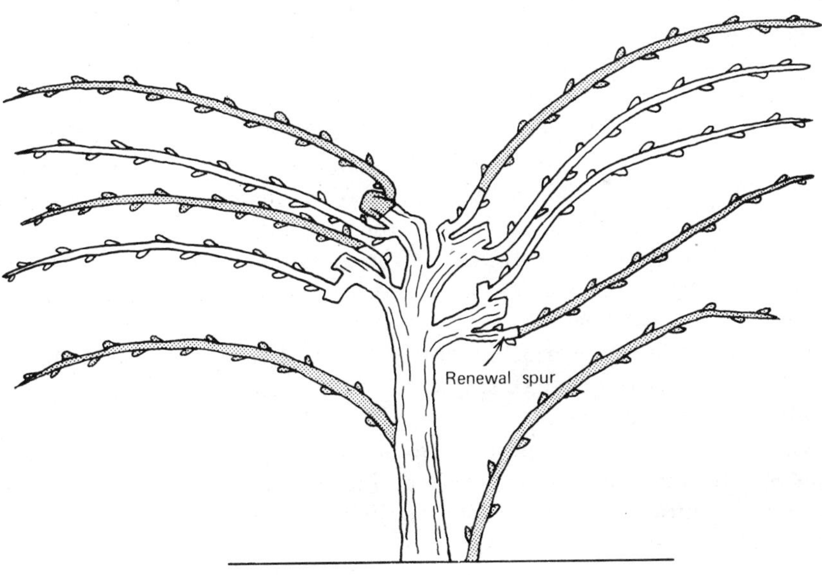

Fig. 11-2. Diagrammatic drawing showing method of pruning a head-trained spur-pruned vine (upper) and a head-trained cane-pruned vine (lower). Dark-shaded water sprouts, suckers, and portions of canes are removed at pruning.

Head Training–Cane Pruning

In the dormant season there are several canes that have usually arisen from renewal spurs, from basal buds, or from old wood (Fig. 11-2). Select strong canes and prune off the weak ones. If two good canes arise from a renewal spur, use the apical one for the cane, and cut the lower one back to a renewal spur. Large canes of ½ in. (12.7 mm) or greater diameter can be pruned to 15 buds, but smaller canes must be pruned shorter. For table grape production where thinning is used, one can leave a standard number of canes. Renewal spurs, usually one per cane, should be placed in the best positions to maintain proper shape of the vine; however, some growers find renewal spurs are unnecessary.

SUMMER PRUNING

Summer pruning consists of the removal of green vegetative tissues during the growing season and includes suckering, pinching, topping, and leaf removal. *Suckering* is the removal of water sprouts from the trunk and below the soil surface. It should be performed several times each year when the vines are young and at least once a year on older vines. *Crown suckering,* the removal of water sprouts from the branches and arms, should be done early when shoots are about 3–6 in. (7.62–15.2 cm) long. Its purpose is to improve the quality of the fruit or to direct growth into other vine parts, and it can cause considerable fruit thinning. Crown suckering is useful on shoots of varieties that produce many water sprouts, such as Muscat of Alexandria. Sufficient sprouts must be retained to shade the head of the vine and its clusters.

Pinching refers to the removal of the growing tip, (usually 3–6 in.) (7.62–15.2 cm) and is used to decrease wind damage and to train young vines.

Topping is the removal of the apical 1–2 ft (30.5–60.9 cm) of a shoot. It provides protection against the wind, but the loss of foliage can greatly weaken the vine.

Leaf removal is often performed on varieties such as Tokay and Emperor to enhance fruit coloration. Leaves are usually removed in June in the cluster area, to allow more light to enter, prevent rubbing of the fruit by the leaves, and to facilitate harvest.

Cardinal and Ribier are quite similar in growth and fruiting characteristics and require extensive hand labor for crown suckering, flower-cluster trimming, lateral shoot removal, removal of shoulders on retained clusters, and removal of leaves for fruit exposure. This must be done at early prebloom when lateral shoots are still tender and easy to remove.

There are usually one, two, or three clusters per shoot on Cardinal and Ribier. Where there are three clusters, growers usually remove the basal cluster, take off the wing on the middle cluster, and remove the tendril of the apical cluster if present. It is common practice to remove the large basal leaves; lateral shoots must also be removed while they are small. Some growers take off the leaves and lateral shoots from the apical cluster to the base of the shoot.

Two basal leaves can be removed without adverse effects, but removal of all leaves from the apical cluster downward greatly reduces the yield of packable fruit, because of a poor set of berries that result in straggly clusters with many shot berries (Jensen et al., 1975).

TOOLS FOR PRUNING

Pruning shears are usually used for pruning (Fig. 11-3). One-handed shears are suitable for smaller vines, but two-handed shears are more

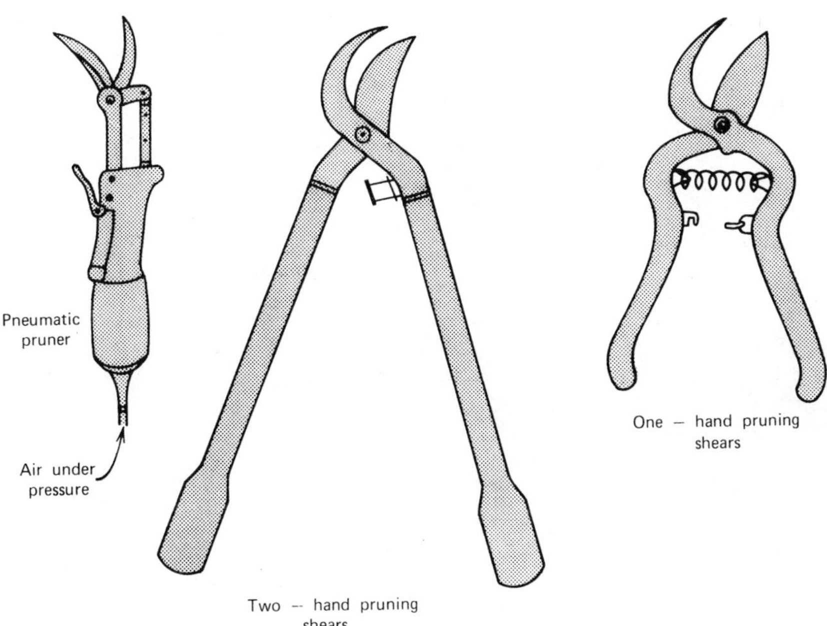

Fig. 11-3. Pruning tools.

appropriate for large ones. Pneumatic air systems, a simpler and more rapid method, are also used for pruning. With this technique, an air reservoir with air under pressure is connected to the pruning shears of the pruners by air lines on a boom. By this method several rows can be pruned at the same time from one power source.

Another pruning aid is a machine called a hedger, which can trim off many canes on the tops and sides of the vines, and thus simplify the rest of the hand pruning. Another machine can trim off a portion of the green shoots on the sides of the vines in summer, a task often performed before harvest to facilitate picking and make entrance into the vineyard easier.

Chapter 12
SPACING AND TRAINING

SPACING OF VINES

Wide spacing of vines (distance between rows is relatively long, often 12 ft, has the advantage of reducing harvesting labor because grapes can be hauled out from between the rows. It also cuts the cost of pruning brush disposal, facilitates the use of power-dusting or spraying equipment, and makes irrigation easier (Jacob 1950; Winkler 1959). The best spacing is a compromise and is the widest feasible, without reducing the crop of the mature vineyard, that is compatible with vineyard operations. The advantage of closer planting is that larger crops can usually be obtained, especially when the vines are young.

Spacing in the Interior Valley

Nearly all table grapes and the most vigorous wine varieties develop well at a spacing of 8 × 12 ft (2.44 × 3.66 m), i.e., 8 ft (2.44 m) between vines and 12 ft (3.66 m) between rows. Moderate-growing varieties should be planted at 6 or 7 × 12 ft (1.83 or 2.13 × 3.66 m), and the most vigorous varieties under highly favorable conditions can be planted at 8 × 12 ft. (2.44 × 3.66 m).

Spacing in Coastal Areas

Moderately vigorous varieties should be spaced at 6 × 12 ft (1.83 × 3.66 m), and vigorous varieties at 7 or 8 × 12 ft. (2.13 or 2.44 × 3.66 m).

Vine Spacing for Machine Harvest

In California, although row widths have generally been 12 ft. (3.66 m), narrower row spacing is being considered to produce optimum returns. With the introduction of over-row harvesters and other cultural equipment developed for narrow rows, mechanically harvesting and normal cultural operations can be performed in vineyards with row widths of 10–12 ft. (3.05–3.66 m) (Christensen et al., 1973).

In coastal counties, spacings of 8 × 12 ft. and 6 × 10 ft. (2.44 × 3.66 and 1.83 × 3.05 m) can be used according to the soil type and variety. In heavier soils and with vigorous varieties wider spacing is required.

PLANTING THE VINEYARD (ADAPTED FROM WINKLER, 1959)

Use 1-year-old rootings except when phylloxera or nematodes are present and require resistant rootstocks. Strong cuttings can sometimes make a successful stand if there is plenty of water. Use certified stock free of known virus disease.

Surveying instruments can be used to divide large vineyards accurately into blocks of desirable size. Locate the position of each row with a surveyor's chain or with a special "row chain" made by melting buttons of solder on a galvanized wire at the distance the rows are to be spaced.

Subsoil the rows after determining their position, to save labor in staking and planting. Locate the positions for the individual vines in each row with a planting line with solder buttons spaced according to the distance between vines in the row. You may plant the vines to the buttons on the planting line, but spacing will be more accurate if temporary pegs or permanent stakes driven in at each button are used.

Handling of Plants

Protect the vines from drying out in all handling operations. When stored, they should be partly or completely buried in moist sand or soil in a cool place; while being moved to the vineyard, they must be well covered with moist sacks or canvas or hauled in tubs containing 2 or 3 in. (5.08 or 7.62 cm) of water. Before planting, cut back the tops of rootings or graftings to a single spur of 1 or 2 buds. For convenience in planting, shorten the roots to 3–4 in. (7.6–10.2 cm) if the holes are dug or made with a power-driven auger, and to 1 in. (2.54 cm) if they are made with hydraulic pressure. In this technique a strong stream of water is injected into the soil to make a hole. Remove all roots within 8 in. (20.3 cm) of the top of the pruned vine.

Planting Operation

Dig holes for the vines on the same side of the planting pegs or stakes, in the row, so that the side or corner of the hole at the peg slopes away from the peg 2 or 4 in. (5.08 or 10.2 cm) distant at the bottom. The hole should be slightly deeper than the vine is long. Place the vine into the hole with the top close to the peg and fill the hole to about one-half of its depth with moist topsoil; then raise the vine to the proper height and pack the soil solidly around the roots. Fill the hole almost full and again pack the soil firmly, then fill the hole completely. In some cases cover the top of the vine with pulverized, loose soil. This can be done under dry, windy conditions. The top of the vine should be exactly at the side of the peg and the roots 2–4 in. (5.08–10.2 cm) away from the peg.

Vines planted by hydraulic means also require firming of the soil around the vine after the water has drained out. Covering with loose soil can be done when necessary.

Plant rootings of fruiting varieties so that the 2 buds retained after pruning are just above the general level of the ground on level land. On contour slopes they should be 2–5 in. (5.08–12.7 cm) above the ground, depending on the slope.

Bench-grafted vines are planted with the union about 3–4 in. (7.62–10.2 cm) aboveground. Rootstock rootings that are to be budded or grafted in the field should have 4 or 5 in. (10.2 or 12.7 cm) of the main stem aboveground, so that the graft union will be above the surface of the soil to prevent the development of scion roots. Cover all tops of dormant stocks to a depth of ½–2 in. (1.27–5.08 cm) with a mound of loose soil to prevent drying before growth starts.

Planting with Preirrigation

A vineyardist (J. Cederquist) in the Salinas Valley uses a successful method of planting after irrigation. After subsoiling strips about 20 in. (50.8 cm) deep, the soil is irrigated. As soon as the soil is dry enough to get onto the land, a pointed steel probe [about 3 feet (91.4 cm)] long and 1.5 in. (3.81 cm) in diameter) is pushed about 18 in. (45.7 cm) deep into the soil. The hole is then enlarged by wiggling the top of the probe. Rootings with roots trimmed back to ½ to ¾ inch (1.27–1.91 cm) long are pushed into the hole, and the soil is tramped. The probe is then inserted into hole and pushed against the cutting so it is tight and cannot be pulled up easily. Then the rooting is covered with soil and tramped again, and the top of the rooting is covered with soil. No irrigation is made for 2 or 3 weeks, until a new shoot has pushed through the soil surface.

TRAINING

There are many different training systems throughout the world, some of which are shown in Figure 12-1. In California the main types are

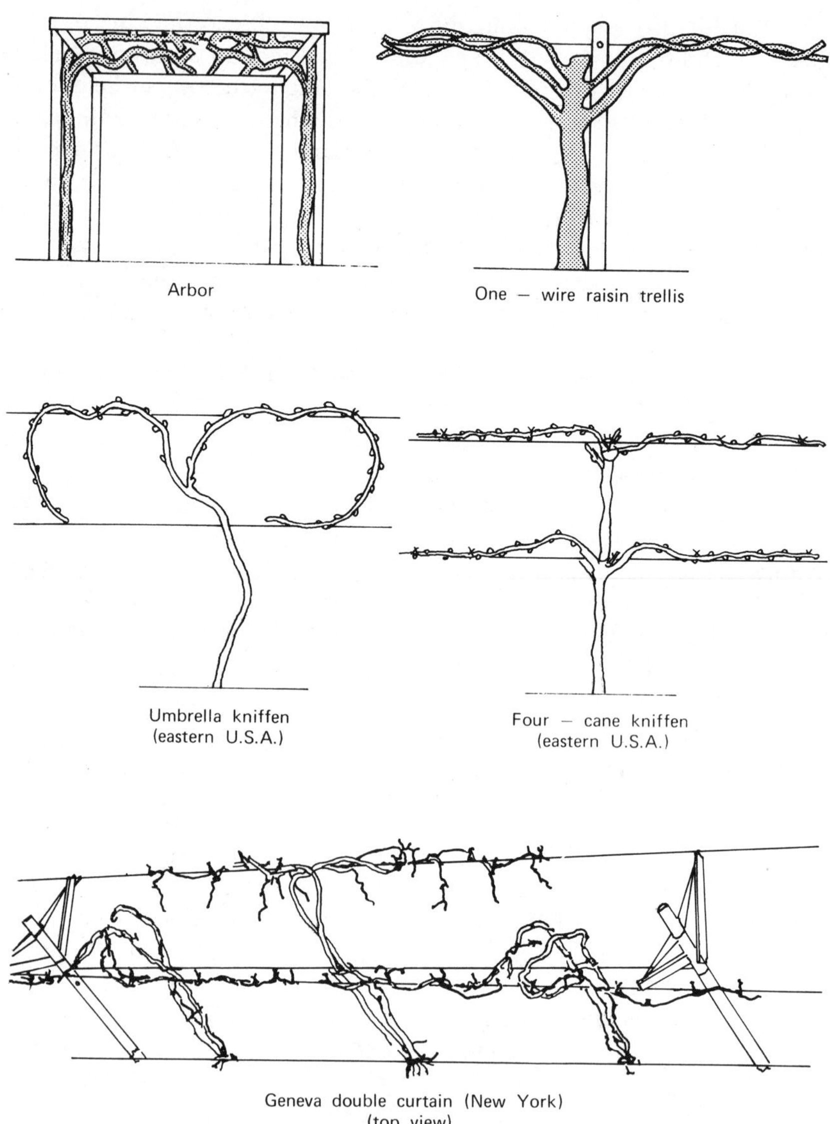

Fig. 12-1. Diagrammatic drawings of various types of vine training found in countries throughout the world. (Nos. 3, 4, after Anonymous, 1973; No. 5 after Bulletin 811, "The Geneva Double Curtain for Vigorous Grape Vines". New York Agricultural Experimental Station, Geneva, N.Y.)

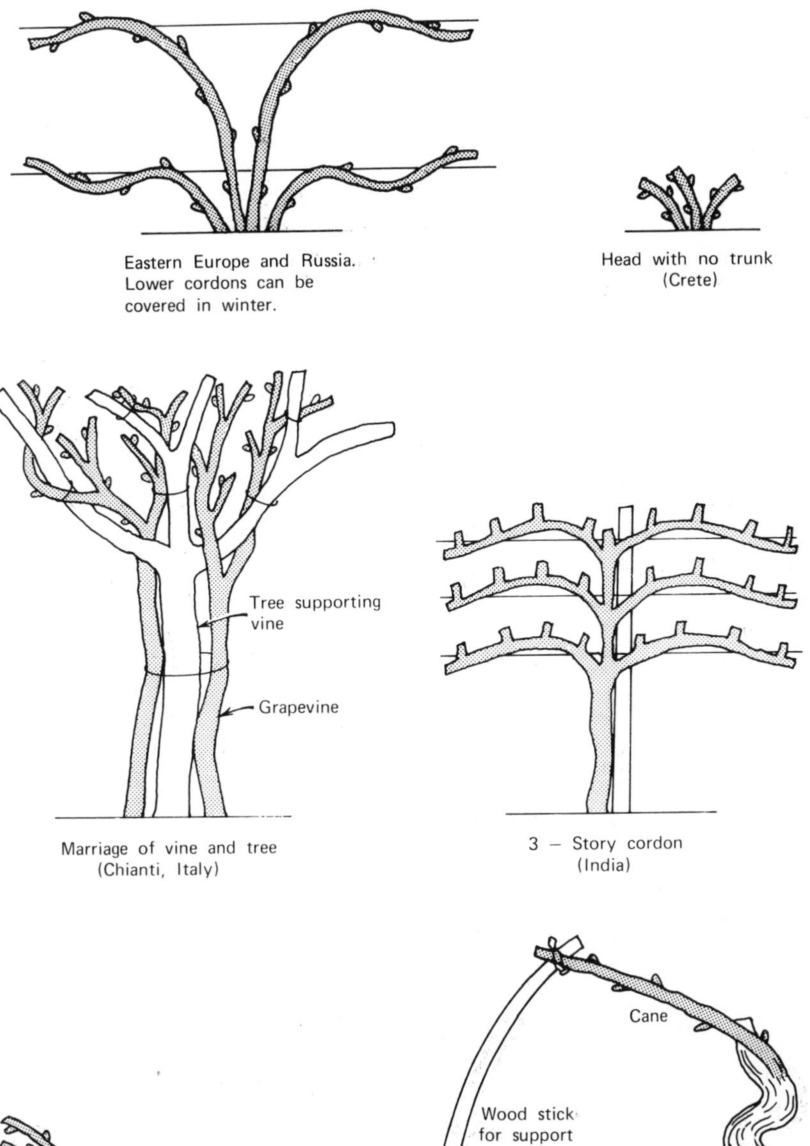

Fig. 12-1. (Continued)

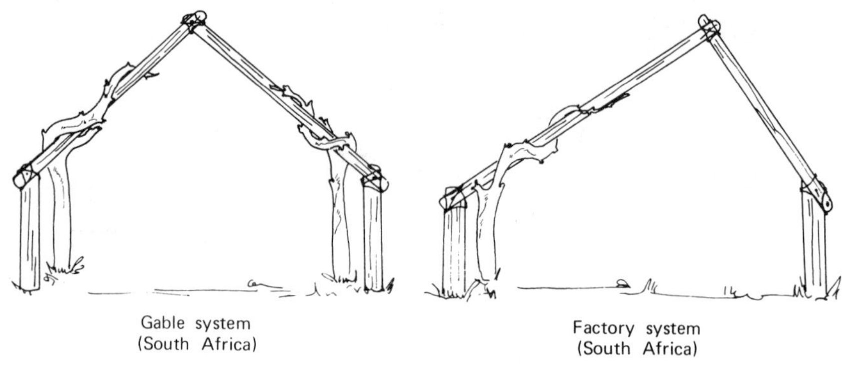

Gable system
(South Africa)

Factory system
(South Africa)

Fig. 12-1. (Continued)

head, cordon, and cane (Fig. 11-1). Usually four seasons of growth or more are required to train a young vine properly. Soon after rootings are planted in the vineyard, the first growing season commences.

Head Training (Adapted from Winkler 1959)

First Growing Season. The main goal is to develop a root system. Usually, at least one irrigation in late spring or early summer is necessary to promote growth of the limited root system of the young vines. In hot climates two, three, or more applications may be required. Late summer or early fall irrigation should be avoided as this may stimulate shoot growth that may fail to mature and thus be injured by

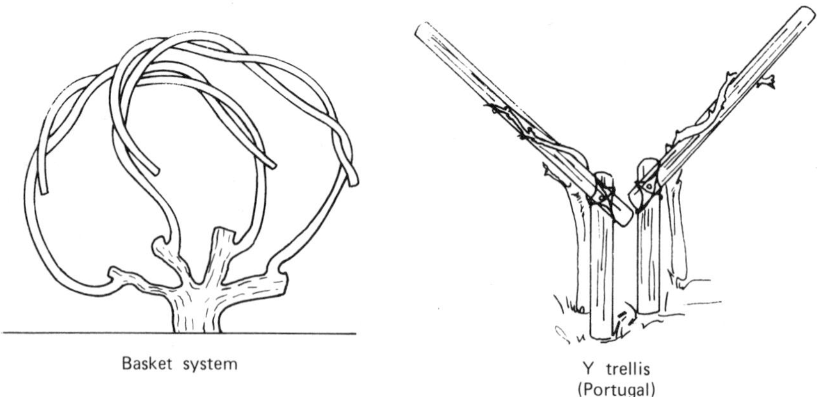

Basket system

Y trellis
(Portugal)

Fig. 12-1. (Continued)

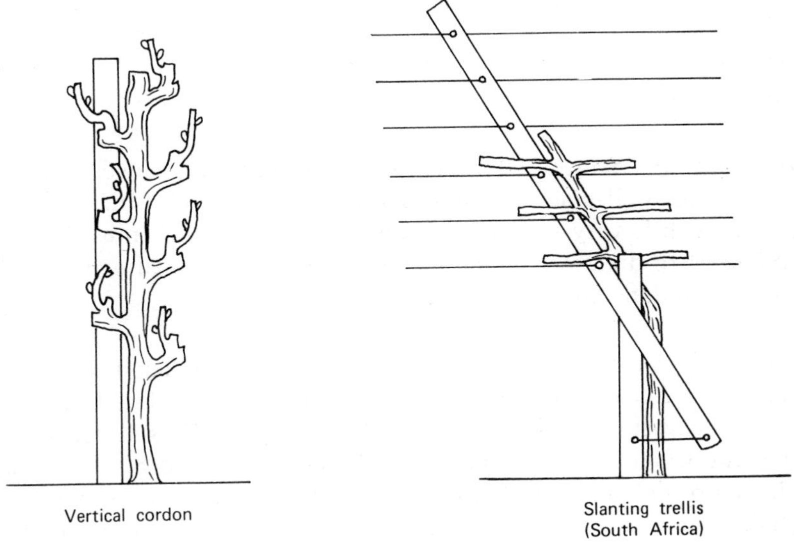

Vertical cordon

Slanting trellis
(South Africa)

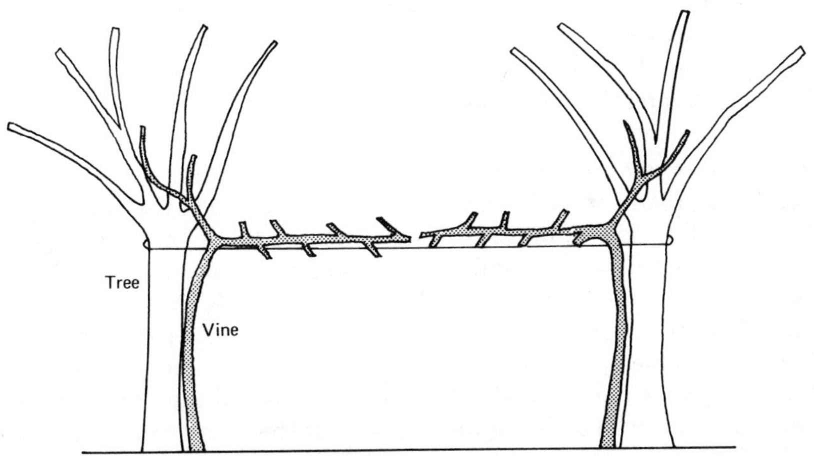

Wires strung between trees
(Italy)

Fig. 12-1. (Continued)

early winter frost. Generally no pruning or training should be done in the first growing season except in very hot regions, where with very vigorous growth the vines may often be trained during the first summer in the manner described for the second summer.

First Dormant Season. The top growth should be pruned except for the strongest well-placed cane, which is pruned to 3 or 4 buds (Fig. 12-2). If growth is very vigorous a cane may be left, but it must be $\frac{5}{16}$ in. (7.87 mm) in diameter or more. Vines should be staked or trellised if this was not done before planting.

Second Growing Season. As soon as new shoots are long enough to tie (about 8 in; 20.3 cm), rub off all but one that is strong and well positioned for growing vertically along the stake (Fig. 12-2). Tie the selected shoot to the stake when it is 8–12 in. (20.3–30.5 cm) long, and retie it once or twice more until it reaches the height at which the trunk divides; this will develop a straight trunk. Remove all other shoots from the old wood as they begin to develop, and break off laterals that grow on the lower half of the reserved shoot.

The main shoot should not be topped until 12–18 in. (30.5–45.7 cm) of shoot apex can be removed. Top above the point at which the trunk will divide to develop arms, to stimulate lateral branch development. Retain only laterals on the upper third to half of the main shoot.

Second Dormant Season. Cut off the trunk at the first node above the area where the head is desired. Make the cut through the node so that the bud will be killed, but leave the swollen portion to make tying easy. Remove all small laterals and all laterals on the lower half of the trunk. On large vines, two or more laterals over $\frac{5}{16}$ in. (7.87 mm) thick on the upper half of the cane can be cut back to 1–3 buds depending on their diameter. These spurs may produce fruit and are necessary for rapid development of the head. On vigorous cane-pruned vines a single fruit cane may be left in addition to the spurs. If the trunk cane is less than $\frac{5}{16}$ in. (7.87 mm) thick at the desired height for the head, it should usually be cut back to 2 buds, and a strong trunk developed the following growing season. Tie the vine to the stake just below the enlargement of the apical node, and place a loose tie around the stake and the trunk cane at about the middle.

Third Growing Season. Some fruit is usually produced, but the main objective is to develop the permanent framework of the vine. While they are small, break off all shoots that grow from the lower half of the

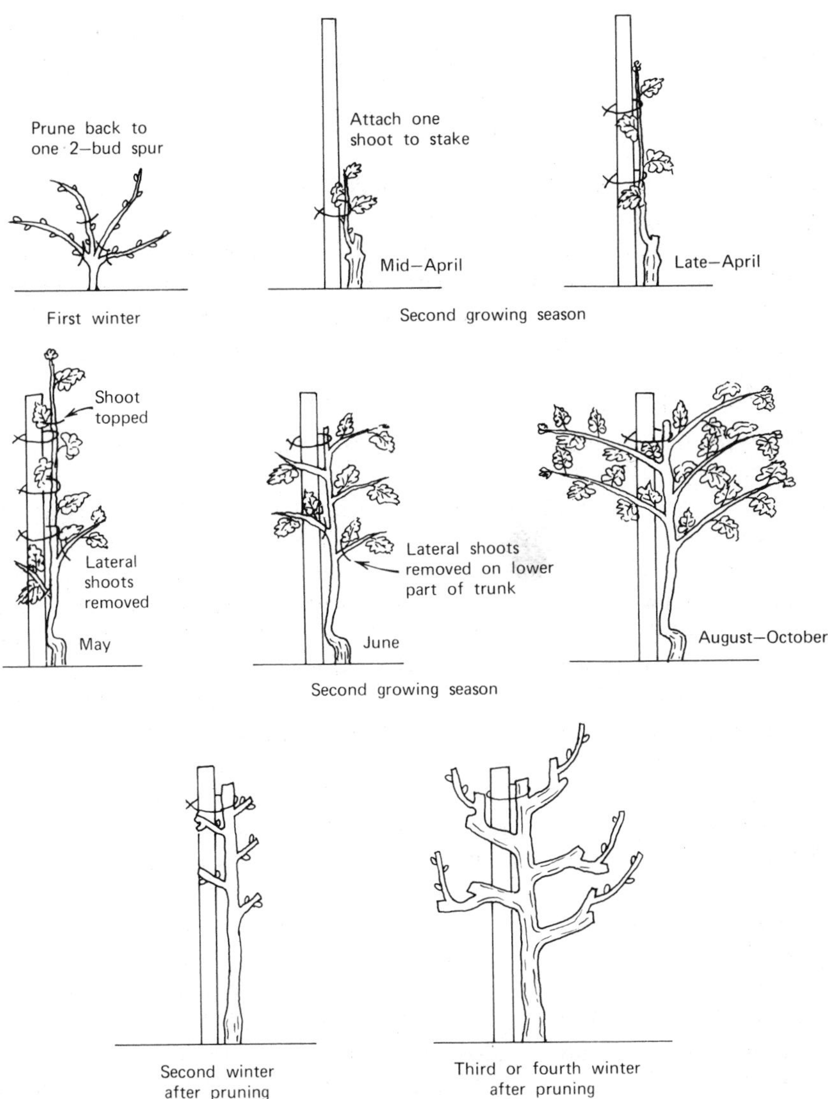

Fig. 12-2. Steps in developing a head-trained vine.

trunks. Shoots on the upper half of the vine are allowed to develop, and the most vigorous ones should be tied or pinched back if there is danger of wind breakage.

Third Dormant Season. Leave 3 to 6 spurs 2–4 buds long depending on vine vigor. Spurs should be as near the top of the vine as is practicable. On cane-pruned vines, leave one or two fruit canes, each 8 to 15 buds long, and tie them to the trellis. Leave two to four renewal spurs to furnish canes for the following year, all as nearly as possible the desired height of the permanent arms into which they will develop.

Fourth and Subsequent Seasons. The goal is to perfect the structure of the vines so that they will bear a large crop of high-quality fruit and cultural operations will be facilitated. Develop vines gradually into symmetrical shapes. Train the heads of cane-pruned vines fan-shaped in the plane of the trellis so they will not interfere with cultivation. During summer, remove all water sprouts from the trunk below the lowest arm on head-trained and cane-pruned vines.

Cordon Training (Bilateral Cordon) (adapted from Winkler, 1959)

Second Growing Season. Cordon training is similar to head training until after topping (Fig. 12-3), then select two laterals to form the cordons. (Occasionally the main shoot and one lateral are used, but this technique is not recommended.) Remove all other laterals. The point where the trunk divides should be 8–12 in. (20.3–30.5 cm) below the wire of the trellis that will support the cordon. After laterals are 18–24 in. (45.7–60.9 cm) long, tie them on the wire on opposite sides of the head of the vine. As growth proceeds keep them straight by tying them loosely to the wire, but do not tie the portion of the shoot that is elongating, consisting of about the apical foot. Occasionally the laterals are pinched after they have grown about 18 in. (45.7 cm) beyond the halfway point to the adjacent vine, but this is usually not done.

Second Dormant-Season. Prune off branches to a point where they are at least 3/8 in. (9.65 mm) thick. If they are thick enough, cut them at a point halfway to the adjacent vines. If the canes are not large enough to reach at least 6 in. (15.2 cm) along the wire beyond the bend, cut them back to a bud on the lower side of the branch within a few inches of the point where the trunk was divided so that a more vigorous cane can be developed the following growing season. Remove all laterals on the trunk below the point of branching and, except in extremely vigorous vines, leave no spurs on the branches. When the cordon canes are full length, turn them around the wire one to 1½ times to keep them straight. This also places the weight of the cordon and subsequently developing shoots on the wire.

Third Growing Season. Rub off shoots on the underside of the

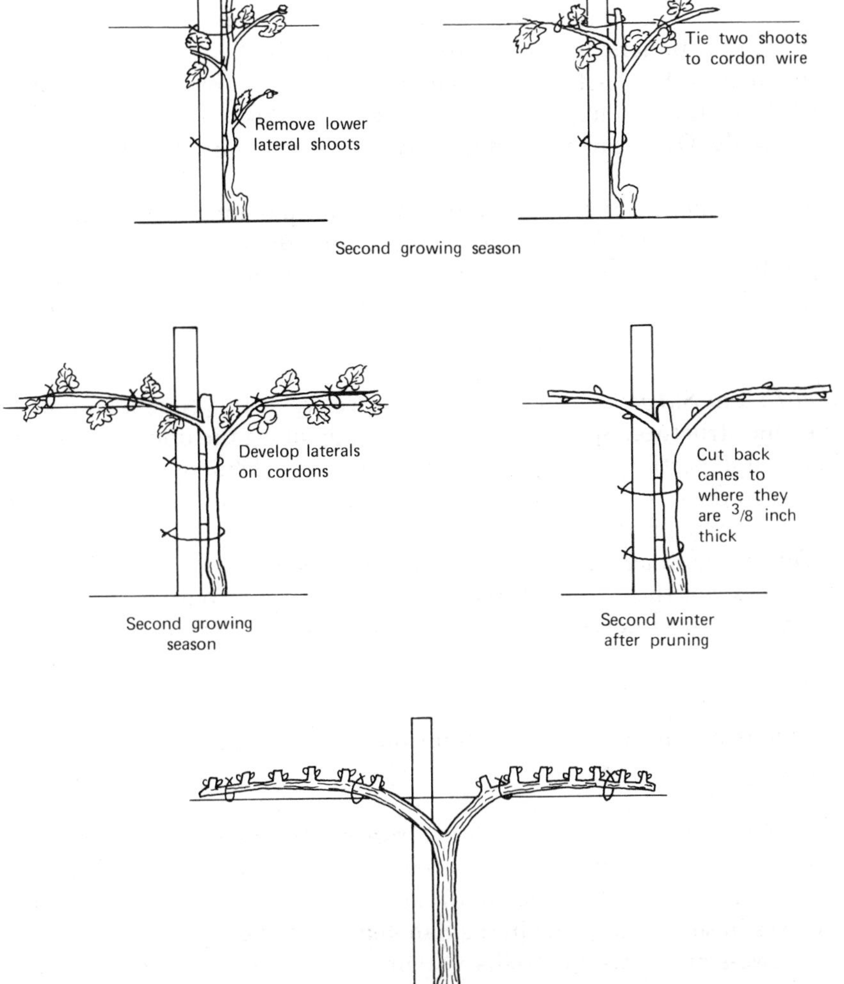

Fig. 12-3. Steps in developing a cordon-trained vine.

branches when the vines begin to grow. Leave shoots on the upper side of the branches, spaced 6–10 in. (15.2–25.4 cm) apart. Also remove all shoots on the trunk or on the bends of the branches several times.

Occasionally the shoots are pinched as soon as five or six leaves have formed to check their growth and allow the weaker shoots to develop evenly with the longer ones; however, this operation is usually not done. With cordons that are not the desired length, develop a shoot from near the end of the incomplete branch and tie it to the wire. A shoot developing from the underside of the cordon is usually best.

As soon as the shoots are long enough, tie several from near the end of each branch to an upper wire to prevent the shoots from turning over completely. Overbearing must usually be prevented by thinning.

Third Dormant Season. Retain spurs 1–3 buds long at regular intervals of 8–12 in. (20.3–30.5 cm), spaced along the upper side of the cordons. Remove all other canes, cut all old ties on the trunk and branches, and retie the vine to make the horizontal portions of the branches as straight as possible.

Fourth and Subsequent Years. The objective at this time is to perfect the vine structure. Space the arms uniformly along the upper side of the horizontal portions of the branches and keep them upright by tying shoots to the top wire. Maintain the arms at the same height and at uniform vigor by pruning and pinching.

During summer remove all water sprouts from the branches and trunks except those needed in developing new arms. With very vigorous vines, the 4-year development work may be accomplished in 3 years.

Arbors

This is usually considered a multiple cordon system with spur pruning (Fig. 12-1).

Two-Story Trellises for Table Grapes (Winkler and Kasimatis, 1964). In the early 1950s, a second crossarm with three wires was added 12–15 in. (30.5–38.1 cm) above the first, above sloping-top trellises in some vineyards in the San Joaquin Valley.

A two-story flat-top trellis, originally designed for cane-pruned varieties grown for table fruit, is now also used for both cane- and cordon-pruned table grape varieties. This is constructed in the same manner as the one-arm sloping-top or flat-top trellis. The crossarms are attached and held in position in the same way, and when vines are spaced 8 ft (2.44 m) or more apart in the row, there should be a lower and upper crossarm at each vine.

Cane-pruned vines should be headed just below the lower crossarm and all fruit canes are tied to the wires on the lower crossarm. The cordons of cordon-trained, spur-pruned vines should be supported below the lower crossarm.

These trellises provide better and more uniform exposure of fruit to the light and also greater foliage cover than do single crossarm trellises. There is also a greater foliage display to the light because of better distribution of shoots and canes and greater trellis height that can result in small increases in yield.

TRAINING AND PRUNING FOR MECHANICAL HARVEST (CHRISTENSEN ET AL., 1973)

Although the principles involved are the same as those previously described, there are some important differences. Adequate training and trellising are necessary for successful mechanical harvesting (p. 200). Cordon training with spur pruning and head training with cane pruning are recommended.

Cordon-Trained, Spur–Pruned Vines

In training one should, if possible, retain only those spurs that are in a vertical line with the vine row, since spurs pointing outward are easily broken or interfere with the harvester rods or fingers. Although most wine grape varieties are usually best suited for spur pruning, harvesting may gradually become more difficult as the cordon-trained, spur-pruned vines increase in age and size. This problem can be minimized in new plantings by better training methods to eliminate poorly directed or positioned spurs.

Cane-Pruned Vines

Cane pruning is usually recommended for the small-clustered varieties, such as Chardonnay, Cabernet Sauvignon, Sauvignon blanc, and Black Corinth. It is also useful for such varieties as Thompson Seedless and Sultana, whose basal buds are not fruitful or of low fertility. Wine varieties that do not fit into these categories usually bear full crops if sufficient numbers of spurs are retained on the cordons.

The main adverse effects in cane pruning large-clustered varieties are overcropping and a decline in vigor. Most wine varieties are easily overcropped on canes because of a high percentage of fruitful buds (e.g., Rubired, Royalty, Muscat of Alexandria, and others) and/or large clusters (e.g., Chenin blanc, Carignane, Grenache, and others). Pruners often fail to prune severely enough by leaving too many canes for the vine capacity.

VINE SUPPORTS FOR MECHANICAL HARVEST (CHRISTENSEN, ET AL., 1973)

Stakes

Stake breakage can be minimized by using douglas fir (sawn), malaysian hardwoods (sawn) (kempas and keruing species), redwood (split), and steel. Sawn stakes with defects such as knots should be avoided, and stakes should be given a preservative treatment. Excessive breakage occurs in sawn redwood and cedar.

Lengthening the Life of Grape Supports (Winkler and Kasimatis, 1974)

In California split-coast redwood stakes may last 20 years or more, but they may rot and break off in about 10 years unless treated. Stakes from young trees or sapwood are especially subject to rot.

Stakes should be treated with a suitable wood preservative. Soak the stakes in undiluted wood-preservative creosote or in a 5 percent solution of pentachlorophenol in diesel oil for at least 24 hr. Only that portion of the stake that goes into the ground plus 6 in. need be treated. Some stakes are now treated with preservatives under pressure.

Stakes treated with either of these preservatives must not be placed close to a young vine until the preservative has completely dried, which may require a month or more. A good practice is to treat the stakes during late summer prior to the time they will be used. At this season they are usually completely dry, insuring a good penetration and drying of the preservative before the stakes are used.

Wire. Wire breakage usually occurs from pruning shear nicks and not from the machine harvesting itself. To minimize wire breakage, the cordon or cane wire can consist of 10- or 11-gauge, low-carbon wire or 12- or 13-gauge, high-carbon wire. Twelve-gauge low-carbon wire or 13- or 14-gauge, high-carbon wire may be used for the upper foliage support wires. Table 12-1 compares low- and high-carbon wires of various gauges using domestic wire data.

Stapling on Rolling Terrain. When vines are planted on rolling lands and the slope change in the direction of the rows is variable, severe strain is placed on the staples holding the trellis wires. The low and high points in the row are always under tension from the vine load. Staples can become loose with continued exposure to weather, and the problem of staple loss during machine harvesting is increased. Attaching the wire

Table 12-1. Comparison of Low- and High-Carbon Wires of Various Gauges*

Wire Type (tensile strength)	Gauge	Wire Diameter (inches)	Wire Diameter (cm)	Feet per Pound	Meters per kg	Minimum Breaking Strength (pounds)	Minimum Breaking Strength (kg)	Pounds/Acre (1-wire)	kg/hectare (1-wire)
Low carbon	10	.1350	.3429	20.6	2.85	1144	518.9	176.2	197.5
High carbon	10	.1350	.3429	20.6	2.85	2863	1298.6	176.2	197.5
Low carbon	11	.1205	.3061	25.8	3.57	912	413.7	140.7	157.7
High carbon	11	.1205	.3061	25.8	3.57	2281	1034.6	140.7	157.7
Low carbon	12	.1055	.2680	33.7	4.66	699	317.1	107.7	120.7
High carbon	12	.1055	.2680	33.7	4.66	1748	792.9	107.7	120.7
Low carbon	13	.0915	.2413	44.8	6.19	526	238.6	80.85	90.62
High carbon	13	.0915	.2413	44.8	6.19	1315	596.4	80.85	90.62
Low carbon	14	.080	.2032	58.6	8.10	402	182.3	61.95	69.44
High carbon	14	.080	.2032	58.6	8.10	1005	455.9	61.95	69.44

* After Christensen et al., 1973.

to the stakes with wire straps, or using slotted or perforated metal stakes may correct this problem.

Trellis Design

There are several general systems that can be used with slapper or beater type machines for trellising for mechanical harvest, which include the following methods: (1) A 5.5-ft. (1.68 m) stake (from soil surface), 2- or 3-wire vertical system; (2) 5.5-ft. (1.68 m) stake with crossarm; (3) 4.5-ft. (1.37 m) stake, 2-wire vertical system; (4) 4.5-ft. (1.37 m) stake, 1-wire system; (5) drying option for raisin varieties (Fig. 12-4). Use either of the two 4.5-ft. (1.37 m) stake trellises, with or without a 2-ft. (0.61 m) wire crossarm on or near the top of the stake, or (6) alternating stake heights (Christensen et al., 1973). see p. 207 for duplex system.)

Method 1 [5.5-ft. (1.68 m) stake]. The most common trellis used for wine varieties. The cordon wire is located 40–46 in. (1.02–1.17 m) above the soil surface. A foliage support wire is usually attached 12–14 in. (30.5–35.6 cm) above the cordon wire, but only 10–12 in. (25.4–30.5 cm) above under windy conditions. To prevent the cordon from twisting, this wire should be attached the year the vine is trained.

The top wire can be moved upward for better foliage support, as the spur positions rise with age. A third wire may be added at the top of the stake for additional support, and be installed along with the other two (Fig. 12-4).

Method 2 [5.5-ft. (1.68 m) stake]. A crossarm may be added to the 2-wire vertical trellis described. Once this trellis is installed, the decision to add a crossarm can be made at a later date. Crossarms are used principally for additional foliage support in very vigorous vines. Spur breakage and foliage interference during harvest is minimized and much of the foliage can be retained beyond harvest.

Very vigorous French Colombard, Sauvignon blanc, Grenache, or Chenin blanc vines can use up to a 36-in. (91.4 cm) crossarm width, whereas a 24–30-in. (60.9–76.2 cm) crossarm will suffice for most others. The harvester to be used must accommodate the crossarm width and design.

Method 3 [4.5-ft. (1.37 m) stake]. This technique can be used for low- to moderate-vigor vineyards, spur-pruned wine varieties, or cane-

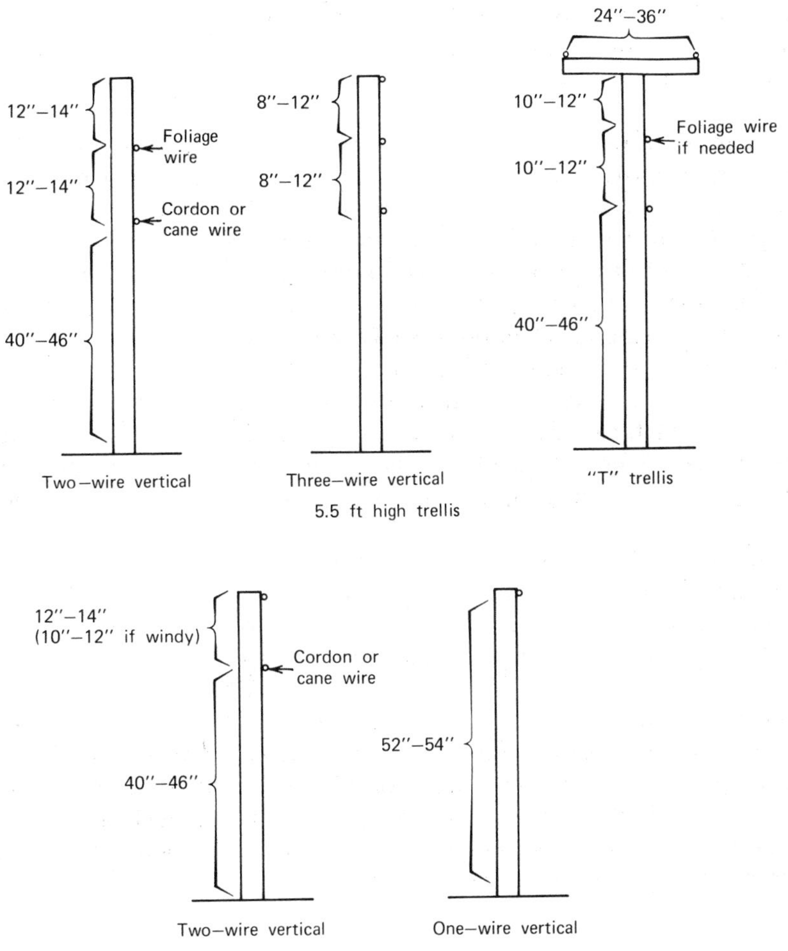

Fig. 12-4. Several trellis systems used with slapper or beater type mechanical harvesters (After Christensen et al., 1973.)

pruned varieties. The lower cordon or cane wire should be at 40–46 in. (1.02–1.17 m). The second or overhead foliage support wire is attached to the top of the stake. The distance between the two wires should be 12–14 in. (30.5–35.6 cm) under average conditions and 10–12 in. (25.4–30.5 cm) in windy areas. The top wire is especially important for catching and supporting upright shoots to prevent cordon twisting in young vines. It should be installed before training the vines.

Method 4 [4.5-ft. (1.37 m) stake]. Used primarily for the cane-pruned Thompson Seedless variety. A single wire is attached to the top of the stake, usually 54 in. (1.37 m) from the ground.

Method 5. Allows for drying of raisins. Either of the two 4.5-ft. (1.37 m) trellises can be used. In addition, a 2-ft. crossarm can be attached with a wire at each end on or near the top of a 4.5-ft. (1.37 m) stake. Raisin drying and machine harvesting are still possible with a crossarm, but exact specifications on crossarm width and careful harvesting procedures are needed. The crossarm provides additional foliage support for vigorous Thompson Seedless and is desirable in Muscat of Alexandria, which is susceptible to fruit sunburn.

The lower wire should be attached on the stake 16 in. (40.6 cm) below the crossarm for cordon-trained vines and 18 in. (45.7 cm) below for cane-pruned vines. This places the cordon approximately 36–38 in. (91.4–96.5 cm) from ground level, the minimum distance for efficient machine harvesting.

The raisin-type varieties, Thompson Seedless, Muscat of Alexandria, and Zante Currant, can also be trellised on 5.5-ft. (1.68 m) stakes as described above. However, row shading from these vines with or without a "T" trellis, eliminates the option to dry the raisins.

Method 6. Alternating 3.5- or 4.5-ft. (1.07–1.37 cm) stakes with 5.5-ft. (1.68 m) stakes is satisfactory in vertical wire support systems. Attaching crossarms only to alternating stakes is also possible with 4.5- or 5.5-ft. (1.37 or 1.68 m) vine spacings in the row, although attaching a crossarm on every stake assures good support and should be used in vines and heavy foliage.

Tying of Vine Parts to Supports

Adequate tying of vines for fruit support is required, especially in cane-pruned and young cordon-trained vines. Loose cane ends may hang too low for harvester recovery and tend to swing with the harvest mechanism, which reduces fruit removal and may cause an increase in vine breakage.

Coated wire ties can be used when the fruit load is light. Twine, vinyl tape, or plastic ties should be used where heavy fruit loads are anticipated. Aluminum cordon ties should not be used on cordons or trunks, as they are often harvested along with the fruit (Christensen et al., 1973).

End Posts and Anchors. Several good end-post materials include steel, wood, and reinforced concrete. Post widths or diameters up to 6 in. (15.2 cm) have caused no difficulty with harvesters entering or leaving rows.

The first vine should be 8 ft. (2.44 m) or more from the end post; this allows complete fruit removal from end vines. Vines placed closer to the end posts should be trained away from them.

End anchors or deadmen avoid harvester interference problems often caused by end posts. The anchor may be a concrete deadman, a buried steel plate or steel pipe, or a short post protruding 12–18 in. (30.4–45.7 cm) above the soil (Fig. 12-5). The angle or incline of the wires from the anchor to the first stake should be more than 45°; if it is less the closure on the catching frame may not open properly and will bind on the wire when entering or leaving the row. Also, it is best to attach the wire to a double stake assembly, since this allows for a more secure attachment for the bottom wire and absorbs the additional wire strain on the first stake. A short marker stake outside of the deadman is recommended to make the assembly boundary visible to equipment operators.

Headland Width

To provide adequate turning space, harvesters should have at least 20 ft. (6.01 m) of headland width. In addition, transfer of fruit from

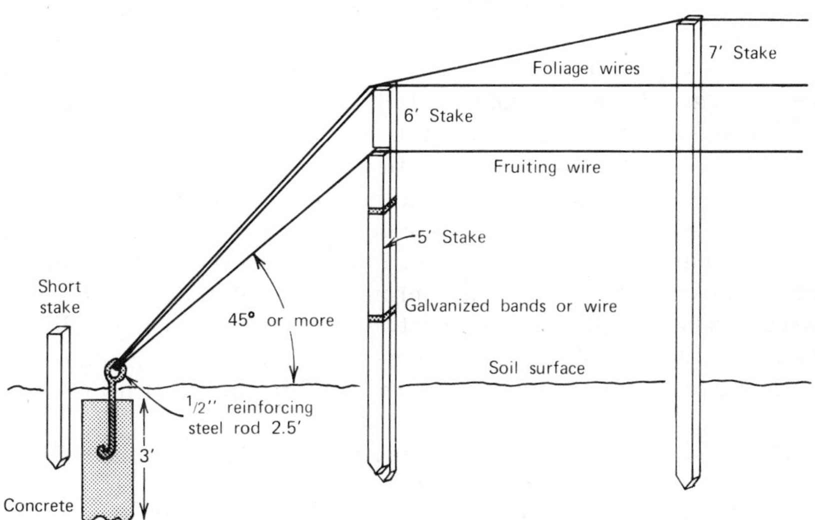

Fig. 12-5. Concrete deadman and end stake assembly used for mechanically harvested grapes. (Redrawn from Christensen et al., 1973.)

vineyard gondolas to highway trucks requires a wider avenue of 36–40 ft. (11.0–12.2 m) spaced every quarter or half mile, (0.40–0.80 km) or the provision for other suitable loading areas.

Row Direction

In windy areas rows should be oriented to be perpendicular to the wind. Then the outside rows will receive the most damage, protecting the remainder of the vineyard. If raisin production is desired, orient the rows east and west.

VINE CONVERSION (ADAPTED FROM CHRISTENSEN ET AL., 1973)

Head-trained or low cordon-trained vines cannot be harvested mechanically with present machines unless they are first retrained.

Conversion of Head-Trained Vines to Bilateral Cordons

Vigorous vines less than 10 years old can be more easily converted than old ones. Straight trunks in line with the row are necessary for conversion; crooked trunks that extend more than 8 in. (20.3 cm) into the row must be cut off about 12 in. (30.5 cm) above the ground and a new trunk formed from a shoot arising as a water sprout or sucker (Fig. 12-6).

Head Height. If the original head is 6–12 in. (15.2–30.5 cm) lower than the proposed cordon wire, select two strong canes from the top of the vine in line with the row. Wrap and tie them to the cordon wire, and cut off all the other arms extending from the trunk (Fig. 12-6).

For low-headed vines, train a cane up to within 4–6 in. (10.2–15.2 cm) of the anticipated cordon wire height 1 year before you plan to complete the conversion (Fig. 12-6). In the spring, remove all the shoots from the selected cane except the terminal two or three, and remove all crop from the retained shoots. In the next dormant pruning, place two strong, mature canes on the cordon wire and remove all arms (Fig. 12-6).

High-headed vines (vertical cordons) require lowering of the height of the head. Cut the head back 1 year before conversion is to be completed. The height of the head should be 6–10 in. (15.2–25.4 cm) below the anticipated cordon wire. Cutting back the head induces strong cane

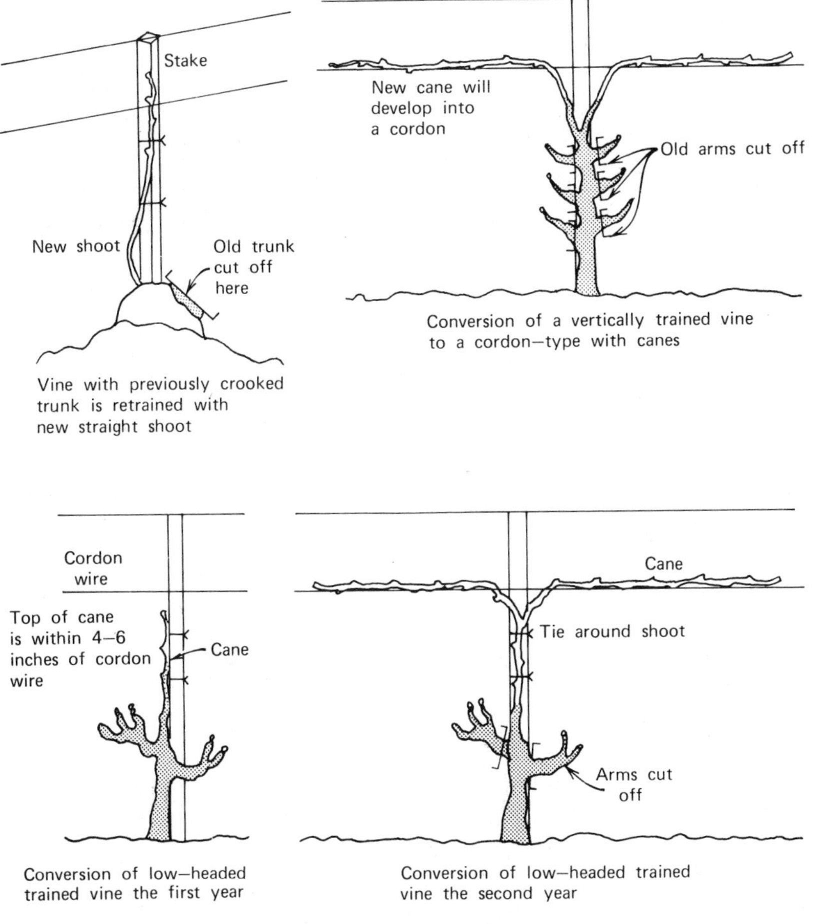

Fig. 12-6. Techniques for conversion of vines to a shape suitable for mechanical harvesting. (Redrawn from Christensen et al., 1973.)

development, so that the conversion can be completed the following dormant season.

Cropping Effects. It is important not to overcrop a vine the first year after conversion. On most medium- to large-clustered varieties, 8 buds on the portion of the cane tied on the wire are sufficient for a good crop. Shoot growth is poor on long canes that are overcropped.

Large-clustered varieties on a converted vine with 8 buds per cane

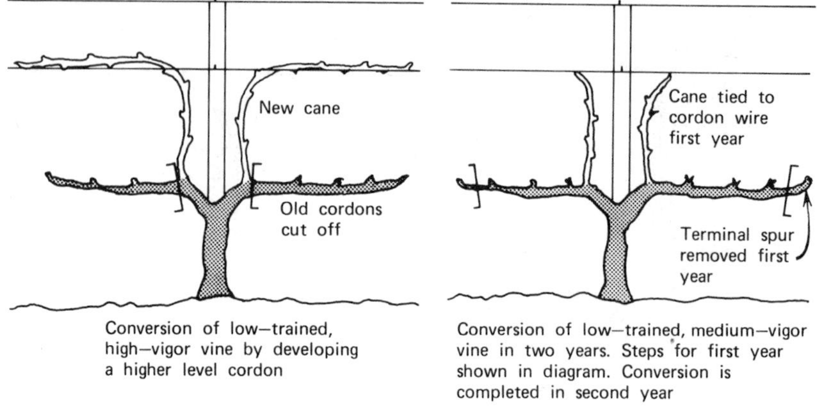

Fig. 12-6. (Continued)

(16 buds per vine) will usually produce as much as the head-trained vines did originally. If vines are very vigorous and high yields are needed to reduce vegetative growth, retain a spur or two on the lower portion of the trunks. However, this will delay full conversion an extra year. Leave fewer fruiting buds the year of conversion than were retained the year before conversion (Christensen, 1973).

Raising a Low Cordon-Trained Vine

Although most harvesters are effective as low as 36–40 in. (0.91–1.02 m), machine operators prefer the cordon to be 40–46 in. (1.02–1.17 m) above the soil surface. Large-clustered varieties should have the cordon at a minimum height of 42 in. (1.07 m). Low cordons should be retrained to improve machine efficiency.

Highly Vigorous Cordon Vines. Select one cane on each side of the stake that originates near the bend of the original cordon (Fig. 12-6). Tie the canes on the new cordon wire, retaining approximately 8 buds per cane on the wire.

Cut off old cordons just apical to the new canes retained. The following spring, break out all shoots that develop from buds located below the wire and retain the 8 buds on the wire. Then follow the practices outlined for a newly developed cordon-trained vine (p. 194).

Medium-Vigor, Low Cordon-Trained Vines. Select two 4–6 bud canes, one on each side of the stake, and tie them vertically to the stake (Fig.

12-6). To improve vigor in the shoots that will produce new cordon canes, remove the terminal two spurs of the old cordons and prune the remainder of the vine normally.

In the second spring, select two strong shoots from each of the short canes and break out the remaining shoots. Remove the crop from the retained shoots to increase vigor.

During the following winter, select two new canes that are well positioned and tie them onto the new cordon wire, retaining 8 buds per cane on the wire. Then cut off the old cordons just beyond the new cordon canes to complete the conversion.

Conversion of Cane-Pruned Vines

Cane-pruned vines are more easily converted for machine harvesting than are older spur-pruned vines. Poor stakes must be replaced in a single-wire, cane-pruned vineyard to straighten the row, and/or to increase stake height. Machine operators prefer the fruiting wires to be a minimum height of 44 in. (1.12 m), particularly if no overhead foliage support is used.

Crooked vines should be straightened as much as possible by heavy pruning. Large arms and multiple trunks out of the vine row should be eliminated. Misalignment of vines should not exceed 8 in. on either side of the stake or vine row.

Vines headed at 30 in. (76.2 cm) or less should have their heads raised to a height of 40–46 in. (1.02–1.17 m), which eliminates low clusters in the vine head and supports more fruit on the wire. Raising the arms of the vines can be accomplished in 2 years. Some yield loss may occur during retraining since some fruiting canes may have to be sacrified to reshape the vine.

In existing trellised table grape vineyards, canes are usually tied to wires on a horizontal or slanted crossarm. For conversion for machine harvesting the crossarm must be removed, pivoted, or reattached vertically to the stake. The canes can then be tied to a single wire on top of the stake or to wires attached vertically on the repositioned crossarm.

DUPLEX SYSTEM FOR MECHANICAL HARVEST

In this system the vine is divided into two parts, the fruit-bearing and the replacement zones (Fig. 12-7). The fruit-bearing zone consists of mature canes wrapped on two parallel wires 3 ft. (91.4 cm) apart (Olmo

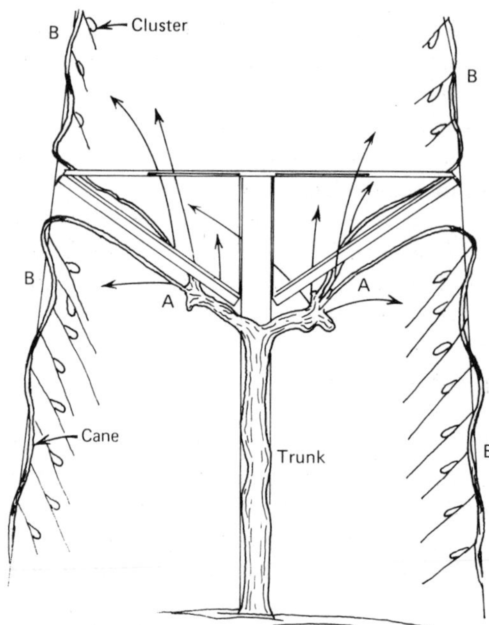

Fig. 12-7. Diagrammatic drawing of the duplex system for mechanical harvesting. (*A*) Replacement zone where new fruiting canes are developed. Arrows indicate positions of selected shoots remaining after spring deshooting and deflorating. (*B*) Fruit-bearing zone is on the trellis wires. Pendent shoots bearing flower clusters of flowers and fruit arise from these fruiting canes (Redrawn from Olmo, 1968).

et al., 1968). The replacement zone furnishes fruiting canes for the following year. In this zone some shoot removal and defloration are performed.

The wires are suspended in a saddle on either side of the trellis. An impacter-type harvest machine developed at the University of California employs a horizontal steel rod that rotates in a curvilinear fashion through a narrow slot in a vertical sheet-metal shield. The upward movement of the rod and sharp collision with the trellis wire snap cluster stems and detach the fruit from the shoots.

This duplex system has the advantage of using a wide-top trellis that can increase yield. Although it is presently little used in California, Australia has shown much interest in the impacter-type machine.

Chapter

13 TECHNIQUES TO IMPROVE GRAPE QUALITY

Girdling, thinning, and applications of plant regulators can cause various beneficial effects on the vine by affecting berry-set, size, color, ripening, cluster compactness, and other parameters.

GIRDLING (ADAPTED FROM WEAVER AND NELSON, 1959)

Girdling (ringing), is the removal of a complete ring of bark $\frac{1}{8}$ to $\frac{1}{4}$ in. (3.3–6.4 mm) wide from the trunk or from an arm or a cane basal to the fruit. As a result, the carbohydrate materials produced in the leaves accumulate in the parts above the wound, including the clusters of blossoms or fruit, and influence their development. Girdling of Black Corinth also can change a bound form of gibberellin to an active form of the hormone, which can enlarge berries (Weaver and Pool, 1965).

Improvement of Berry Set

Girdling at bloom increases the number of seedless berries that set and occasionally the number of seeded berries. It increases the yield of Black Corinth which normally produce small, straggly clusters of tiny seedless berries. Black Corinth should be girdled during bloom, and it increases the number and size of seedless berries. Trunk girdling is favored over girdling the arms or fruit canes because it affects the whole vine uniformly. Girdling was replaced by hormone sprays in the late 1950s (p. 216).

Increase in Berry Size

A complete girdle that is open during the rapid growth of the berries following bloom increases the size of seedless berries 30 to 100 percent, and of seeded berries usually less than 20 percent. Girdle immediately after the normal shatter of berry following bloom (fruit-set) to obtain the greatest increase in berry size with little or no change in number.

For production of table grapes, girdle Thompson Seedless vines as soon as possible after fruit-set. If girdling is done too early, before the normal berry drop is complete, the clusters become too compact. If girdling is delayed more than a few days after fruit-set, less increase in berry size results; if it is delayed 3 weeks or more, there is little increase in berry size. Girdle either the trunks or the fruit canes. When Thompson Seedless are girdled for table use, the girdles are usually made $\frac{3}{16}$ or $\frac{1}{4}$ in. (4.76 or 6.35 mm) wide, and heal over in 3–4 weeks.

Thinning is always necessary when Thompson Seedless for table are girdled. The increase in total crop, without thinning, is roughly proportional to the increase in berry size. Thompson Seedless clusters from ungirdled vines are normally well filled but are not compact. Since girdling increases the size of the berries but not the length of the stem parts, clusters often become too compact. On Thompson Seedless for table fruit a bloom spray of GA_3 is used for thinning (p. 217).

At fruit-set, cluster thinning should be done to eliminate clusters that are too compact, too small or too large, misshapen, or otherwise defective, leaving the required number of the best clusters. Any of the remainder that are too compact must be berry-thinned. Thinning usually combines the cluster and berry methods, plus spraying with gibberellin.

Although the number of shot berries (small seedless berries) of seeded varieties increase in the same manner as the berries of seedless varieties, the normal-seeded berries increase only slightly in size. Girdling to enlarge berries of normal-seeded varieties is not recommended.

Hastening Coloration and Maturation

Girdles should be made at the beginning of coloration and they must be open during the early part of the ripening period in Stage III. At the proper time of girdling of Red Malaga and Ribier the °Brix usually ranges from 5 to 6 and total acid from 2.3 to 2.6 (Weaver, 1955). Girdling hastens the maturity of such varieties as Cardinal, Red Malaga, and Ribier by improving coloration and accumulation of sugar. Best results are obtained from vigorous vines having only a light crop; with a normal to heavy crop, there will often be no response.

Girdling to hasten ripening is used in early districts where a few days' advance in maturity can result in a higher price for the crop.

Girdling Technique

Various types of double-bladed knives are used to girdle the trunks (Fig. 13-1). Cane girdling is best performed with girdling pliers that

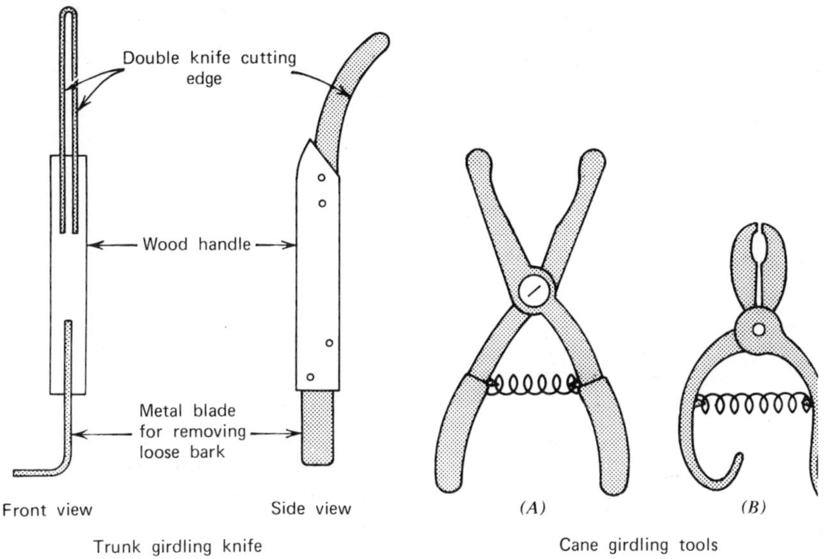

Fig. 13-1. Tools used for girdling grapevines.

have double blades on each side. With one type (A), cut through the bark by pressing on the handles, then release the pressure, rotate the pliers on the cane, and cut another section of the ring by squeezing the handles. After the ring of bark is cut completely around the cane, remove it by rotating the pliers, under slight pressure, around the cane. With another type (B), put the girdler on the cane and rotate it around the cane without pressure on the handles until the bark is removed. Almost all cane girdling done in California is performed with the latter type of girdling plier. In nonirrigated areas in Greece, only a simple cut with a knife blade is made on Black Corinth.

In trunk girdling the loose bark is first removed with the end of a trunk-girdling knife (Fig. 13-1). Then a strip of bark is removed completely around the vine with the double blade (Fig. 13-2).

Adverse Effects of Girdling

Girdling stops the downward movement of food materials past the wound until after healing. Thus the roots of a girdled vine live on reserve carbohydrates until the wound is healed. If the food reserve below the girdle is small, the vine may have difficulty in maintaining

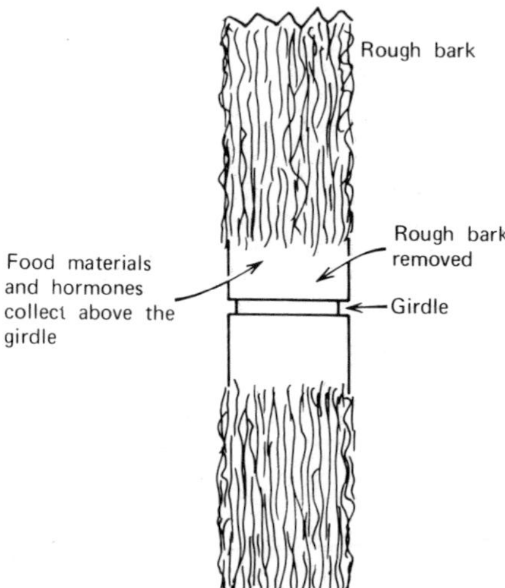

Fig. 13-2. Diagrammatic drawing of a trunk girdle.

adequate root growth essential for the absorption of water and nutrient materials. The top growth is retarded and the leaves may turn yellow. The longer the wounds remain open, the more serious is the weakening effect. Trunk girdles that do not heal during the growing season can kill the vine. Cane girdles that fail to heal are less serious because only one cane is killed. Girdles made during or soon after blooming, not more than ¼ in. (6.4 mm) wide, will usually heal in 3–4 weeks (Jensen et al., 1975); girdles ⅛, 1/16, or ¼ in. (3.1, 1.5, or 6.4 mm) in width are equally effective. Girdles made later in the season or reopened to influence ripening will heal more slowly and cause greater weakening. Irrigate immediately after girdling and keep the soil moist until girdles are healed.

Avoid overbearing of girdled vines. This is done with Black Corinth by pruning more severely, and with Thompson Seedless and other table varieties by thinning the girdled vines so that the final crop will be reduced about 25–33 percent compared to the crop of ungirdled vines. Many Thompson Seedless vines have been girdled for 20 or 30 years with no harmful effects.

In making the girdles remove only the bark; if even a narrow strip of bark remains, the beneficial effect of the girdle is greatly reduced. Cutting too deep into the wood is harmful because it destroys many active conducting vessels in the outer wood layers and causes a lack of water above the girdle.

THINNING*

The purpose of thinning is to reduce the grape crop to a normal load so that high-quality fruit is produced, and to produce clusters that are less susceptible to rot and are easier pack in boxes for shipment. The main types of hand thinning are flower-cluster, cluster, and berry thinning. Much thinning is done by gibberellin sprays (pp. 220).

Flower-Cluster Thinning

The clusters of rudimentary flowers appear with the leaves in early spring. The individual flower parts usually develop for 6–8 weeks until blooming. If some flower clusters or their parts are removed soon after they emerge, without removing leaves, the nutrition of retained clusters is improved and there may be a better set of normal berries.

With highly fruitful varieties such as Cardinal and Ribier, which produce two or three clusters per shoot, flower cluster thinning is the cheapest way to reduce the crop. Occasionally, growers also flower-cluster thin Thompson Seedless grapes for table use when the crop is excessive. Usually the retained clusters do not become too compact.

Flower-cluster thinning is much cheaper to perform than cluster thinning, since at this early thinning time the clusters are not obscured by the foliage. This is sometimes done in Emperor, which may be thinned at either time without producing compactness or other adverse effects. Usually however, the problem with Emperor is to obtain a large enough crop. The advantage of flower-cluster thinning is offset somewhat by the greater difficulty of selecting the best potential clusters. Also, late-developing clusters may be missed and the crop will thus be heavier than planned.

For many years it was too time-consuming and expensive to flower-cluster thin wine grapes. About 1974, however, some growers started using this technique on wine grapes. Many others have used GA_3 to thin their wine grapes (p. 220).

Cluster Thinning

This is the removal of entire clusters after the berries have set following bloom. This method can be used to reduce the crop on overloaded vines of high-producing wine- and raisin-grape vineyards, and for table grapes. Leave sufficient fruiting wood at pruning time to produce a good

* This section is adapted from a leaflet by Weaver and Nelson, 1959.

crop in years of poor set, and reduce the overload in good years by cluster thinning. Clusters that are undersized, misshapen, or oversized are removed. This favors the nutrition of the retained clusters, improves berry size and coloring, and hastens maturity. It is adapted to such varieties as Thompson Seedless. At thinning time, disentangle the clusters of table varieties from one another or from shoots or trellis wires to prevent damage to clusters at harvest.

Berry Thinning

This is the removal of cluster parts soon after the berriers have set. The apical end of the main stem is cut off so that 4–8 branches are retained at the base of the cluster, depending on their size (Fig. 13-3). Berry thinning improves quality only when too many berries make the cluster too compact or when very large cluster parts interfere with proper coloring and maturation. Varieties such as Tokay and Thompson Seedless are berry-thinned to keep the clusters from becoming too compact and to increase berry size. Thinning reduces cluster size and alters the shape.

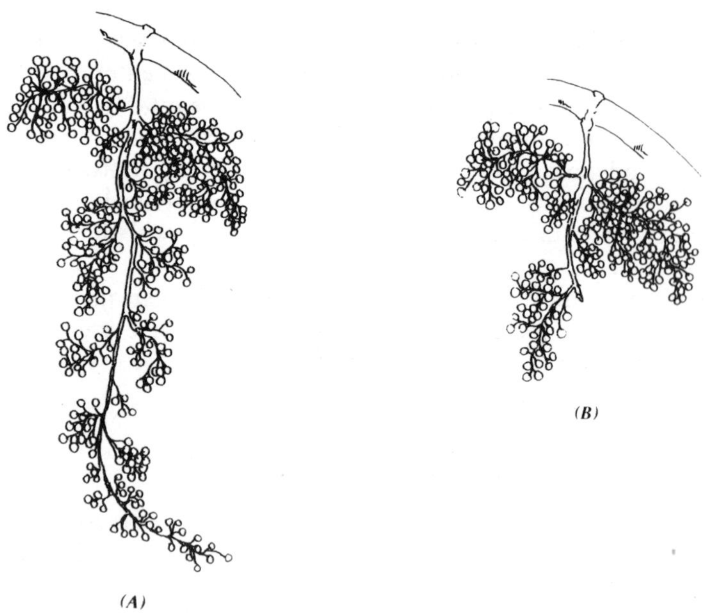

Fig. 13-3. Diagrammatic sketch showing how to berry-thin a cluster. (*A*) Whole cluster, not berry-thinned; (*B*) cluster after thinning.

Thinning of GA_3-treated Thompson Seedless grapes should be done during fruit-set or as soon after as possible for maximum increase in berry size.

The Perlette variety is exceedingly compact so that further hand thinning must be done on the retained portion of the cluster. Picking shears are used to snip out individual berries or laterals in a very expensive operation. Some success has been obtained with GA_3 sprays.

PLANT REGULATORS

Increasing Fruit-set in Grapes

Auxins on Black Corinth. Fruit-set in Black Corinth is induced by stimulative parthenocarpy, but berries are small and never grow to full size. The set varies from year to year. Girdling vines at anthesis has been used for over a century to improve set and increase berry size.

Beginning in the early 1950s, sprays of 4-CPA at 2–10 ppm were used to replace girdling (Weaver, 1956), and produced a set equal to or greater than girdling. The clusters should be thoroughly sprayed. Since vines on heavy soil may produce compact clusters subject to rotting, in the latter part of the 1960s gibberellin replaced 4-CPA. Gibberellin either has no effect on set of Black Corinth, or it decreases set.

Gibberellin on American Varieties. Applications of GA_3 to the Concord variety increases set by about 16 percent (Bukovac et al., 1960). These marginal benefits on crop increase may be useful in vineyards where early berry fall is a problem. GA_3 should be sprayed at 100 ppm about 11 days after bloom.

Inhibitors. SADH, Succinic acid-2, 2-dimethylhydrazide (abbreviated SADH; trade names Alar, B-nine, and B-995) can increase fruit-set in Himrod and Concord grapes. For example, SADH at 2000 ppm just before anthesis increased set 100 percent (Tukey and Flemming, 1967).

In some Eastern states, such as New York and Ohio, a total of about 10,000 acres (4,047 hectares) per year are sprayed with SADH from beginning of bloom to 50 percent bloom to increase yield (Shaulis, private communication). A concentration of 500 ppm which increases set about 20 percent is recommended. Although berries are smaller, total yield is increased. A Federal label allows use on selected varieties only: Concord, Niagara, Aurora (Seibel, 5279), Chancellor (Seibel 7053), and De Chunac (Seibel 9549), all of which are grown primarily in the eastern United States except for the Concords in the Northwest.

(2-Chloroethyl)trimethylammonium chloride (CCC, chlormequat), sold under the trade name Cycocel, increased set of berries on Thompson Seedless in New York (Barritt, 1970).

Inhibitors on *vinifera* grapes

In California, CCC and SADH have been tested on several varieties of small-berried wine grapes including Cabernet Sauvignon, Pinot noir, and Chardonnay. Although sprays of CCC at 1000 ppm, or SADH at 2000 ppm, applied 2 or 3 weeks before bloom resulted in a larger set of smaller berries, the yield was about the same. Usage is still in the experimental stage, and has not been officially cleared for general use. Possibly grapes with smaller berries produce better wine than those with larger berries due to the larger skin-to-pulp ratio in a given weight of small berries.

Increasing Berry Size of Grapes

Black Corinth. To improve berry size of this variety, GA_3 at 2.5–5 ppm should be applied during the interval 95 percent capfall to 3 days later. The clusters must be thoroughly wetted. It is not necessary to girdle the vines. Sprays of 4-CPA (2–10 ppm) applied 1 to 3 days after bloom are effective, as is 2,4-D at 1–5 ppm. However, the latter compounds are not cleared for use in California, and the 2,4-D can be especially injurious to the vine.

Thompson Seedless and Other Seedless Varieties. Thompson Seedless grapes have been girdled in California since the 1920s. In the 1950s, the plant regulator 4-CPA at concentrations of 5-15 ppm was applied to Thompson Seedless on a limited scale to increase berry size either as a replacement for or as a supplement to girdling. Occasionally however, there was a delay in maturation, and it often failed to produce as uniform a berry size as did girdling.

Within 3 or 4 years after the first experiments in which GA_3 was applied to grape in 1957 (Weaver, 1957; Stewart et al., 1958), practically all grapes of the Thompson Seedless variety intended for table use were being sprayed with GA_3. They were sprayed with the compound at 20–40 ppm at the fruit-set stage to increase berry size (Fig. 13-4). Some clusters were very compact, subject to bunch rot, and

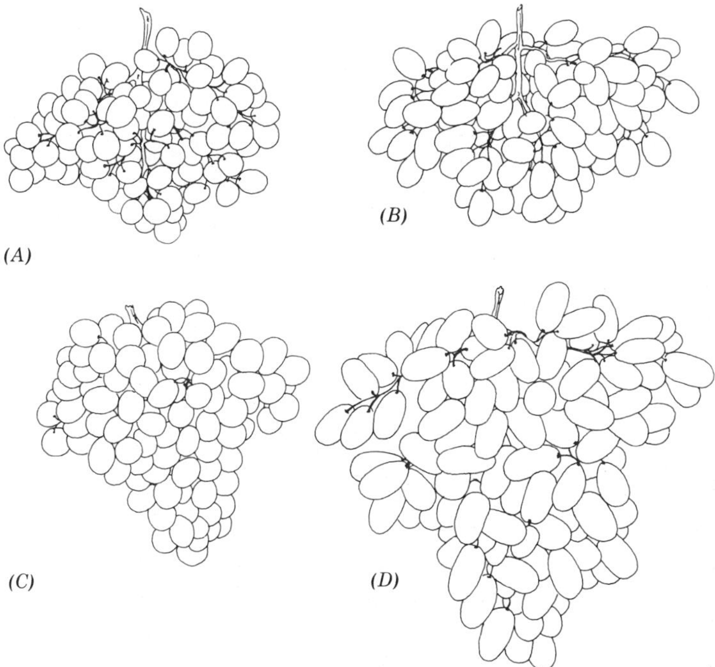

Fig. 13-4. Berry-thinned Thompson Seedless clusters at harvest that received different GA_3 treatments at bloom and/or fruit set. (*A*) unsprayed cluster; (*B*) application of GA_3 at 10 ppm at bloom; (*C*) application of GA_3 at 40 ppm at fruit-set; (*D*) application of GA_3 at 10 ppm at bloom, and at 40 ppm at fruit-set. Sprays at bloom loosen the clusters; repeat spray results in the largest berries.

difficult to pack in shipping boxes, since sufficient berry thinning usually had not been performed to compensate for the greatly enlarged berries.

The present recommendation is to use two applications of the compound. The first application of GA_3 is made with the compound at 2.5–20 ppm when 30–80 percent of the caps have fallen (Fig. 13-4). Concentrations higher than 20 ppm may result in shot berries. The purpose is to decrease set resulting in less compact clusters, and to increase berry size. The percent of berries that set is often reduced by 30–50 percent. The bloom-time thinning spray tends to change the shape of the berry from the typical oval to a longer configuration (Fig. 13-5).

A second application at concentrations of 20–40 ppm is made on the same vines at the fruit-set (normal girdling time) stage to further increase the size of the berry (Fig. 13-4). Girdling is also performed at

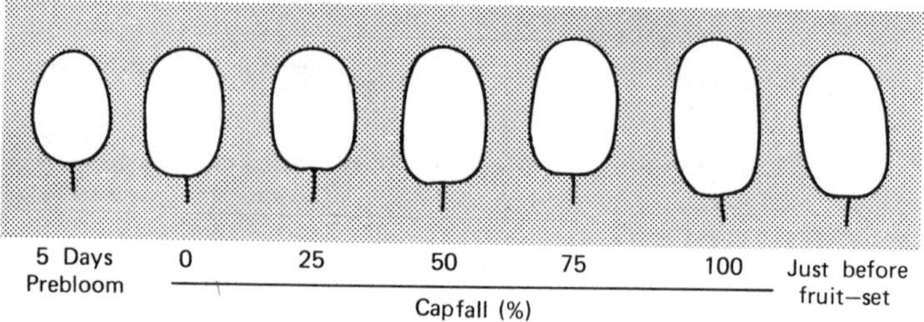

Fig. 13-5. Effect of stage of flower and/or berry development at time of treatment with GA_3 on shape and size of mature Thompson Seedless berries at harvest.

fruit-set stage to obtain more uniform berry size, a larger berry, and to increase the attachment of the pedicel to the berry. The cluster area should be thoroughly wetted and GA_3 should be applied as soon as possible after the shatter of berries.

Sprays of GA_3 at bloom can also be used on Thompson Seedless utilized for raisins or canning, when clusters are compact and subject to rotting. Higher concentrations of GA_3 may be required at fruit-set stage in desert regions for size increase because of the small berries growing in such environments.

The response to GA_3 of other seedless varieties such as Black Monukka, Perlette, Delight, Beauty Seedless, and Seedless Concord is generally similar to that of Thompson Seedless. The proper timing of sprays is more critical with Perlette than with Thompson Seedless (Kasimatis et al., 1971).

Seeded Table Grapes. These grapes usually increase only slightly in size, if at all, as a result of GA_3 application. Dattier (Waltham Cross) is a notable exception. In the Republic of South Africa, clusters are dipped at fruit-set into GA_3 at 10–20 ppm to increase berry size, and an increase of 50–70 percent is obtained both in berry size and crop yield. Begun in about 1970, this method is presently practiced on about 50 percent of the total acreage of Waltham Cross, and its use increases each year.

Delaying Ripening of Vitis vinifera

The auxin BOA–benzothiazole-2-oxyacetic acid (BOA; BTOA)—could be useful in viticulture because it greatly delays maturation (Weaver, 1956, 1962), from a few days to several weeks, in both seedless and seeded grape varieties. The compound should be applied at concentrations of 5–40 ppm, depending on length of delay desired, 4 or 5 weeks after fruit-set. At present, few growers are interested in inducing late maturation of grapes, so there has been no commercial use of BOA. The compound does however, offer possibilities for staggering harvest dates or meeting late market demands, although it is not currently cleared for use on grapes by the Environmental Protection Agency.

Promotion of Grape Ripening

For many years grape growers have universally sought a chemical means to hasten grape ripening. Some of the advantages of an earlier grape crop are that early market demand could be met, harvest dates could be staggered, and fall rains might be avoided, especially in regions with a Mediterranean climate such as California and Greece.

Ethylene occurs naturally in plants and is frequently referred to as a ripening hormone. Ethephon is an ethylene-releasing agent. Ethephon spray applied at about 15 percent berry color enhances berry color. It sometimes increases the sugar-to-acid ratio, and this change is due mainly to lower acidity (Hale et al., 1971; Weaver and Pool, 1971). In some of the colder growing areas where acidity is often excessive for optimum wine quality, use of ethephon might beneficially initiate and maximize loss of acids at an early date.

With table grapes such as Tokay and Emperor, ethephon is applied at 100–200 ppm at 5–15 percent berry color (Jensen et al., 1975). Both clusters and foliage should be sprayed for best results. Ethephon produces some softening of the berries and may occasionally produce excessively dark fruit. The main beneficial effect on table grapes is to hasten coloration, but berry size, acid and sugar are not changed. Ethephon can greatly increase the amount of grapes harvested at early pickings.

Ethephon can increase anthocyanin pigment in wine grapes. Vines may be sprayed with ethephon at 300–500 ppm at about 15 percent berry color. Spray both clusters and foliage. Usually there is no change in soluble solids, but in some cases total acidity decreases. An increase in content of skin pigment would be advantageous for red wine production. Ethephon is especially useful in varieties and locations where natural color development is poor.

Induction of Seedlessness in Delaware and Other Varieties

Gibberellin treatment can induce seedlessness and advanced maturity in Delaware, a seeded variety of *V. labrusca* and the leading table grape in Japan. Clusters that usually grow on overhead arbors are dipped in

GA_3 at a concentration of 100 ppm, 10 days before anthesis and again 2 weeks after bloom. The first dip results in seedlessness and the second in berry enlargement. Treated clusters produce large seedless berries somewhat smaller than seeded berries that color and ripen 2–3 weeks before untreated berries. Almost 100 percent seedlessness can be obtained. Similar results have been obtained in the northwestern United States (Clore, 1965). Gibberellin sprays at 100 ppm should not be used on vegetative vine parts, or phytotoxic effects on the vine will occur.

The seedlessness induced by GA_3 is caused mainly by the compound's injurious effect on the ovules (Ito et al., 1969) and not by its reduction of pollen germinability, an effect of GA_3 that has also been demonstrated (Weaver and McCune, 1960).

Use of GA_3 to produce seedlessness in other grape varieties has proved commercially unsuccessful, as a mixture of seedless and seeded berries reduces marketability.

Berry Shrivel in Emperor

In the San Joaquin Valley of California, the Emperor variety is sometimes affected by a condition known as "berry shrivel," in which the fruit loses turgidity approximately one month before harvest. If more than a few berries per cluster are affected, the cluster must be discarded, causing some growers to suffer an appreciable crop loss annually. Jensen (1970) found that applications of GA_3 to Emperor reduce the amount of berry shrivel. GA_3 should be applied at 20 ppm from 1 to 2 weeks after shatter at 200–250 gal/acre (1,870–2,337 liters per hectare). The predominant berry diameter during this period should be in the range of 0.39–0.59 in. (10–15 mm).

Another beneficial effect of this spray was that average berry size increased, mainly as a result of increased size of one-seeded berries in the clusters.

Loosening Compact Bunches of Seeded Varieties

Wine Grapes. The incidence of rot in seeded wine grapes is often excessive in varieties that produce compact clusters. Vines should be sprayed about 3 weeks before bloom when longer shoots are about 15–20 in. (38.1–50.8 cm) long, and clusters average 3–4 in. (7.6–10.2 cm) in length and range from 2 to 5 in. (5.08–12.7 cm). Thorough coverage of the cluster area is necessary, and will require about 100 gal of spray per acre (935 liters per hectare).

Loosening of the cluster is a result of reduced set, cluster elongation, and/or production of shot berries (Fig. 13-6). Clusters with rot contribute little to the total yield because of their light weight. Probably any reduction in crop weight as a result of GA_3 application is more than compensated for by the elimination of or decrease in amount of rotting. Since the severity of bunch rot varies from year to year, a prebloom application of GA_3 is considered to be a form of insurance.

In California GA_3 treatments are cleared for the following varieties, with proper concentrations in ppm in parentheses: Tinta Madeira (1–2.5), Palomino (1–2.5), Carignane (2.5–5), Valdepeñas (2.5–5), Aleatico (2.5–5), Zinfandel (5–10), Petite Sirah (5–10), and Chenin blanc (5–10).

The recommended dosage of GA_3 must not be exceeded since seeded vines may be injured by the hormone. Often, if too high a concentration is applied to seeded varieties, bud growth the following spring will be sharply decreased and fruitfulness reduced.

Table Grapes. Generally, gibberellin should not be applied to seeded table grapes for thinning purposes because the shot berries that develop from the application detract from the appearance of the cluster. In India however, clusters of "Anab-e-Shahi" table grape are dipped in

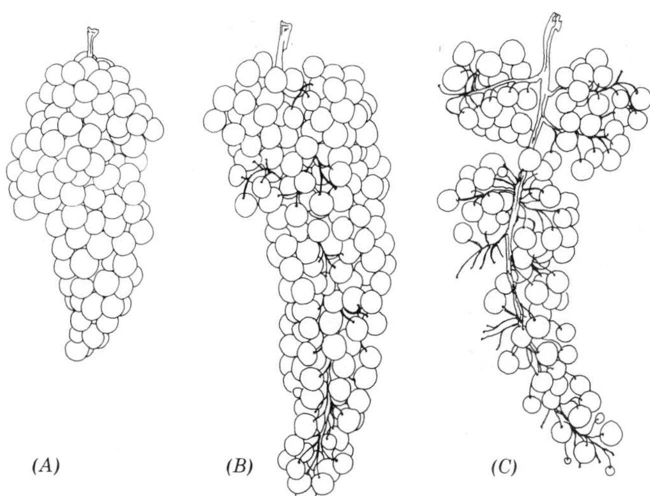

Fig. 13-6. Diagrammatic drawings of Zinfandel, a compact-clustered wine grape, sprayed two weeks before bloom with GA_3 at 0 ppm (*A*), 7.5 ppm (*B*), and 20 ppm (*C*). Satisfactory loosening of cluster resulted with GA_3 at 7.5 ppm (*b*), but 20 ppm (*c*) reduces crop.

low concentrations of GA_3 at bloom to reduce set and increase berry size.

Bloom Sprays of GA_3 to Thin Thompson Seedless

Application of GA_3 to Thompson Seedless during bloom produces a very loose cluster. Application of the compound at concentrations of 2.5–20 ppm at 30–80 percent capfall is most effective. The percent of berries set is reduced by about 30–50 percent. Loose clusters are less subject to summer bunch rot and easier to pack. Application of GA_3 during the early stages of bloom increased berry size to a lesser degree than when applied at later bloom stages. Concentrations higher than 20 ppm may cause shot berries.

Application of GA_3 at the bloom stage results in striking longitudinal berry elongation; later treatments also cause radial expansion.

Thinning Perlette Grape with GA_3

Perlette has overly compact clusters and extensive removal of berries by hand thinning is required following fruit-set. In many vineyards, application of GA_3 at concentrations of 10–15 ppm during the bloom period has resulted in satisfactory thinning (Kasimatis et al., 1971). In other vineyards, however, satisfactory thinning of clusters was not attained. Gibberellin applied at 40–80 ppm increases berry size.

Chapter 14
HARVEST AND POSTHARVEST OPERATIONS

In this chapter we discuss table, wine, and raisin grapes separately because different techniques are involved.

TABLE GRAPES

Proper Picking Time

Table grapes should be picked when they are attractive and have good eating quality, when they keep and carry well and, if possible, when they can reach the market and be sold for high prices (Jacob, 1950). There may also be legal requirements regulating the proper picking conditions.

Ripening consists mainly of an increase in sugar, a decrease in acidity, and the development of characteristic color, texture, and flavor. These changes occur only as long as the grapes remain on the vine and practically cease after picking. At véraison (the stage when berries begin to soften and color), the ripening rate increases sharply. Generally there is a gradual improvement in quality until the optimum stage is reached for the intended use of the grapes; afterward gradual deterioration occurs.

The acid and sugar content are the best measurements of maturity and occasionally color estimates or measurements are also useful.

Sugar Test

The sugar content of ripe grapes can be measured with a hydrometer or a refractometer. These instruments are generally calibrated in the Balling or Brix scale and read directly in percent sugar by weight (Fig. 14-1). The hydrometer measures specific gravity, the weight per unit volume as compared with that of pure water. Since sugar is the chief

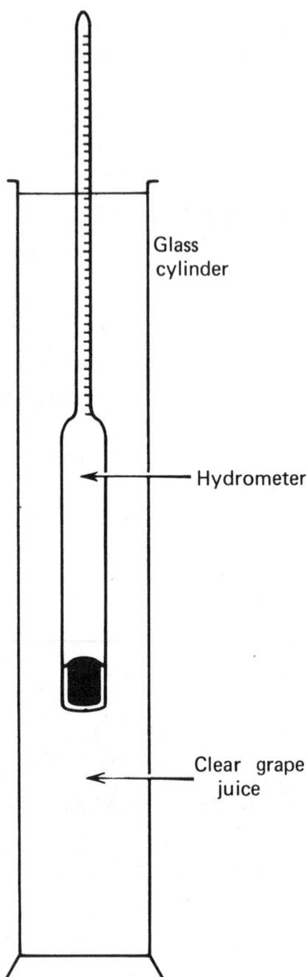

Fig. 14-1. Brix hydrometer used for measuring soluble solids.

substance affecting specific gravity in ripe grapes, one may estimate the quantity of sugar present by measuring the specific gravity.

The system of graduating hydrometers on the basis of percentage by weights of sucrose in a pure sucrose solution was first made by Mr. Balling. The Balling scale was subsequently recalculated by Mr. Brix. The term Brix hydrometer is now usually used to designate these instruments, and degree Brix is used for the values. Most Brix hydrometers are standard at 20°C. The percent soluble solids and refractometer reading is the same value as the °Brix.

To make the test, take a representative sample of berries picked from clusters at random from various vines, mash the grapes, and express the juice from the pulp. The extracted juice is poured into a cylinder filled to overflowing, and the hydrometer is lowered into the juice until it comes to rest. The temperature of the juice is taken, and the temperature of calibration shown on the hydrometer is used in measuring the result to obtain the approximate percentage of sugar.

A refractometer will also give results in percent of sugar weight and is a more convenient and faster method (Fig. 14-2). The refractive index of a sucrose solution is an accurate measure of the concentration of dissolved substance. One can squeeze juice from a berry onto the glass prism of the refractometer and make a reading. For a more representative reading it is best to get a reading on the extracted juice of 100–150 randomly sampled berries. Using a food blender is an effective and convenient method for juice extraction. Field readings should be taken with a temperature-compensating refractometer or one with a built-in thermometer for temperature correction of readings. Otherwise readings should be taken indoors at room temperature to minimize temperature variations.

Acid Titration

To measure acid, pipette 10 ml of clear juice into a small flask, then add 50–100 ml distilled water and 2 or 3 drops of phenolphthalein indicator solution. A standard solution of sodium hydroxide (0.133 *N*) is slowly run from a burette into the flask containing the diluted juice, with a constant stirring of the flask, until the color turns pink. A solu-

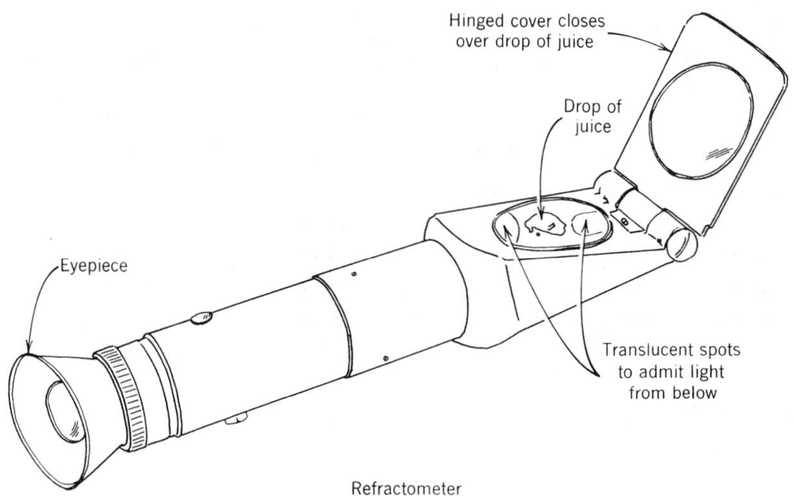

Fig. 14-2. Refractometer used for measuring soluble solids.

tion standardized to 0.133 N is equivalent to 0.01 g tartaric acid per ml. The milliliters of sodium hydroxide solution used divided by 10 equals the tartaric acid in the juice in g/100 ml.

Degree Brix Acid Ratio

The degree Brix divided by the acidity in grams per 100 ml gives the °Brix acid ratio (often referred to as sugar-acid ratio). This is a better measurement of palatability than sugar content or acidity used alone. The minimum desirable °Brix acid ratio varies with different varieties: The ratio for Thompson Seedless, Malaga, and Ribier should be about 25:1, the Ohanez, Cornichon, Muscat, and Emperor 30:1, and the Tokay, Olivette blanche, and Red Malaga 35:1. In addition, most other varieties should have a °Brix of 17 or higher. Ribier, Red Malaga, and Emperor should be at least 16°Brix (Jacob, 1950).

Climatic conditions affect the °Brix acid ratio. If ripening occurs during hot weather, the ratio will be high and the grapes palatable at relatively low sugar; in cool weather the acid will be higher, and more sugar will be necessary for equal tastiness.

Estimating Maturation

By the use of several criteria one can judge when a cluster is ripe for picking. If the peduncle is brown and woody, or if the stem framework of the cluster is of light straw or yellow color, the grapes are probably mature. Although tasting the berries can indicate if the cluster is ripe, the sense of taste is quickly dulled by frequent use, so the picker must rely principally on other characteristics such as the fruit color or the condition of the stems. Often the vineyardist may correlate these visual criteria with the results of °Brix-acid measurements to provide a suitable guide for pickers. Red or black grapes develop their characteristic color as they ripen; the best colored grapes are usually the ripest. Green or white varieties become more nearly yellow or white as they ripen.

Since all clusters do not ripen at the same time, even those on the same vine, picking is often done twice or more to harvest the grapes at the proper stage.

Criteria of Quality

Clusters should be well developed, well filled, and have fresh, straw-colored or yellow cluster stems and cap stems. The berries should be firm, plump, and have the typical shape and uniform color for the cultivar. Clusters should be free of sunburned or shot berries and berries scarred by thrips, powdery mildew, or wind damage. Firm berry attachment to the cap stems is also important. The berries should not be shriveled, crushed, or decayed. Undesirable aging is shown by brown, dry stems, shriveled and dull berries, and visible mold on stems or berries.

Packing

Table fruit is all hand-picked. In harvesting and packing, fruit is moved as rapidly as possible from the vines to coolers in refrigerator cars or cold storage. Some grapes are packed in the field, and others are house-packed after they are brought in from the field.

Field Packing. The picker should hold the cluster by the stem to remove it from the vine. The stem should then be cut with picking shears or a sharp knife. All defective berries, particularly those broken or decayed, should be carefully removed by clipping with shears the stem that attaches the berries to the cluster. Berries should not be pulled off with the finger, leaving the wet brush attached to the cluster, as this will give fungi and other microorganisms a medium on which to begin growth. The cluster can sometimes be improved by removing the undersized or insufficiently colored berries. Often however, such clusters are better packed in a separate, lower quality pack. Crushing grapes or breaking them loose from the stems should be avoided because any break in the skin permits easy entrance for molds, yeasts, and other decay-causing organisms.

Table grapes in lugs are usually packed by the "stems up" method. To make this pack, the box can be tilted by placing it crosswise in another box or on a special stand. Lay one or more clusters horizontally in the lower end of the box, and continue to fill the box from this end, with all clusters placed nearly upright except those needed to make the bottom of the pack solid. Occasionally press the fruit toward the low end of the box so that a firm pack is made (Jacob, 1950).

House Packing. Grapes are picked into shallow boxes and carried to packing houses or sheds where clusters are trimmed, and carefully packed into standard size boxes of wood, corrugated board, or plastic (Fig. 14-3). The movement of full and empty boxes and of the cull fruit is usually mechanized with conveyors, belts, and forklifts. Sometimes

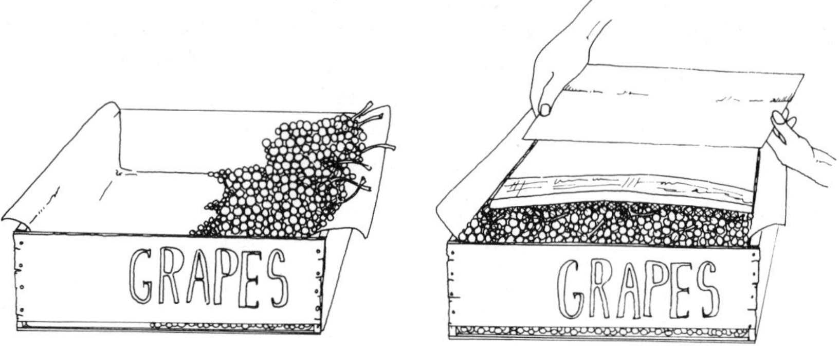

Fig. 14-3. Packing grapes in a lug box, using the stems-up method. Lug box on right is ready for lidding.

grapes are vibrated into position in the shipping container, and this may prevent much shattering of berries. Cushion pads at the tops and bottoms of boxes help prevent bruising of fruit. The bottom pad must also be absorptive to prevent fruit "wetting" from juice released from damaged berries. Other variations from the standard hand-place packs include the South African cluster-wrap pack and the sawdust export chest pack.

Cooling and Storage

It is important to cool grapes as soon as possible after removal from the vine, to minimize deterioration before they reach the market. Since mature grapes have a higher sugar content than most fruit, they freeze at a lower point [26.5°–28°F (minus 3.1 to −2°C)] than other fruit. Table grapes can be safely stored at 30°–32°F if the temperature is closely controlled. A relative humidity of 90–95 percent is recommended (Ryall and Pentzer, 1974).

Cooling of grapes is usually performed in special rooms or in forced air coolers. *Vinifera* table grapes vary greatly in storability. A few cultivars, such as Emperor, Ohanez, and Calmeria, can be stored for as long as 6–7 months. Other cultivars will remain in marketable condition for only 1 or 2 months.

In room cooling, grapes in containers are exposed to cold air in a refrigerated space. The air sweeps by the containers (usually boxes with vents), which are stacked so that uniform channels are left between rows of the lugs. Heat transfer from the grapes is primarily by conduction to the outer surface of the lugs. A uniform air distribution with at least 200–400 fpm (foot per minute) equivalent to 61.0–122.0 mpm air circulation is required as are well-vented containers.

In forced air or pressure cooling, the grapes are cooled rapidly by producing a difference in air pressure on opposite faces of stacks or pallet loads of vented containers. This pressure difference forces air through the containers and removes heat primarily by flow around the individual clusters. Speed of cooling is regulated by adjusting air volume. The time required for this type of cooling is $\frac{1}{6}$ to $\frac{1}{10}$ of the time required for room cooling.

Sulfur Dioxide Fumigation

Modern grape storage would be impossible without the decay control provided by SO_2 fumigation, which also prevents browning of the cluster stem structure.

Vinifera grapes can be given an initial fumigation for 20 minutes with about 1 percent concentration of SO_2, depending on the amount of space to be treated (Ryall and Harvey, 1959). Treat the fruit after it is packed so that crushed or cracked berries or those pulled loose from the pedicels during packing will be exposed to the SO_2. Gassing is usually done before the grapes are precooled. Since warm fruit absorbs more gas than cold fruit, the latter requires longer exposure to obtain the same effects in killing the spores of fruit rot. The toxicity of SO_2 to spores of *Botrytis* increases about 1.5 times for each 18°F (10°C) rise in temperature between 32°F and 86°F (0° and 30°C) (Couey and Uota, 1961). A fumigation treatment that causes the grapes to absorb 5–18 ppm of SO_2 provides good decay control (Pentzer and Asbury, 1934). Although the initial treatment with about 1 percent SO_2 kills most of the spores on the fruit, it does not destroy infections already caused by the pathogen.

Formula for Calculating SO_2 Dosage

The absorptive capacity of the fruit, lugs, and the space they occupy must be accounted for when the amount of SO_2 is determined. The following formula was devised by Nelson and Baker (1963):

$$A = \frac{B \times C}{D} + (E \times F)$$

where A = the number of pounds of SO_2 required;
B = the percent concentration of SO_2 desired;
C = the number of cubic feet of free space of the room;
D = the volume occupied by 1 lb of SO_2 gas (5.5 cu ft at 32°F (0°C);
E = the number of pounds of SO_2 absorbed by each carload (1000 28-lb lugs of fruit);
F = the number of carloads of fruit.

Factors B and E are related, and the values used depend on (1) the interval between storage treatments, (2) the condition of the fruit (3) the uniformity of air distribution and velocity during treatment and (4) accessibility of fruit, including container alignment and openness of the packs.

Fumigation Rooms. Precooling rooms, refrigerator cars, or trucks can be used for initial treatment with SO_2. However, because of the corrosiveness of SO_2, special fumigation rooms are often used for the initial fumigation. Air circulation is required in the fumigated space to obtain uniform concentrations. A fan at the brace of the refrigerator car that blows upward past the outlet of the liquefied gas is effective. Usually the SO_2 gas is released from small cylinders containing the amount required for the space to be fumigated.

If grapes are stored in well-ventilated display lugs used in California, they can be fumigated at intervals of 7–10 days with about 0.25 percent SO_2 for 20 minutes. Repeated fumigations kill any spores that escaped the first fumigation before they can infect the grapes. Air circulation during fumigation is needed so that the SO_2 can penetrate the containers and to attain uniform distribution of the gas throughout the storage room.

Sulfur dioxide is colorless, has a pungent odor, and is very irritating to the mouth, nose, and eyes; it also corrodes pipes and other equipment.

Bisulfite Application. Release of SO_2 from bisulfite within the package of grapes is an effective practice. Five grams of sodium bisulfite can be placed in the bottom excelsior pad of display lugs holding 28 lb (12.7 kg) of grapes. The chemical may also be mixed with sawdust used in export packs of grapes (Pentzer, 1939; Ryall and Harvey, 1939). If the bisulfite is mixed with sawdust before packing the fruit, a more even distribution of the chemical is obtained than by adding it to the top of the packed chest as it is vibrated (Nelson and Gentry, 1966).

A two-stage method of releasing SO_2 has been developed by Gentry and Nelson (1968) for use in tight containers that retard moisture loss from the grapes but cannot be fumigated by adding SO_2 to the ambient air. There are two packets containing bisulfite, one with a thin paper cover and the other with a thick cover. The thin-covered packet releases SO_2 quickly and the other provides a slow release. Water vapor contacting the bisulfite gives off SO_2 and enters the thin-covered packet faster than the thick-covered one. Although the two-stage method is useful in overseas shipments, it cannot be used for long storage.

Berry Injuries

Bruising can occur after picking and during packing operations. In boxes where grapes are packed too tightly, berries on inner surfaces of containers may become flattened and discolored. High absorption of SO_2 by these injured areas may bleach them. Pads in field boxes and shipping containers can prevent excessive bruising.

Cracking and Splitting. Some grape berries can crack at the stylar end or on the side, and sometimes partial or complete circles form around the cap stems. Rains occurring late in the season can also cause cracks, and such injury may lead to decay during shipment and storage. Several varieties susceptible to skin cracking include Thompson Seedless, Red Malaga, Tokay, Muscat of Alexandria, Ribier, Emperor, Almeria, Petite Sirah, and Mission.

Internal turgor pressure exerted against the skin probably causes cracking or splitting of the berry. Good aeration of fruit in the vineyard along with proper thinning techniques and tight fill of grapes in the package help to reduce cracking.

Freezing Injury. This can occur if the storage temperature is allowed to drop too low. Frozen grapes are soft, flabby, and have a dull appearance.

Internal Browning. Thompson Seedless berries sometimes undergo internal browing after 2–3 months of storage, and is visible through the fruit skin. In Thompson Seedless it begins in the vascular areas and extends into the flesh (Pool and Weaver, 1970); Malaga also can develop browning. Since there is a direct correlation between browning and maturation of fruit, harvested fruit should not be too mature for storage.

Sulfur Dioxide. This can cause bleaching of the skin, especially around the cap stems or in cracks in the skin. Sometimes pits appear as scattered bleached areas occur; this is quite common in Tokay. When grapes injured by SO_2 are warmed to room temperature, they have a cooked sulfurous taste. To prevent injury use proper gassing techniques, consider different varietal sensitivity to SO_2, age of fruit (younger fruit is more susceptible to injury), and grape temperature at time of gassing (Ryall and Pentzer, 1974).

Ammonia. Leaks of this gas in refrigeration systems must be avoided. The gas can turn red berries blue, and green-colored fruit takes on a bluish tinge; as the NH_3 leaves the grapes, the pigments return to normal. The gas damages color because the skin becomes less acid. Prolonged exposure to NH_3 will brown the grapes permanently and make then unattractive.

WINE GRAPES

Wine grapes are picked by hand or mechanical harvesters. The proper picking time for wine grapes depends mainly on the kind of wine to be made. Grapes for dry wines should have high acidity and moderate sugar content. Therefore such grapes are usually harvested from 20–24°Brix. Grapes for sweet wines should be high in sugar and moderately low in acid, and should attain as high a sugar content as possible without raisining, usually around 24°Brix or higher (Jacob, 1950).

Hand Picking

The entire crop is usually harvested at a single picking. For fine wines, one may make several hand pickings to obtain uniform fruit at the optimum stage of maturity. Even when the crop is all harvested at a single picking, it is desirable to remove clusters that have waterberry (soft, flabby berries) and those that are very green, or raisined, decayed, or moldy since such berries can detract from wine quality.

The grapes can be picked into field lug boxes and hauled in them or in bulk loads to the winery. A better technique is the pull gondolas through the rows, so the fruit can be picked into pans or buckets and then dropped directly into these containers. Small gondolas (1 or 2 tons) are usually hoisted mechanically and the grapes are dumped into larger gondolas near the edge of a road for hauling to the winery.

The fruit should be taken to a winery immediately for crushing and placing into fermentation vats, since harvested grapes can spoil quickly and should not be allowed to set for more than a few hours.

Mechanical Harvest (Adapted from Christensen et al., 1973)

Several methods for mechanical harvest of grapes have been tried. An early concept of training grapes was to use a "T" trellis so that clusters hang downward at the same level. At harvest, a machine harvester fitted with a cutter bar cut off the clusters (Winkler et al., 1957). This technique had to be abandoned partly because the popular wine grapes had peduncles that were too short for the cutter bar to remove the fruit effectively.

A second method was to remove the clusters or berries with a suction tube with a strong vacuum. One problem encountered with this technique was that leaves as well as clusters were collected. Fresh leaves produced a green suspension in the wine, and dry leaves produced wine with a tealike flavor.

The mechanical harvesters currently in use are nearly all "beater" or "slapper" types that strike and shake the fruit from the vines (Fig. 14-4). Machines that shake off clusters by vertically striking the trellis wires are sometimes used, employing flexible, horizontal, two-wire

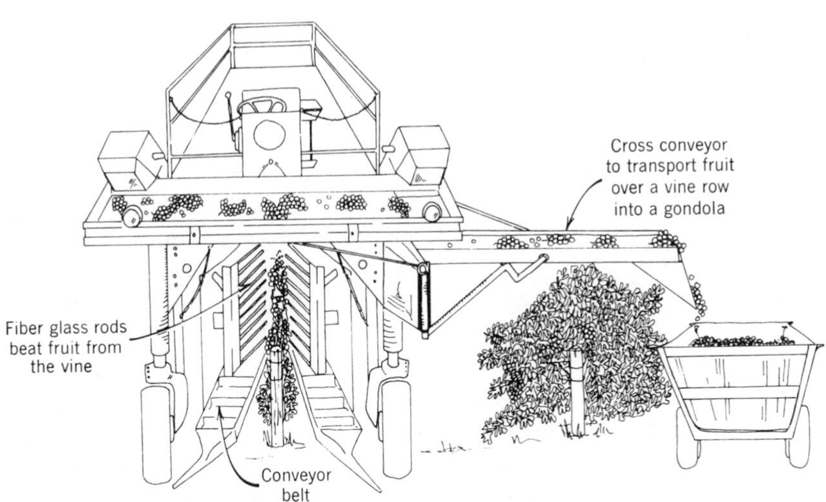

Fig. 14-4. Diagram of a slapper-type mechanical harvester.

trellises which support and position the fruit for ease of machine removal. The vines are trained by the duplex system (p. 207).

Slapper Type Harvester. These harvesters have double banks of flexible horizontal rods that strike and shake the vines, removing cluster parts, individual berries, and juice. With difficult-to-harvest varieties the harvester delivers single berries and juice, but with easily harvested varieties the fruit is delivered as cluster parts and single berries, and sometimes a few whole clusters.

With easy-to-harvest, low juicing varieties, mechanical harvesters can probably deliver tonnage to the winery equivalent to that picked by hand crews. With difficult-to-harvest, heavy juicing (berries that lose juice easily) varieties, mechanical harvesters may deliver less fruit to the winery than do hand-picking crews.

Fruit Characteristics. Fruit of most varieties is removed mainly as single berries, particularly in Thompson Seedless, Rubired, Royalty, and other varieties with a fairly loose berry attachment.

Varieties such as French Colombard, Chenin blanc, Grenache, Barbera, and Carignane have a firmer berry attachment and thus require more energy for fruit removal. In such varieties more of the fruit removal is associated with berry rupturing and breakage of cluster parts, and weakness in the peduncle or framework helps in the process. For example, the green stems of Aramon and Barbera clusters often fall off or break into pieces as single berries are detached.

Varieties with a firm berry attachment and a tough or wiry cluster framework, such as Emerald Riesling, are the most difficult to harvest. After harvesting, the cluster stem structure and the large wet brushes still adhere to the vine.

The soft juicy berries of such varieties as Sémillon, Muscat Canelli, and Burger lose much juice during mechanical harvest and transit, and conversely the very firm berries of the easily harvested Tokay and White Malaga varieties lose practically no juice during machine harvesting.

Vine Growth Habit. Dense foliage can interfere with the harvester, making access to fruit difficult. Excessive leaf removal during harvesting can also slow down fruit conveying and leaf removal operations in the harvester. French Colombard, Sauvignon blanc, Sylvaner (Franken Riesling), Mission, and Gray Riesling are examples of varieties with dense foliage.

Varieties with a large vine framework and heavy cane growth can also cause harvester interference (e.g., Grenache and French Colombard). The brittle wood of French Colombard often causes spur breakage and clogs fruit conveyors with canes and shoots.

Mechanical trimming of shoots before harvest, the use of foliage support trellising, and suitable pruning practices can minimize vine growth problems. Mechanical preharvest shoot trimming should be practiced on very vigorous vines such as French Colombard.

Varietal Adaptability for Mechanical Harvest. Grape varieties vary widely in the ease and mode with which they are removed from the vine. The greatest contributing factors to machine harvestability involving fruit characteristics, vegetative growth habits, and type of training and trellising have been discussed (training and trellising for machine harvest vines were discussed on p. 197). Varieties that prove impractical for mechanical harvesting will either be hand-harvested or eliminated.

Difficulty in harvesting and juicing amount of various varieties are shown in Table 14-1. The ratings are general averages obtained from different growers, vintners, and University of California Agricultural Extension personnel (Christensen et al., 1973).

Condition of the Soil. The ridge of soil in the vine row must not be too high or wide or it can interfere with the catching frame of the harvester. Also, the area under the vines should be kept free of trash and excessive weed growth.

The soil between the rows should be flat, smooth, and firm to aid in steering and facilitate the efficiency of the harvester. If the vines are hedged (shoots cut back) before harvest, the trimmings should be kept out of the area where they might interfere with the catching frame.

Time of Day to Harvest. Grapes are usually easier to harvest at night from about 10 P.M. to 10 A.M, possibly because berries are more turgid at night. For the same reason grapes are more difficult to harvest if the vines are stressed for water. One should irrigate close enough before harvest to avoid stressed vines and fruit wilting.

Slope of Land. Although standard over-the-row harvesters are designed to work where there is no slope, successful harvesting can be accomplished without machine conversion on slopes up to 7 percent. Some models can be used in vineyards with slopes up to 20 percent. Other machines have a hydraulic ram on each wheel that can self-level the machine on sloping terrain.

Table 14-1. Adaptability of Various Wine Varieties to Mechanical Harvest*

Variety	Ease of Harvest	Amount of Juicing
White Wine Varieties		
Burger	Hard	Heavy
Chardonnay	Medium	Medium
Chenin blanc	Medium	Medium
Emerald Riesling	Very hard	Very heavy
Flora	Easy	Light
French Colombard	Medium	Medium
Gewürztraminer	Easy-medium	Light
Gray Riesling	Easy-medium	Light
Malvasia bianca	Medium-hard	Medium
Muscat Canelli	Hard	Heavy
Palomino	Easy-medium	Medium
Pedro Ximinez	Easy-medium	Heavy
Pinot blanc	Medium	Medium
Saint Emilion (Ugni blanc)	Medium	Heavy
Sauvignon blanc	Medium	Medium
Sémillon	Medium-hard	Heavy
Sylvaner (Franken Riesling)	Medium	Light
White Riesling	Medium	Medium
Black and Red Wine Varieties		
Aleatico	Hard	Heavy
Alicante Bouschet	Medium-hard	Medium-heavy
Aramon	Medium	Light-medium
Barbera	Medium	Medium
Cabernet Sauvignon	Easy-medium	Light-medium
Carignane	Medium-hard	Medium-heavy
Gamay	Medium	Medium
Grenache	Hard	Medium-heavy
Mission	Medium-hard	Medium
Petite Sirah	Medium	Medium-heavy
Pinot noir	Medium	Medium
Royalty	Easy	Medium
Rubired	Easy	Medium
Ruby Cabernet	Medium-hard	Medium
Saint Macaire	Medium-hard	Medium-heavy

Table 14-1. (Continued)

Variety	Ease of Harvest	Amount of Juicing
Salvador	Easy	Medium
Souzão	Medium	Medium
Tinta Madeira	Medium-hard	Light-medium
Valdepeñas	Medium-hard	Medium-heavy
Zinfandel	Hard	Medium-heavy
Raisin and Table Grape Varieties		
Black Corinth (Zante Currant)	Medium	Medium
Calmeria	Easy	None
Emperor	Easy	None
Muscat of Alexandria	Easy-medium	Light
Niabell	Easy	Light
Ribier	Easy	None
Sultana	Easy	Medium-heavy
Thompson Seedless	Easy	Very light
Tokay	Easy	Very light
White Malaga	Very easy	None

* After Christensen et al., 1973.

Sprinklers. The internal height clearance of over-the-row harvesters varies from 6.75 to 7.5 ft. (2.06–2.29 m) to clear the risers and heads of permanently installed sprinkler systems.

Since sprinkler heads can be broken off due to the whipping action imparted to loosely held risers (tubes to conduct water to sprinkler head), risers are often strapped to a vine-supporting stake. Channel-shaped steel or grooved wooden stakes can be used to hold the risers tight by strapping. Rigid plastic risers can be protected by using a larger diameter polyvinylchloride (PVC) tubing as a collar, which provides rigidity, mechanical protection, and shields against weather.

Some growers use galvanized steel pipe for risers that are connected to the underground lateral by a short section of flexible polyethylene tubing. Another possible technique is to remove the risers and lay them on the ground parallel to the row close to the vines before harvesting. This can be done by extending the polyethylene connector to about 6 in. (15.2 cm) aboveground to provide a bending point. Plastic clips may also be used to facilitate attachment and removal of plastic risers.

Field Crushing. After harvesting, grapes are usually hauled immediately to the winery in gondolas on trucks. However, some crushing in the field is done which may be considered an extension of mechanical harvesting. During mechanical harvesting a few staples, nails, and the like are removed from the vine trellis when wooden stakes are used. Staples and nails can jam the must pump on the field crusher. However, a magnet placed on the discharge conveyor or mechanical harvester, as demonstrated by Petrucci and Siegfried (1975) at the California State University at Fresno, practically eliminates this problem. Field crushing and stemming (removal of stems and cap stems or clusters) is used mainly by wineries that produce varietal table wines.

As soon as grapes are harvested, broken and bruised berries are subject to spoilage by enzyme activity, leakage of cellular constituents into the juice, and fungal and bacterial contamination. The main advantage of field crushing and stemming is that it allows the immediate introduction of an antioxidant fungistat, such as SO_2, to prevent juice browning and possible spoilage. Also, the early removal of stem parts from the must (crushed berries and juice) will minimize excessive tannin.

The must can be put into stainless steel tanks mounted on the harvester, or it can be pumped directly into a closed, stainless steel gondola pulled by a tractor in the adjacent vine row. With the first method, the tanks are discharged into a trailer-tanker for transport to the winery; with the second method, the closed gondolas are transferred to a flatbed truck by a forklift and hauled to the winery.

To obtain the best quality wine, grapes should be delivered to the winery soon after picking, the juice separated, and fermentation begun. Field crushing and the introduction of 100 mg/liter SO_2 can delay browning for 8 hr. A layer of CO_2 over the crushed grapes reduces wine color slightly as compared to an air layer. The lower the handling temperature, the better the wine quality will be. Long holding times (over 12 hr) without the addition of SO_2 can result in bacterial and yeast growths and thus decrease wine quality.

Postharvest Treatment and Delivery to Winery. Growers may help protect white table wines from browning and oxidation by the application of SO_2 at harvest time. This must be done following directions of the receiving winery. One technique is to apply potassium metabisulfite at a rate of 5–6 oz (142–170 gm) per ton on the bottom of the gondola at the beginning of harvest. Dispersion of this relatively stable compound occurs as the chemical dissolves in juice in the bottom of the tank, and the mixing results from the motion of the gondola. One can also meter the potassium metabisulfite powder onto the harvested grapes as they pass to the gondola on the cross-conveyor. Another technique is to drip a 100 ppm SO_2 water solution on the harvested grapes in the conveyer. There may, however, be excess on loss of SO_2, and crushed berries will fix more SO_2 than sound berries.

Grapes for red table wine production can also benefit from SO_2 applications in the field by decreasing bacterial spoilage that can occur with must temperatures above 80°F (27°C) especially if winery delivery is delayed for 24 hr.

RAISIN GRAPES

The higher the degree Brix at harvest time, the better will be the quality of raisins produced. It is best if the berries are at least 20°–22°Brix. Picking time is determined by ripeness of fruit and the risk of unfavorable drying conditions if the grapes are allowed to remain on the vine too long.

For grapes that will be dehydrated, weather conditions are only a minor factor in the drying; the influence of maturity on the quality of the raisins is also less marked than with the natural sundried product. Even for dehydrated raisins, however, the grapes should be at the highest possible degree Brix. Harvesting must be completed before the early rains cause deterioration of quality.

Sundrying of Thompson Seedless (Adapted from Jacob, 1950; Kasimatis and Lynn, 1967)

Sundrying is performed in the vineyard, where raisins are dried between the rows of vines. Around 90 percent of California raisins are dried in this manner.

Preparation for Drying. This consists of smoothing and leveling the loose soil between the vine rows to provide flat surface for the trays. If the vineyard rows run in an east-west direction, the bed should be prepared so that it slopes to the south and exposes the grapes more directly to the sun, to hasten drying and provide drainage. For good drainage, a minimum of 5 percent slope (about 2 in. per 3 ft or 5.08 cm per 91.4 cm) should be used. Drainage outlets at the ends of the rows eliminate excess water from the vineyard. For north and south rows, the drying area requires only smoothing and leveling. Some growers, however, prefer a slope to the west to provide drainage in case of rain.

Grape maturity affects yield, quality, and drying ratio of the raisins.

Grapes with higher sugar content mean more mature fruit, better quality raisins, lower drying ratio, and higher yields.

The higher the degree Brix at harvest, the greater the weight of raisins produced. For example, grapes harvested in from 16–22°Brix, with pounds of raisins produced from one ton of grapes given in parentheses, were: 16° (394), 17° (416), 18° (438), 19° (460), 20° (481), 21° (503), and 22° (524). (The respective weights in kilograms were 179, 189, 199, 209, 218, 228, and 238.) Over this range there was a 130 lb (59 kg) increase in raisins (Kasimatis and Lynn, 1967). Eating quality of raisins also increases with a high degree Brix.

Drying ratio is the number of lb (.45 kg) of fresh grapes required to make 1 lb of raisins. This ratio may be used to compare the price offered for raisins with that bid for fresh grapes. One must consider differences in harvesting costs between selling fresh grapes and drying, and the additional risk involved in making raisins.

As the amount of sugar at picking time becomes higher, the drying ratio decreases. Less grapes are required to make raisins when the degree Brix is high. The decrease in drying ratio as the degree Brix increases from 16° to 22°, with the drying ratio in parentheses, is: 16° (5.08), 17° (4.81), 18° (4.57), 19° (4.35), 20° (4.16), 21° (3.98), and 22° (3.82) (Kasimatis and Lynn, 1967). Sugar readings should be made from berries taken from various cluster positions and from approximately 100 vines. Separate sampling in a number of areas in each vineyard block is necessary, particularly if there are weak spots and areas of vigorously growing vines.

Time of Harvest. For high-quality raisins, grapes for natural sundrying should be 19° Brix or higher. In California picking should start soon

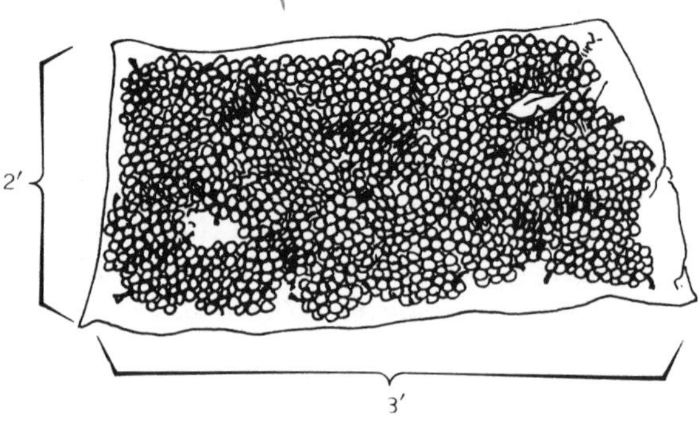

Fig. 14-5. Clusters of grapes drying on a paper tray.

Fig. 14-6. In vineyards where rows run east to west, trays should be placed in the sun near the north row. Trays should slope downward toward the middle of the row to enhance drying, and also for drainage in case of rain.

after the grapes reach 19° Brix, or on September 1, whichever is earlier. Grapes of less than 17° Brix should not be dried for raisins because of low raisin quality and greater harvesting and handling costs as a result of the poor drying ratio. All sundried raisins should be on trays by September 20. If grapes cannot be harvested for raisins by this date, it is usually best to sell them green or perhaps run them through a dehydrator.

Drying Trays. The wood tray has been largely replaced by the 2-×3-ft (61.0 × 91.4 cm) paper tray as a drying surface (Fig. 14-5).

Picking Raisin Grapes. In east-west rows, trays should be placed high on the terrace to prevent shading by the vines and kicking sand on the trays, and to provide for adequate drainage (Fig. 14-6). In vineyards with north and south rows, place trays near the middle of the smoothed area. Do not put bunches with rot and insects on the trays. Avoid cut and crushed berries that can be sticky, accumulate sand and dust, and attract insects such as vinegar flies and dried fruit beetles. Use knives to remove the bunches and spread them evenly on the trays. Very large bunches dry more rapidly if cut into smaller units. Overloading of trays results in slower drying and increases the probability of rot and serious mold contamination if it rains before drying is completed. Lighter trays of 20–22 lb (9.07–9.98 kg) of grapes will facilitate drying, reduce the need for turning, and retain less water on the trays following rain. Do not leave foot tracks in the drying area or kick sand on the drying trays.

Turning, Rolling, and Boxing. When the top layer of berries has browned and shriveled, the grapes are usually turned over, insuring rapid even drying of large clusters. Roll the paper trays when the grapes are dry or almost dry. Removal of off-grade bunches prior to rolling is recommended, since the entire cluster can be discarded before off-grade berries loosen and mingle with the sound berries.

Rolling the paper trays into a biscuit shape is preferable to other methods (Fig. 14-7). Biscuit rolls protect raisins from rains and reduce raisin-moth infestation, sand accumulation, and excessive drying. Cigarette (folded) rolls reduce costs in boxing and rolling, but do not offer as much protection from insects and weather as biscuit rolls (Fig. 14-7).

Since raisin-moth eggs are laid on fruit at night, rolling done in the afternoon avoids enclosing live eggs in the rolls because many eggs are killed by direct exposure to the sun.

After the raisins are cured in the rolls, they are picked up for boxing in the field or hauled to a central vineyard boxing area. Special machines designed for pickup and boxing in the vineyard accelerate this operation.

For more thorough cleaning of raisins, the rolls should be taken to a central boxing area and dumped onto a mechanical shaker. Undesirable bunches can be eliminated by hand sorting. Shaking removes excess sand and eliminates as much as 90 percent of raisin-moth and dried fruit beetle larvae and eggs. Use clean, dry boxes for storing raisins. Most growers dry raisins to about 14–15 percent moisture before delivery to the raisin plant.

Rain Damage. If it rains before the trays are rolled, as soon as they are dry enough to handle, pull the trays slightly higher up the terrace slope to tighten the paper and allow the water accumulated in pockets to drain. Turn the trays as soon as practical to allow raisins to dry on the undersides, and replace all deteriorated trays.

None of the permissible fungicides applied to raisins that have been

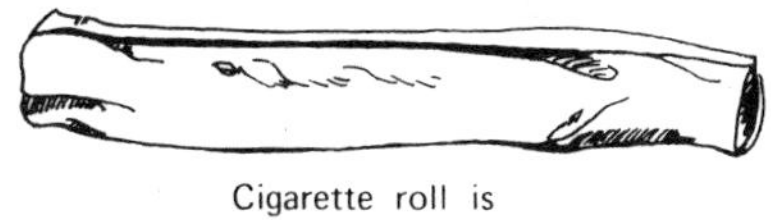

Cigarette roll is open at both ends

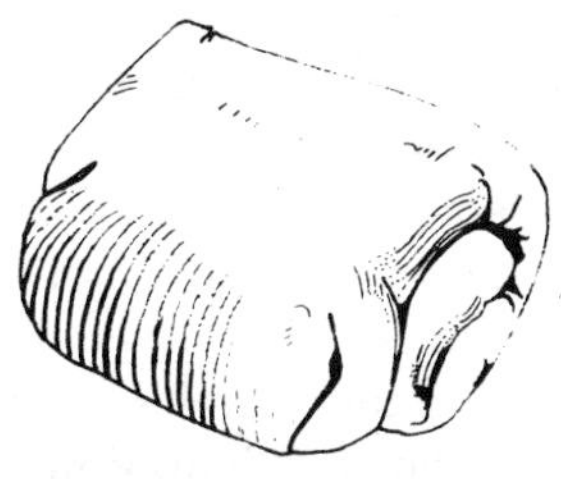

Biscuit roll. Grapes are completely enclosed

Fig. 14-7. Paper trays are rolled into biscuit or cigarette rolls for further cluster drying.

rained on will prevent mold damage. Discard moldy bunches as the trays are rolled. If rains occur in late season before drying is completed on late-picked grapes, it usually is best to box the raisins and complete the drying in a commercial dehydrator or in an improvised drier on the farm. The latter type consists of partially filled sweat boxed stacked in a row with a plastic covering. The boxes are spaced for air channels between them so that the stack serves as a tunnel. A portable grain drier or a space heater is the heat source at the intake end, and a squirrel cage fan at the exhaust end provides the air flow through the covered stack.

Raisin Storage on the Farm. If the raisins are not delivered directly to the packing plant, they must be protected from weather and insect and rodent infestation. Fumigate with methyl bromide to control different raisin-storage insects. A gastight structure for fumigation can be made by covering a stack of boxes with a gastight material such as Sisal Kraft or polyethylene sheeting; use soil to seal the bottom edges of the covering. Properly applied and with a sloping top, this covering will provide adequate weather protection.

Rodents are a constant menace. During the winter months, baiting and trapping are very effective since other natural foods are scarce. During the spring and summer months, weeds should be eradicated near the stacks because they may harbor rodents. Properly covered stacks offer rodent protection as long as the paper is intact. In the spring, check and fumigate stacks of raisins in storage, as warm weather increases insect activity.

Sundrying of Black Corinth, Muscat of Alexandria, and Sultana

Black Corinth are sundried without pretreatment. These raisins are only about one-fourth the size of Thompson Seedless, have a tangy flavor, and are useful for flavoring breads, mince pies, and fruitcakes.

Muscats are natural sundried raisins of Muscat of Alexandria. These large sweet raisins can be obtained with seeds, or the seeds can be removed by a special machine. The raisins are very large, grayish-black or grayish-brown, with a strong muscat-raisin flavor.

Sultana is a nearly seedless variety with round berries that are sundried like Thompson Seedless. They are inferior to Thompson Seedless for raisins as they are less meaty, have a higher acid content, and occasional hard seeds. Only a small acreage is grown in California. The tart flavor of these raisins make them useful in many baked products such as mince pies. The Sultana of Australia and the Republic of South Africa are the same as Thompson Seedless in California.

Other Drying Techniques (Adapted from Jacob, 1944)

Golden-Bleached Raisins. Thompson Seedless grapes are dipped in a solution of 0.2–0.5 percent sodium hydroxide (caustic soda, lye) at 190°–200°F (88–93°C) until the skins show slight checking or cracking, after which the grapes are cooled and washed in a rinse or spray of cold water. They are then exposed to the fumes of burning sulfur in large ovens for 2–4 hr. The quantity of sulfur varies from two to four pounds (0.91–1.81 kg) per ton of grapes. After sulfuring, the grapes are dehydrated at up to 140°F (60°C) for about 18 hr.

The raisins are a brilliant lemon to golden yellow. Although their sulfur dioxide taste may be objectionable to some, it disappears completely during cooking or baking. The raisins are very popular for use in fruitcakes during the holiday season.

Soda Dipped. This method involves dipping Thompson Seedless in hot lye, or a mixture of sodium hydroxide and sodium carbonate. The grapes are not sulfured, but are tunnel dehydrated. The raisins are amber to brown.

Oil Dipped. Thompson Seedless grapes are treated in one of two dips.

One solution is cold [100°F (38°C) or less] and contains about 4 percent sodium carbonate and a little caustic soda, with a thin film of olive oil floating on the surface. The other is a hot dip [170°F (77°C) or higher], but a weaker solution of sodium carbonate is used. A film of oil is also used. The grapes are dried on trays in direct sunlight. Oil-dipped raisins are medium to dark brown, fairly tender, and slightly oily but not sticky. They are not made commercially in California.

Valencias (Lexias). In Spain and Australia, some muscats are dipped before drying. The caustic soda dip used for golden-bleached raisins, or a hot dip containing sodium carbonate or potassium carbonate with lye is often used. Sometimes olive oil is added. The Spanish raisins are dried in direct sunlight, stemmed, and sold as Valencias or lexias. The Australian rack-dried product is called lexia.

Dehydrators

To avoid field contamination of raisins, grapes can be taken directly from the vineyard to a tunnel dehydrator. The objections to this method are that color and flavor of the dehydrated fruit are different from those of sundried raisins and the process is more expensive. About 50 hr are required to produce raisins from fresh grapes. While dipping fruit in hot alkaline solutions cuts drying time in the dehydrator to about half, production costs are still much higher than for sundrying.

Drying Raisins on the Vine

Drying raisins on the vine enhances mechanical harvesting for raisin production. For many centuries the finest Black Corinth raisins in Greece were produced by cutting off the clusters and hanging them to dry in the shade on shoots and arms of the vine. To aid mechanical harvest, severing of canes to aid in grape drying has been successful with and without the use of drying sprays applied to the vines.

Black Corinth. In this variety, on-the-vine drying is initiated by severing the fruiting canes from the vines just below the trellis wire (Christensen et al., 1970). The canes and fruit are left on the trellis wire supports to complete the drying process. After drying, the raisins can be machine-harvested and taken to packing houses. Usually fruit is adequately dried within 3½–5 weeks after cane cutting.

These results were obtained using the 2-wire duplex system (p. 207), and a 3-wire system (fruit canes wrapped on a single wire 15 in. (38.1

cm) below a 24-in. (61.0 cm) wide T trellis). Harvesting on the duplex system was done with a vertical impactor system developed at the University of California (p. 207). The T trellis was harvested with a machine using the horizontal side beater system.

Berries dried on the vine are a uniform dark blue and retain their even natural shape, with the bloom left intact. Tray-dried raisins have a reddish cast and are often misshapen, with less uniform wrinkling.

Thompson Seedless. On-the-vine drying with cane severing has not proved completely successful with Thompson Seedless. However, moisture content of the fruit can be reduced 25–35 percent over a 6–9 week drying period (Studer and Olmo, 1973). Under these slow drying conditions the raisins develop a typical sundried color. A final moisture content of 15 percent or less can be attained by drying the fruit for about 12 hr in a tunnel dehydrator, or under deep-bed drying conditions.

Another approach to hasten drying is to apply emulsion sprays to fruit on the vine. Work on this aspect was initiated in Australia (May and Kerridge, 1967).

In California experiments, Petrucci and Canata (1974) pruned vines with fruit at 20–23° Brix by cutting fruit canes at the same location as during dormant pruning. This procedure facilitates even drying and efficient removal of the finished fruit product.

The next step is to spray the fruit with a 2 percent mixture of ethyl oleate and potassium carbonate [17 lb (7.71 kg)] of potassium carbonate and 2 gal (3.79 liters) ethyl oleate/100 gal (379 liters) water]. A second spray is applied 5 days later at 1 percent.

In vineyards of moderate to heavy vigor, 600 gal of spray per acre (5,609 liters per hectare) is required to obtain satisfactory drying of berries on the vine. Petrucci and Siegfried (private communication) found that in a good drying year the fruit basal to the cut cane (10–12 percent of the total crop), which dries at a slower rate than the fruit from the cut portion of the cane and has a greater moisture content, may be mechanically harvested along with the drier fruit on the cut portion of the cane containing approximately 14 percent moisture.

The harvested fruit is stored in 180-lb (81.6 kg) boxes or 1000-lb (453.5 kg) bins and allowed to "sweat." This is a moisture equalizing process, and fruit with higher moisture content can thus be mixed with drier fruit with no apparent hazard.

Bolin et al. (14) showed that using ester emulsions of fatty acid to accelerate drying in conjunction with mechanical harvesting can reduce labor, energy, and processing requirements for raisins.

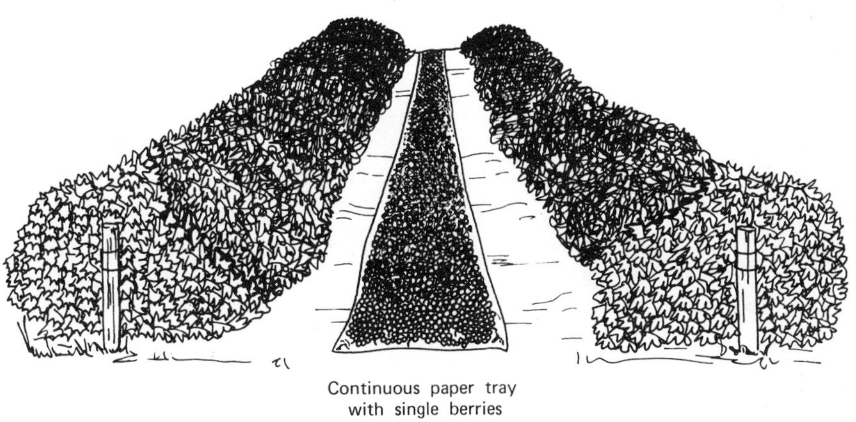

Fig. 14-8. Continuous paper tray in vineyard, used for mechanical harvest.

Mechanical Harvest

Thompson Seedless that have been cane cut can be mechanically harvested for raisins, although the work thus far has been done on an experimental basis. A machine to lay a continuous paper tray in the row has been developed (Studer and Olmo, 1971). The fruiting canes are cut 4–6 days before harvest at a point near the juncture of the vine head and the cane, to induce easy separation of the fruit as single berries from the vine. Without cane cutting the cap stems of the berries are usually pulled from the berry, leaving an open wound from which the juice exudes. With cane cutting, cap stems dry and become brittle and are generally retained by the berry, so that damage to the berry is minimized. The berries can be easily conveyed and spread in a uniform fashion on continuous paper trays for sun drying (Fig. 14-8). Single berries dry more rapidly and uniformly than do grape clusters. Drying time on the tray usually ranges from 11 to 14 days. An experimental machine has been developed to pick up the raisins from the continuous paper tray. Flavor and appearance of machine-harvested raisins is enhanced by cane cutting (Studer and Olmo, 1974). Product evaluations indicated that most consumers could not tell the difference between mechanically harvested and hand-picked raisins (White and Petrucci, 1974).

Part 4
CROP HAZARDS

Chapter

15

CROP HAZARDS—VIRAL, FUNGAL, AND BACTERIAL DISEASES

Vine hazards include viral, fungal and bacterial diseases, insects, noxious weeds, birds, rodents and deer, and frost. All are discussed in this section; except for frost, discussed in Chapter 5.

VIRUS DISEASES

Viruses are small infectious particles (about one-millionth of an inch) composed of a core of nucleic acids surrounded by a protein sheath that reproduce only in living cells of the host. Virus symptoms vary with the type of virus and vine.

Grape Leafroll (White Emperor Disease)

Grape leafroll is the most widespread virus disease in California, and it results in great economic losses due to decreased vine vigor, productivity, and delayed fruit maturity. In this disease leaves roll downward and turn progressively red toward the cane tips. The veins remain green (Goheen, 1970). Symptoms appear in early June in nonirrigated vineyards and in August in irrigated vineyards. Vines have small clusters, few clusters per vine, and berries have less sugar than normal vines. Red fruit varieties such as Cardinal, Emperor, Mission, Red Malaga, Queen, and Tokay develop fruit lacking normal color. The name White Emperor arose from the lack of color in the Emperor variety. Berries of diseased white grapes such as Thompson Seedless, Riesling, Sylvaner, and Melon develop a yellowish-white color at harvest instead of the normal greenish-white. The disease spreads mainly by propagation from infected mother vines.

Leafroll delays ripening and decreases fruit coloring. However, a leafroll infected vine of Burger with a 50 percent crop reduction will have about the same rate of sugar accumulation as that of a fully-cropped healthy vine (Lider et al., 1975). The acidity of the fruit of leafroll-affected vines is higher than that of healthy vines regardless of crop or sugar levels.

Fanleaf

Three strains of the fanleaf virus produce three closely related diseases: fanleaf, yellow mosaic, and veinbanding. All are transmitted from diseased to healthy shoots in soil by the nematode, *Xiphinema index*, and the three diseases may sometimes be found in the same vineyard.

Fanleaf. The veins of infected leaves are spread so that the leaf resembles a fan (Vuittenez, 1970). Leaf symptoms are very conspicuous on Mission and French Colombard. In some cultivars of grapes enations develop on the undersides of basal leaves under poorly defined circumstances. Infected vines have shorter and more irregular internodes than healthy ones. There are double nodes and stem fasciations, and lateral sprout development gives a bushy appearance. Often crop is sharply reduced by a shattering of berries and formation of many shot berries.

Yellow Mosaic. This disease causes a chrome-yellow mottling of leaves in early spring, in which the shoots may become partially or completely yellow (Dias, 1970). Degree of injury can vary from occurrence of yellow net veins, yellow veinbanding, yellow spots, or blotches to complete yellowing. When infected leaves age the yellow color becomes paler and occasionally cream-colored, and these areas may bleach, die, and result in leaf drop. In late spring and summer new leaves may also show symptoms. Clusters are small and contain many shot berries. The disease is spread by grafting.

Veinbanding. This disease causes chrome-yellow bands along the principal veins of mature leaves. Usually the symptoms are not visible until late spring or early summer. Many flowers do not set but shell from the cluster, resulting in straggly looking clusters when the fruit matures. Both leaf and fruit symptoms vary and are more severe some seasons than others. Growth of infected vines is little affected, and may be slightly less than that of healthy vines. The disease is spread by rooting diseased cuttings, by grafting, and by nematodes.

Yellow Vein

This disease occurs in Carignane, Grenache, and Valdepeñas, and is caused by infection with the grape strain of tomato ringspot virus. Fruit-set can be reduced from slight to 100 percent, but second-crop clusters are normal (Gooding and Teliz, 1970). First-crop clusters have a straggly appearance. Leaves showing symptoms can vary from one to many per vine. A faint chlorotic mottling first appears. The commonest leaf symptoms are scattered chrome-yellow flecks and yellow vein-bands; by late summer the yellowing fades to a bleached appearance. Diseased vines are often larger than normal ones, partly because of their low crop. In Emperor, the yellow flecks along the veins merge to form a continuous band. The disease is spread mainly by cuttings rooted from mother vines that are carrying the virus, and is transmitted by nematodes in some areas.

Corky Bark

This disease is more common in coastal wine vineyards than in other areas. Varieties such as Palomino, Petite Sirah, Mondeuse, Gamay, and Cabernet franc show cane and leaf symptoms (Beukman and Goheen, 1970). Bud break may be delayed in spring and vines may have one or more dead spurs; canes bend downward with upright tips and the wood is rubbery or limber; the bark at the base of the canes sometimes splits, and may have a purplish cast in late season.

On most *vinifera* varieties corky bark shows no characteristic symptoms. On Carignane, only leaf symptoms are visible. Young leaf tips of affected canes are light yellow, similar to yellow mosaic. The mosaic symptoms fade during the summer but corky bark symptoms persist. The disease can be carried by infected rootstocks that show no discernible symptoms.

Yellow Speckle

Grapevine yellow speckle usually causes small irregular yellow speckles ranging in size from a pinpoint to about 1 mm on exposed mature leaves (Taylor and Woodham, 1972). The speckles can be along main veins or veinlets, or may be scattered in interveinal areas near smaller veinlets. Symptoms may be masked completely or may occur on one or several leaves per vine. The virus has no obvious effects on growth, yield, or fruit-set. Symptoms of the disease are similar to those of fanleaf and yellow vein.

Asteroid Mosaic

Leaf symptoms on *vinifera* resemble star-shaped spots irregularly distributed over the blade (Refatti, 1970). The spots can coalesce, and occur more frequently between primary and secondary veins. Sometimes a netlike vein bordering occurs. Leaves are asymmetric, twisted, and puckered along the veins. Normal green blisters occur on some varieties. Usually symptoms are less severe in summer. Symptoms have been noted in inoculated Zinfandel, Merlot, Mission, Colombard, Emperor, Carignane, Thompson Seedless, and Valdepeñas vines. The disease is rare in commercial vineyards.

Viruslike Conditions

Spindle Shoot of Colombard (Hewitt, 1970). Infected shoots elongate rapidly and have small leaves one-fourth to one-half the size of normal ones. Leaf margins are chlorotic, and leaves are variously cupped and twisted. By midseason, however, growth appears normal. This disease has been found only in a few vineyards of French Colombard.

Viruslike symptoms can be induced on vines by growth regulators, herbicides, insects, and by chimeras. The latter arise from mutations, and refer to a portion of a plant composed of two or more genetically distinct tissues growing adjacent to each other as parts of the composite organ such as a leaf. (For a discussion of these effects and of grape viruses in general see Frazier et al., 1970.)

FUNGUS DISEASES IN CALIFORNIA

Several types of fungi cause various diseases in vines. Some diseases affect mainly the roots and base of the trunk near the soil, or the trunk and mature canes; others harm the green tissues of young canes, leaves and fruit forms, and still others damage the fruit. For convenience in diagnosis, the fungus diseases will be grouped arbitrarily by the principal plant parts affected. Some fungus diseases are difficult to group under this system because they affect several different tissues.

Soil Fungi

These organisms attack the roots or base of the trunk primarily. The diseases produced are root and crown rots, which usually kill vines. Fortunately these diseases are limited by environmental conditions and are usually confined to small, discrete areas in vineyards.

Armillaria Root Rot (Oak Root Fungus). This fungus (*Armillaria mellea*) causes rotting of the roots and trunks of many trees and shrubs as well as vines. There is usually a decline in vigor, growth ceases, and foliage turns yellow (O'Reilly, 1963); in subsequent seasons cane growth is often weak and the leaves are small. The vines may die suddenly after showing some of these symptoms. White, fan-shaped plaques of the fungus often occur, spreading between the layers of bark and between the bark and the wood. Infected roots and underground trunks may have small, black, smooth and shiny threadlike strands on the surface and penetrating into the bark. After rains in the fall and winter, clumps of mushrooms may also appear at the base of affected vines.

Control. Infected roots remain in soils after grapevines and trees have been removed. Soil treatment with carbon bisulfide or methyl bromide are the most promising remedies. After vine roots are removed, cultivate soil to a depth of 24 in., (61.0 cm) disc, and make a smooth surface. Carbon bisulfide applications can be made by hand or by tractor-drawn applicators.

For methyl bromide, inject at least 400 lb/acre (448 kg per hectare) by chisel applicator to a depth of 36 in. (91.4 cm) and cover with a polyethylene film seal. The seal should not be removed for at least 7 days and the treated soils or crown gall should be aired for 7 days or longer before planting.

Collar Rot. Collar rot is caused by soil fungi (Jacob, 1950). The disease is seldom found in vineyards older than 5 years. The fungi thrive in wet soil and usually enter the grape tissues near the crown of the vine in early spring. They kill the cambium and the bark, forming dead cankers that often girdle the trunk near the soil surface. A large calluslike overgrowth usually develops above the canker. Sometimes the callus tissues will bridge over and heal a small canker.

The tops of the diseased vines usually wilt and the vines dry up some time during the summer. Some vines grow all summer but show signs of early maturation during the fall.

Control. Remove 3 or 4 in. (7.62—10.16 cm) of soil from around the vine to let the bark dry out. The soil can be plowed back around the vine after the rains cease.

Cankers Caused by Fungi Attacking the Trunks

These organisms become established in the mature wood of the vine trunks or cordons and produce canker diseases. They are difficult to control because the time of infection is not well understood. The diseases, often overlooked in early investigation, are extremely important and are the cause of considerable economic loss in vines that are 8 years or older.

Eutypa Dieback. This disease has also been called dying arm disease. It is probably caused by the fungus *Cytosporina* (Moller et al., 1974), the imperfect stage of *Eutypa armeniaceae.* In the spring affected vines show weak, stunted shoot growth bearing small leaves, which exhibit varying degrees of yellowing, speckling, distortion, and necrosis but do not wilt. Such shoots contrast strikingly with healthy shoots nearby. If the apical end of a cordon or branch is affected, the disease progresses basally until the whole cordon, branch, or vine is dead. Spring is a good time to observe the effects on the foliage. The same fungus has been isolated from apricot trees that show dieback (English, 1963).

Black Measles (Spanish Measles, Black Mildew). This disease is associated with fungi that produce heart rot in the vine trunks (Jacob, 1950). The principal symptom, however, is a speckling and mottling of white or red grape berries with reddish-brown or purplish spots when the fruits are maturing. The fruit of severely affected vines often cracks and splits open and may dry on the vine before maturing. Severe fruit symptoms are accompanied by discoloration and dropping of leaves, and dying back of the shoot tips. The leaves develop various degrees of mottling, bronzing, spotting, and necrosis of tissue between the veins. The discolored leaf areas may enlarge into yellow spots that later dry up and turn brown or red.

An entire vine may be affected, but usually the symptoms are found only on a single arm or branch. Measles may show in some vines one season, and not the next. Occasionally the disease appears suddenly, and the vines can drop all of their leaves within a few days: some vines die and others start new growth within a few weeks.

Control. Black measles may be caused by wood-rotting fungi. Preventive treatment is to spray vines with sodium arsenite [4 lb. arsenic trioxide/gal. (0.479 kg per liter)] in the dormant season 4 or more weeks after pruning; use 3 qt. per acre (7.02 liters per hectare). The interval between pruning and spraying allows closing of the pruning wounds. For most effective control, all vines should be sprayed instead of spot-spraying the most diseased vines.

Dead-Arm. This disease is caused by a fungus (*Phomopsis viticola*) and occurs most frequently in the table grape area of the northern San Joaquin Valley. The fungus lives through the winter in diseased canes, arms, spurs, and petioles left on the vines. Infection usually occurs in the spring, when the shoots are young. The spores are spread from the old diseased parts of the vine to the young tissues by late spring rains. The disease is most conspicuous in late spring.

Leaves, petioles, canes, and flower-cluster stems develop small angular spots, most of which have yellowish margins with dark centers. These spots often coalesce to form large brown areas with numerous dark spots in them.

The diseased areas may split, forming open diamond-shaped cankers in the older shoots and canes (Jacob, 1950). Some shoots are severely stunted and may be killed. Vigorous shoots usually continue to grow, and later in the summer the diseased portion will appear only at the base of the cane; most of the cankers callus over and there is little indication of further increase in diseased areas during the summer. In autumn, many of the diseased areas resume growth.

Diseased areas on the cluster stems spread into the fruit and cause bunch rot; diseased areas on the canes frequently enlarge and can kill buds. Sometimes the fungus grows back into the wood of the arm and gradually kills the arm.

Control is the same as for Black measles when the vine is dormant. At the bud swell stage Captan 50 percent wettable powder at 1½ lb per 100 gal (0.18 kg per 100 liters) can also be used in severe cases. In mild cases, the Captan spray may be sufficient without a dormant treatment with sodium arsenite. If dead arm is severe and late rains are forecast, a second Captan spray may be applied when shoots are 6–8 in. (15.2—20.3 cm) long.

At pruning time in the dormant season, the diseased parts of the canes, arms, or trunk should be removed.

Fungi Attacking Green Shoots, Leaves, and Immature Fruit

These organisms usually produce new infections on green tissues each season. The diseases produced are leaf, shoot, or cluster blights. Treatments with fungicides at intervals during the growing season or sanitation practices that reduce the resting spore stages effectively control these diseases. If treatments are not made or the inoculum is not reduced the diseases can destroy the crop.

Powdery Mildew (Oidium). This important and widely distributed disease is caused by the fungus *Uncinula necator,* which can grow on all aboveground parts of the vine (Fig. 15-1). Mildew appears on the surface of the canes as grayish-white powdery growth (Hewitt and Jensen, 1973). When rubbed off, weblike black, or dark-brown discolorations

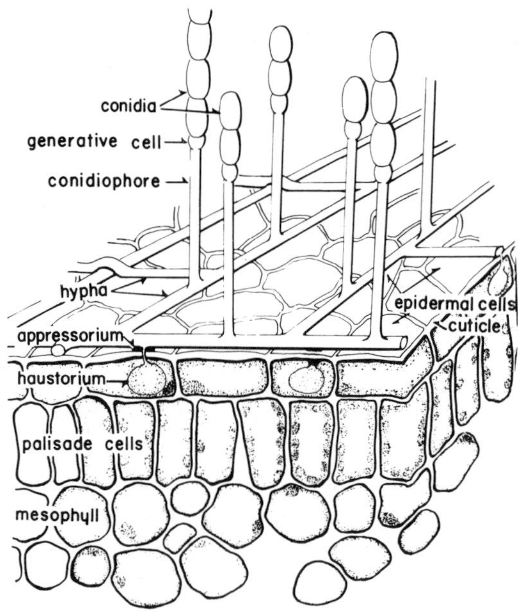

Fig. 15-1. Diagram of a cross section of a grape leaf in which upper epidermis is infected by a fungus colony of grape powdery mildew. Note how fungus spreads by hyphae, and how spores are produced on conidiophores. Fungus obtains food from the leaf by extending the growing haustoria into the epidermal cells. (Redrawn from Hewitt and Jensen, 1973.)

are visible. Mildew causes curling and withering of young leaves and dark staining on the surface of mature leaves. Other symptoms include dropping, discoloration, or splitting of the berries, browning, and poor maturation of canes. Another sign of the disease is a musty, mildewlike smell in the vineyard. The fungus reproduces by spores called conidia.

Control. Prevention is the best means to control powdery mildew. Sulfur dust properly applied to cover the green tissue of vines will prevent powdery mildew from developing. If a sulfur particle contacts a germinating conidium, the fungus is killed before it will infect and establish itself on the plant surface. A dusting machine that mixes the sulfur dust with air and provides a uniform distribution of sulfur particles over the grapevine foliage is essential. If a hand-operated duster is used, the sulfur is applied so that it drifts through the vines. Each application requires from 5 to 10 lb of dusting sulfur per acre (5.6–11.2 kg per hectare).

Sulfur dust should be applied when shoots average about 6, 12, and 18 in. (15.2, 30.5, and 45.7 cm) in length. Additional applications are made every 14 days until the fruit reaches maturity. If the sulfur has been washed off by rain, it should be reapplied. After ripening has begun, the schedule may be reduced for early grapes. Applications of sulfur should be made in the fall to prevent powdery mildew from developing on cluster stems of late-ripening grapes.

If any of the first three regularly scheduled dustings are omitted, the vines should be drenched with a spray including wettable sulfur (see *Eradication*). When the temperature is over 100°F, (38°C) sulfur can burn leaves, shoots and fruit; no dusting should be done at such times. No more than 10 lb (4.54 kg) of sulfur should be applied per acre per treatment. The sulfur program can usually be discontinued in wine grapes if the fruit is free of mildew in midsummer. It must be continued in table grapes until the fruit is mature.

Eradication. Apply wettable sulfur at 1.5 lb/100 gal (180 g per 100 liters) with a wetting agent such as Triton B-1956. Use eradication spray if any of the first three dust applications were not made, or if mildew has not been held in check with sulfur dust. If table grapes are sprayed after berries are more than one-third their full size, the bloom is altered and a spray residue can collect at bottom of the berry.

Diplodia Cane Blight. This disease causes blighted areas of green canes that are gray to brown, with dark-brown streaks projecting at the advancing margin of the blight canker. Gray areas are often speckled with numerous black dots.

Wounds made by girdling and by large pruning cuts on grapevine trunks are sometimes infected by *D. natalensis*. Cankers develop under the bark, then the bark dries, cracks, and peels away.

Fruit Fungi

Many fungi attack the fruit at various stages during its development or even after the fruit has been harvested and shipped and cause fruit rot diseases. Sanitation practices and, in some cases, fungicidal treatments will reduce the inoculum of fungus spores and will help prevent these diseases. Some of the fungi involved are very dependent upon weather conditions, and if rains occur before the fruit is harvested the diseases will cause extensive crop loss.

Botrytis Rot (Slipskin, or Gray mold). The rot caused by *B. cinerea* is one of the most prevalent and widely distributed of the fruit rot diseases of grapes (Hewitt, 1974). The rot is common at harvest time and is associated with rainfall or heavy dew. The fungus can attack in early summer when there are no rains, following rains at harvest, and in stored fruit. *Botrytis* often appears as a gray mold growing on rotten, withered grapes. In some countries, grapes having *Botrytis* rot in the slipskin or early mold stage are processed separately from other grapes. Wines made from grapes partially rotted by *B. cinerea* are known for their distinctive flavor and aroma.

Control. Apply 20 lb per acre (22.4 kg per hectare) of sulfur dust containing 15 percent Captan. Starting July 1 to July 15 apply four dusts at 3–4 week intervals. To prevent infection, dust must be applied before it rains. Although dusting after rainfall has little effect on established disease, it helps protect against new infection if later rains occur.

Summer Bunch Rot. Summer bunch rot is a special type of sour rot found mainly in Thompson Seedless and to some extent in Perlette, Delight, Black Monukka, Red Malaga, and occasionally in Emperor (Hewitt, 1974). Development of summer bunch rot disease is started by *Diplodia natalensis.* This fungus also causes a cane blight on the same varieties, a canker disease originating from infection through pruning and girdling wounds.

Summer bunch rot is characterized by excessive dripping of juice from rotting grapes and the presence of numerous fruit flies, dried fruit beetles, and larvae. One can distinguish summer bunch rot initiated by *D. natalensis* from other sour rots by its association with *Diplodia* cane blight. The two diseases occur together.

Phomopsis Rot. This disease is caused by the fungus *Phomopsis viticola,* the same fungus that causes dead-arm disease (Hewitt, 1974). Phomopsis rot of grapes generally develops from infected cluster stems or pedicels, or both. The fungus grows from the stems into the grape and causes rotting. The disease has been observed in Emperor, Khandahar, Malaga, Molinera, Olivette blanc, Olivette noir, Thompson Seedless, and Tokay.

Sour rots. These rots are caused by microorganisms that may enter grapes by direct penetration or through wounds. They have a pungent, sour, vinegar odor (Hewitt, 1974). There are generally a complex of microorganisms involved in the development of sour rots, including species of *Alternaria, Aspergillus, Cladosporium, Diplodia, Penicillium, Rhizopus, yeast, Acetobacter,* and other bacteria as well as fruit flies and dried fruit beetles.

Sour rots may be grouped into two types based on overall symptoms: those that rot a large proportion of the center of the cluster (a typical bunch rot), and those that rot only a few grapes or clusterettes on the shoulder, side, or tips of clusters.

DISEASES IN THE EASTERN UNITED STATES (McGREW AND STILL, 1972)

Several diseases occur in the eastern states that do not occur in California chiefly because of the greater humidity and summer rains in the former location.

Black Rot

This fungus causes greater loss of grapes in the eastern states than all other diseases combined. It attacks the leaves, young canes, tendrils, and fruit. Only the youngest tissues are attacked except for the fruit, which may become infected until it is almost fully grown. Spotting occurs on the leaves and berries. The decaying berries soon shrivel and become mummified and black in 7 to 10 days.

Downy Mildew

This fungal disease is most severe in cool moist weather. Mainly it attacks grape leaves, beginning with the older ones, then spreads apically on the shoot. The first symptoms are light yellow spots on the leaves that can spread and kill them, causing the leaves to dry and crumple. Early in the season downy mildew can attack and cause malformation of shoots, tendrils, or berries.

Anthracnose (Bird's-eye Rot)

This fungus can attack the berries, young shoots, tendrils, petioles, leaf veins, and fruit stems. Spots sometimes occur on young shoots and can combine and girdle the shoot, which then dies. Spots may also develop on petioles and leaves, especially on the undersides of leaves. Spots on berries are circular, sunken, and ashy gray, and dark-bordered in late stages of the disease. The name "Bird's-eye rot" is derived from the appearance of these spots on the berries.

Ripe Rot and Bitter Rot

These two rots may appear when the fruit begins to mature, and cause a bitter, off-flavor taste. Bitter rot discolors the berry at the cap stem and, as the disease progresses, the spores give a sooty appearance to the infected berries. Berries infected with bitter rot may show concentric zones and usually fall off before they dry.

DISEASES OF MUSCADINE GRAPES

Muscadine grapes, which are adapted to the southeastern United States, are generally less severely injured by disease than bunch grapes. Black rot and bitter rot, which affect these grapes, have been discussed earlier.

Cercospora (Angular Leaf Spot)

This is the most important disease economic attacking muscadine grapes. The berries are not attacked by the fungus. Irregular brown spots appear on the leaves, which enlarge and combine. Defoliation and reduced quality of fruit may result.

BACTERIAL DISEASES

Bacteria can infect grape tissues at various stages of vine development. One disease produced by bacteria is devastating to grapes and has limited their production in the southeastern United States to muscadine or other grape types resistant to the organism. The bacterial diseases are difficult to control once they become established in a grape-producing area.

Pierce's Disease (Hewitt, 1970)

This destructive disease was first observed in 1884 in southern California, where it was called Anaheim disease or California vine disease. It is now present in most of the state, and has assumed epidemic proportions in some local areas. For many years this disease was believed to be caused by a virus, but it has recently been shown to be due to bacteria. The characteristic symptoms in early summer include delayed shoot growth, leaf mottling, and dwarfing of the new shoots. In late summer and fall diseased vines show burning, scalding, or drying of the leaves; wilting or premature coloring of the fruit, and uneven cane maturity. Diseased vines die in 2 or more years, depending on the variety of grape and the age of the vines. Ribier usually dies within 2 years; Emperor, Malaga, Thompson Seedless, and most other *vinifera* varieties die within 2 to 5 years. Some of the American (slipskin) varieties live much longer. No fully effective control of Pierce's disease is known.

To maintain the vineyard in production, remove all diseased vines and replant with healthy vines or with layers from adjacent healthy vines. The disease is not carried over in the soil. The disease is caused by rickettsialike bacteria that have been observed in leaf vessels from infected vines (Goheen et al., 1973; Hopkins et al., 1973). Pierce's disease is spread by several types of leafhoppers, the spittlebug, and by grafting.

Insect Vectors of Pierce's Disease. There have been two serious outbreaks of Pierce's disease in California. The first was between 1880 and 1900, and the second between 1934 and 1947; both periods were associated with above-normal rainfall. Four spittlebugs and 20 sharpshooters, a kind of leafhopper, can transmit the bacterium that carries the disease from plant to plant. The two important vectors for transmitting the disease to grapevines in the San Joaquin Valley are the green sharpshooter (*Draeculacephala minerva*) and the red-headed sharpshooter (*Carneocephala fulgida*). In coastal counties the blue-green sharpshooter (*Hordnia circellata*) is the major vector. Seventy-three plant species are hosts for the vectors and serve as disease reservoirs.

Black Knot or Crown Gall

Black knot is caused by soil-inhabiting bacteria (*Agrobacterium tumefaciens*), and is often a serious problem when vines are propagated by softwood cuttings in the greenhouse if strict sanitation is not observed. It is usually not serious in California vineyards, except when the trunks and arms have been cracked by freezing (Jacob, 1950). Injuries on the arms and trunk can develop tumorlike spongy overgrowth during the spring and summer. These overgrowths on the aerial parts of the plant often turn dark.

The disease also occurs around the crown of the plant, often just below the soil surface. This form is usually called crown gall. The bacteria can be spread to the aerial parts of the vine by rains and may enter the tissues through fresh wounds. The pathogen can move in the vessels, spread to aerial parts of the vine, and cause secondary tumors to develop (Lehoczky, 1968). The bacteria in the plant tissues stimulate the cells to grow rapidly, resulting in the formation of overgrowths. The best prevention is sanitation. Avoid cutting into affected tissue with pruning shears.

Chapter

16

INSECTS

Insects can be classified according to the part of the vine that is attacked.* Some attack flowers and fruit, roots, wood, and so on, and others are pests in wineries. For details on the various insects, their life cycles, injurious effects, habits and enemies, the reader should refer to the California Agricultural Experimental Station Circular 566 (Stafford and Doutt, 1974), and for control programs to the current Pest and Disease Program for Grapes (Barnes et al. 1973).

INSECTS THAT ATTACK FLOWERS AND FRUIT

Consperse Stink Bug (*Euschistus conspersus*)

Adults are brown and 3/8–1/2 in. (9.53–12.70 mm) long and have amber-colored legs with minute black spots. Antennae are pale green or yellow and the body is hard and shield-shaped. Adults can fly considerable distances. The bug inserts its mouth parts into the ripening berries and the punctures result in a rapid breakdown of the berry.

Grape Mealybug (*Pseudococcus maritimus*)

Many clusters of table grapes may be lost because of mealybugs, their egg masses, cast skins, honeydew, and sooty mold that grows on honeydew. Adult females are oval, wingless, flattened insects about 1/8–3/16in. (3.17–4.76 mm) long. They appear to be dusted with a white waxy secretion, and fine wax filaments protrude from the margins of their bodies. Honeydew is secreted by both young and adult insects, and may be found inside bunches of grapes; enough collects to cause running and dripping. Heavy infestations spoil raisins.

* Much information in this chapter was adapted from Stafford and Doutt, 1974.

Grape Thrips (*Drepanothrips reuteri*)

These are tiny insects about 1/32 in. (0.79 mm) long. They can scar berries and spoil cluster appearance, so fruit is unacceptable for the table market. They also feed on tender green shoots and leaves. Injury occurs when shoots begin growth, especially in cool weather. Shoots may be stunted but resume normal growth at onset of warm weather. Most fruit damage is done by the time berries are one-third grown.

Hoplia Beetle (*Hoplia oregona*)

Adults are 1/4–1/3 in. (6.35–8.46 mm) long. On the upper side these beetles are mostly reddish-brown, with darker heads; the underside is shiny and silvery. Beetles fly into vineyards when shoots are about 12–14 in. (30.4–35.6 cm) long and feed on clusters and young leaves.

Vinegar Fly (*Drosophila* sp.)

The yellowish adult flies are about 1/10 in. (2.54 mm) long, and are attracted to all kinds of fermenting fruits. The larvae are shaped like maggots and are about 1/4 in. (6.35 mm) long. During the growing season the number of vinegar flies builds up on waste and culls of several vegetable and fruit crops grown in locations near the vineyards. If berries are pulled from their clusters the fleshy fruit is exposed, and in such locations vinegar flies lay their eggs. Larvae develop and feed on the berries. Flies are attracted to fermenting bunches and carry bunch rot pathogens to previously uninfested clusters, which is their most damaging effect on vineyards. Cultural practices to reduce number of tight bunches and incidence of bunch rot help to control vinegar flies.

Western Flower Thrips (*Frankliniella occidentalis*)

These are commonly found in flower clusters during bloom. They are about 1/24 in. (1.06 mm) long and range in color from yellow to brown. The thrips dwarf and scar young shoots in early spring and may be in the flowers from prebloom to fruit-set. Peak population occurs at 50–60 percent capfall. When eggs are layed in young berries, a small dark scar surrounded by a lightened area occurs; this is called a "halo" spot. These spots can spoil the sppearance of such varieties as Almeria, Calmeria, and Italia. Feeding by nymphs also causes scarring, which can first be observed at shatter stage (fruit-set), and is completed shortly after completion of the shatter.

INSECTS THAT ATTACK ARMS, CANES, AND TRUNKS

Branch and Twig Borer (*Polycaon confertus*)

These beetles can kill half the young shoots of a vineyard. Females are brown or black, cylindrical, and about ⅔ in. (16.9 mm) long; males are about half as large. In the spring when shoots are 8–10 in. (20.3–25.4 cm) long, the adult can bore a hole into the crotch formed by a shoot and a spur. A wind can then break the base of the shoot so it hangs down and wilts. The larvae feed on both living and dead wood, and plug the furrows behind them with frass and dead wood. This material looks like fine, tightly packed sawdust.

Minor Cicada (*Platypedia minor*)

This insect occasionally injures grapes. Cicadas can make loud clicking or buzzing noises. Adults are about ¾ in. (19.1 mm) long, have black or bronze bodies and two pairs of transparent, colorless wings. The noise they produce resembles that of two glass marbles struck together rapidly. Damage is caused when the cicada punctures the cane during egg laying. Slivers of wood protrude from each puncture.

Darkling Ground Beetles (*Blapstinus* sp.)

These beetles feed on trunk wounds occasionally produced by cultivating equipment, and on other tissue exposed by fresh cuts or on callus tissue of older wounds. Eventually they girdle the vine, often making girdles 2–3 in. (5.08–7.62 cm) wide. The larvae that live in the upper, dry 2–3 in. (5.08–7.62 cm) of soil do no harm to the vine.

Scale Insects

Scale insects are not usually of economic importance in California. Some scales feed on more than one part of the vine, and others remain stationary after selecting a place to feed. The brown apricot scale (*Lecanium corni*) and cottony maple scale (*Pulvinaria vitis*) are examples of the first type, and the grape scale (*Diaspidiotus uvae*) is typical of the second group.

Subterranean Termite (*Reticulitermes hesperus*)

The adult sexual forms seen during swarming after the first fall rains are black and have two pairs of long slender wings, and broad waists (ants have a slender, threadlike waist). The males and females dig a small hole in the soil and raise a brood of wingless workers that can enter the vines through old beetle holes and various types of wounds. Termites feed on the heartwood (inner dead wood of trunks and branches) almost exclusively. Older vines show more termite damage than younger ones. Stakes are also destroyed by termites. Cedar heartwood is quite resistant to termites.

INSECTS THAT ATTACK THE LEAVES

Achemon Sphinx Moth (*Pholus achemon*)

This moth is about the size of a hummingbird and hovers near a flower while feeding. When fully grown, the caterpillars are about 4 in. (10.2 cm) long and resemble the green tomato worm. A large worm can eat nine mature grape leaves every 24 hr. After a vine is defoliated, the worms move to other vines with green leaves. Several years of severe damage are usually followed by several years of negligible injury.

Grape Erineum Mite (*Colomerus vitis*)

This mite can be seen only under the microscope, as the adult female is about $^{8}/_{1000}$ of in. (0.203 mm) long and about one-quarter of that in width. Its elongated body has short legs attached near the anterior end. Damage begins in early spring, when pinkish or reddish swellings or galls appear on the upper surfaces. The concave portion of the leaf beneath the gall is lined with a felty mass of curled plant hairs termed "erinia," which can sometimes completely cover the lower surface of the leaf. Later the erinia turn yellow, and in August they turn brown. The swellings on the upper side of the leaf disappear. This has not been a serious problem in vineyards.

Grape Leaf Folder (*Desmia funeralis*)

These moths fly about all night on warm nights and feed on ripe table grapes. The adult moth is dark brown and has a wing expanse of about 1 in. (25.4 mm) (Fig. 16-1). The forewings have two white spots, and there are two white bands across the abdomen in both sexes. There are three broods each year. The first moths of the season emerge from pupae that overwintered. Eggs about 1/32 in. (0.79 mm) long are laid on leaves in places sheltered from the wind. Two weeks after hatching, larvae feed singly in pencil-size leaf rolls made by spinning strands of silk from the edge of the leaf to points near the center (Fig. 16-2). As the strands dry, the leaf edges curl into a roll with the upper leaf surface forming the outside. Larvae are yellowish-green until feeding, when they become bright green. Rolling of leaves restricts food production for vines and exposes berries to sunburn. Larvae can also break berry skins and spoil fruit. Severe damage occurs only with the second and third broods.

Grape Leafhopper (*Erythroneura elegantula*)

This pest feeds on leaves and produces light-colored mottling. These narrow insects are pale yellow with reddish and dark-brown markings in a characteristic pattern (Fig. 16-3). There are distinct dark brown spots on scutellum and wings. When vines leaf out in spring

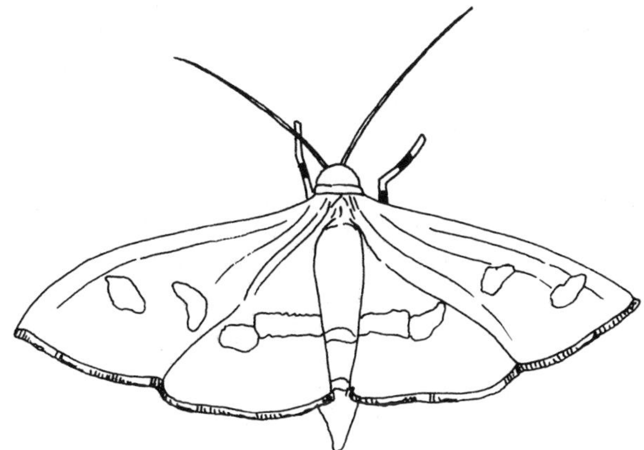

Fig. 16-1. Adult female leaf folder. Wing expanse is about 1 inch (2.54 cm). (After Jensen et al., 1973a.)

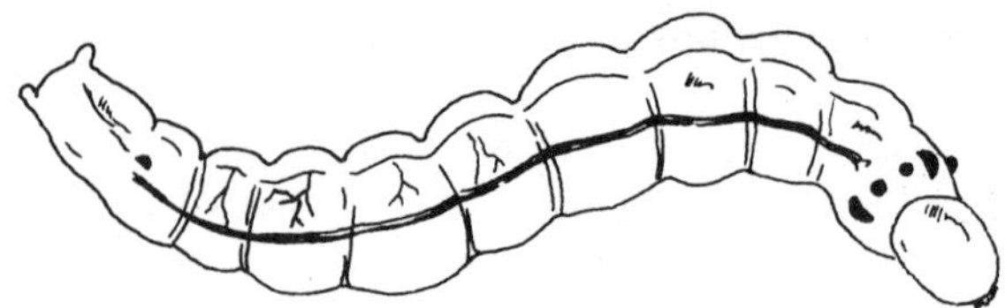

Fig. 16-2. Nearly full-grown larva of grape leaf folder. Actual size, about ¾ inch (1.91 cm) long. (After Jensen et al., 1973a.) (Drawn from photo by J. E. Dibble.)

overwintered adults move into the vineyards and feed on the leaves. Three broods are produced during the season. Both nymphs (Fig. 16-4) and adults feed by sticking their mouthparts into the leaves and sucking out the contents. As more leaf punctures are made, pale areas develop and the leaf becomes mottled or variegated; later, whole leaves turn pale, dry up, and fall. Sticky drops of leafhopper excrement support growth of black fungus.

Grape Rust Mite (*Calepitremeris vitis*)

These mites feed on leaves and cause a yellowing of white grape varieties similar to injury caused by Pacific Spider mite (pp. 273). Injured leaves of dark grapes turn a brilliant red. These microscopic, light-amber mites appear wormlike under a 14-power lens; they are broader at the front end and move slowly. During the season they attack upper and lower surfaces of leaves, although control is not usually necessary.

Grape Whitefly (*Trialeurodes vitatus*)

This mothlike insect is about 1/16 in. (1.59 mm) long. Its white color is due to a dense, white, waxy powder covering the body and wings. Eggs are laid on surfaces of grape leaves, and there are several generations per year. The whitefly soils berries with a sticky excrement on which a

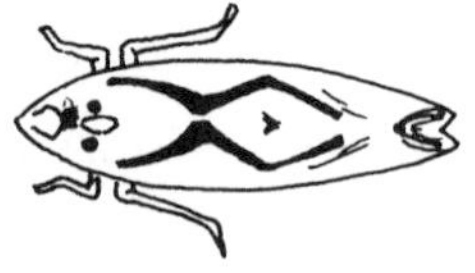

Fig. 16-3. Adult grape leafhopper. Actual size, about ⅛ inch (3.17 mm) long. (After Jensen et al., 1973.) (Photo by J. E. Dibble.)

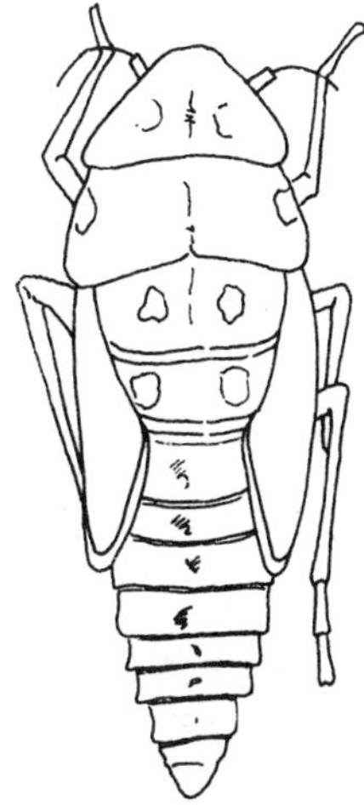

Fig. 16-4. Grape leafhopper nymph. Actual size, 1/10 inch (2.54 mm). (After Jensen et al., 1973b.) (Drawn from photo by J. E. Dibble.)

black sooty fungus can develop. Backyard grapes are more apt to be infested than commercial vineyards. Ornamental shrubs may provide a suitable host on which the insect can overwinter.

False Chinch Bug (*Nysius raphanus*)

The adults are about 1/8 in. (3.17 mm) long and are light or gray. In some years, large numbers of these insects may breed in pastures and grasslands in early spring. When the grass dries up, the insects migrate to find green food, and if a vineyard is attacked serious injury may result. These bugs migrate by walking since they are mainly in the wingless young stage; any adults present also march. A horde of these insects can swarm up the trunks of the vine and suck the juice from the leaves. Thus a vine can wilt and turn brown within 3 hrs. Most destructive migrations occur in May and June, but occasionally they take place in September and October.

Grasshoppers

Commercial damage is generally caused by the green valley grasshopper (*Schistocerca shoshone*), the vagrant grasshopper (*S. vaga*) and the devastating grasshopper (*Melanoplus devastator*). The green valley grasshopper is green, has red hind legs, and a yellow stripe along the midline of the head and thorax. The vagrant grasshopper has a brown body, brown legs, and a tan stripe along the midline of the head and thorax. The front wings are tan with brown mottling. The adult devastating grasshopper is about 1 in. (2.54 cm) long, amber to brown in color, and has dark markings on the thorax and a row of dark spots on the front wings. Grasshoppers may attack in large numbers and injure the vine by feeding on young shoots and defoliating them.

Omnivorous Leaf Roller (*Platynota stultana*)

This insect attacks the grape berries. Yellowish-white or cream-colored larvae drop on a fine thread when the foliage on which they are feeding is disturbed. Adult moths are about $\frac{1}{2}$ in. (12.7 mm) long with a wing expanse of $\frac{9}{16}$–$\frac{11}{16}$ in. (14.3–17.5 mm) (Fig. 16-5). In May and June the larvae attack new foliage and clusters (Fig. 16-6). When ripening begins increasing numbers of larvae are found in the bunches, where they often make webs and feed where two berries touch. This damage attracts vinegar flies, dried fruit beetles, and other secondary feeding insects that carry spoilage organisms which cause bunch rot.

Orange Tortrix (*Argyrotaenia citrana*)

This moth has caused considerable injury to wine grapes in the Salinas Valley since 1968. The adults are about $\frac{3}{8}$ in. (9.53 mm) long, and are brown or buff with a saddle of darker coloring across the folded wings. Full grown larvae are about $\frac{1}{2}$ in. (12.7 mm) long. The head, thoracic plate, and body are straw-colored, but the body may sometimes be green, dark gray, or smoky in color. The larvae wriggle sideways or backward when disturbed, drop to the ground, or are suspended from a

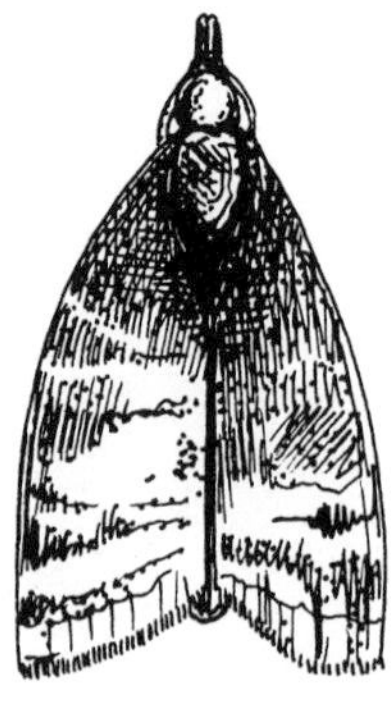

Fig. 16-5. Adult omnivorous leaf roller. Actual size, $\frac{3}{8}$ to $\frac{1}{2}$ inch (9.53 to 12.7 mm) long. (After Jensen et al., 1973c.)

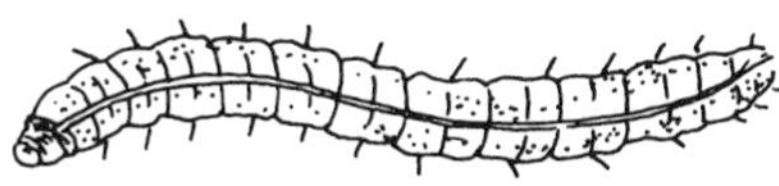

Fig. 16-6. Large larva of omnivorous leaf roller. Actual size, about ⅝ inch (15.9 mm) long. (After Jensen et al., 1973c.)

leaf by a silken thread which they can ascend. They feed on berries and stems and can cause berry drop and stem girdling. Pupae may contaminate the clusters, and larvae feeding can cause spoilage.

Pacific Mite (*Tetranychus pacificus*)

This is a much more destructive red spider mite than the Willamette mite. Under a magnifying glass the Pacific mite appears pale amber to reddish in color, with two, four, or six large black spots on its back (the Willamette is similar, but is usually pale yellow with a row of inconspicuous black dots along each side of the body). Pacific mites are larger and produce more webbing than Willamette mites.

The first signs of Pacific mite injury are yellow spots on the upper leaf surface. Weak areas in the vineyard are favorite locations for the mites. Symptoms first appear in April or May; and in warm June weather mites increase and damage entire leaves. In 10 days a healthy vineyard may turn brown. On Thompson Seedless and other white varieties, yellow leaves turn brown and later dry out and die. On Zinfandel and other black varieties, yellow areas turn red, then purplish red, and then die. (The grape leafhopper causes a pinkish-red color.)

Western Grape Skeletonizer (*Harrisina brillians*)

Caterpillars live in close groups on the lower surfaces of leaves. They are yellow with two purplish transverse bands and several narrow black ones, and have four tufts of poisonous black spines on each body segment that can raise welts on the skin like those caused by nettles.

The small black heads are retractable. Fully grown adults are about ½ in. (12.7 mm) long. The caterpillars line up side by side and eat the lower epidermis and green part of the leaf. The upper epidermis is left intact, although large caterpillars also eat this part. This insect is found mainly in southern California and produces three generations a year. Localized infestations in northern California have usually been eradicated.

White-Lined Sphinx Moth (*Celerio lineata*)

Mature larvae are about 3 in. (7.62 cm) long, bright green (or occasionally black), with a posterior yellow horn (Barnes, 1970). In early spring larvae may migrate into vineyards in the Coachella Valley and defoliate the vines. A preventive measure is to place insecticide dust at the bottom of a ditch across the line of migration.

Willamette Mite (*Eotetryanychus willametti*)

This spider mite is not a serious pest and does little damage north of Fresno. It causes less burning and defoliation than the Pacific mite, and usually results in bronzing and yellowing of the leaves. Sometimes the Willamette mites attack the youngest unfolded leaf on a shoot and then move up to the next leaf, which is also killed. Eight or ten young leaves may be killed in this manner. Sometimes shoots are bronzed and distorted.

Yellow Woolly Bear (*Diacrisia virginica*)

These caterpillars may be found in San Bernardino County in early spring (Barnes, 1970). Mature larvae are 1½–2 inches (3.81–5.09 cm) long, hairy, straw-yellow, have transverse black lines, and damage leaves heavily.

INSECTS THAT ATTACK BUDS AND YOUNG SHOOTS

Cutworms

This is the larva form of a moth that hides during the daytime and fly at night. The three commonest species that attack grapes are variegated cutworm (*Peridroma saucia*), greasy cutworm (*Agrotis ypsilon*), and brassy cutworm (*Orthodes rufula*). Fully grown cutworms are about 1½ in. (3.81 cm) long with smooth bodies marked with faint spots or lines. Vines are attacked by cutworms from swelling of buds until shoots are several inches long. Buds and succulent shoots are eaten partway through so that they are weakened and fall over.

Click Beetle (*Limonius canus*)

The immature larvae are called wireworms. Female, full-grown click beetles attain a length of ½ in. (12.7 mm), have reddish-brown wing covers and a brown head and thorax. They are slender and hard-shelled. If held, a click beetle can arch its body backwards and then straighten out with an audible snap. It damages buds in the same manner as grape bud beetles (see below), but feeds in full daylight (cutworms and grape bud beetles feed at night). Click beetles can fly.

Flea Beetle

Like its namesake, the flea beetle can jump. The important species in California is the steel-blue beetle (*Altica torquata*), about $^3/_{16}$ in. (4.76 mm) long, and metallic blue or purple in color. Adults emerging from hibernation attack and destroy opening grape buds.

Grape Bud Beetle (*Glyptoscelis squamulata*)

This insect is found on grapes in Fresno County and in the vicinity of Lodi. It is about $^1/_4$ in. (6.35 mm) long, hard-shelled, and light gray. It feeds at night and hides in the daytime. The insect feeds on swollen or opening buds in spring, leaving bud scales intact but gouging out the inside of the bud. The shoots grow poorly after they are 1 to 2 in. (2.54–5.08 cm) long.

Little Bear Beetle (*Pocalta ursina*)

This insect is widespread, but damages grapes mainly in the southern San Joaquin Valley. Adult beetles have bodies $^1/_2$ to 1 in. (12.7–25.4 mm) long, with broad trunks, hard bodies, and wing covers that are usually dark reddish-brown, although some may be light-brown or black. The head and thorax are black. Adults sometimes attack young, succulent shoots in late March and April.

INSECTS THAT ATTACK ROOTS

Grape Phylloxera (*Dactylosphaera vitifoliae*)

For thousands of years this pest lived on native wild grapes. It was first seen in California in 1852, and is now generally distributed throughout northern California although some localities probably have none. Occasionally the winged form is present in coastal areas in late summer and fall but cannot establish new colonies as in Europe and the eastern United States.

Phylloxera are pear- or oval-shaped, and their entire life is spent on the vine roots. The adult insect is microscopic in size and is yellowish-green or yellowish-brown. The adult female remains almost stationary on the root and her eggs pile up around her. When the young hatch they begin feeding. Some crawl through cracks in the soil to the surface, travel a short distance, and then crawl down and start new colonies on another vine. Feeding on the roots by adult females and young causes gall formation.

On older roots semispherical swellings are produced, and on young rootlets the galls are hook-shaped. After the galls decay in a month or two, the insects move to other locations on the roots and produce new galls. Decaying galls plus a poisonous saliva injected into the roots are probably the cause of the stunting and decline of the vines. Destruction of fine feeder and other roots is also involved.

Phylloxera requires a soil that will contract and crack when drying, so that there will be pathways for the insect to travel in and infest the whole root system as well as to travel to other vines. Use of rootstocks is the main preventive measure. Care must also be taken not to spread the insect into noninfested areas by grape rootings, bench grafts, irrigation, or tillage tools. To clean nursery stock, all soil should be washed from vine and roots and fumigated. A dip into emulsions of insecticides may also be effective.

Ground Mealybug (*Rhizoecus falcifer*)

This is rarely a serious pest, but sometimes causes damage to backyard vines in coastal areas. Superficially it does not resemble a mealybug because of its smaller size, long slender body, and absence of wax rods and filaments typical of other mealybugs. Its body is covered with white, waxy powder.

Ground Pearls (*Margarodes meridionalis*)

This scale insect is found on roots of grapevines in the Imperial Valley (Barnes, 1970), and is also present on roots of Bermuda grass in that area especially along irrigation canals. It is favored by sandy soils. The young female is a globular cyst about 3/16 in. (4.76 mm) in diameter found attached to rootlets, and has a glossy pearly-colored covering.

Nematodes

The important types include root-knot, root-lesion, daggar, and citrus nematodes.

Root-knot nematodes (Meloidogyne sp). These insects can cause weakening of growth and lowered production. The young are microscopic and worm-shaped. Mature males remain wormlike but females are pear-shaped. They are spread by infested rootings, water, and tillage equipment. Young worms travel through root tissue and soil to reach other roots. The spread of nematodes is favored by porous, sandy, or loam soils; in heavy soils infestations are less serious.

Infested roots develop knots or galls. Swellings usually cause enlargement of whole root; swellings caused by phylloxera occur mostly on one side of the root or cause hooked-shaped galls at tips of rootlets. The most effective means to control nematodes is the use of resistant rootstocks. Preplant treatments of soil with fumigants has provided enough control of root-knot nematodes so that a new vineyard can be established.

Root-lesion nematodes (Pratylenchus vulnus). These pests can cause poor vine growth. Young and adult worms can be found in roots and soil surrounding infested plants. No gall formation occurs on roots. The male and female are elongated worms and can penetrate the roots in both larval and adult stages. Severe infestations can greatly reduce root systems, and lighter ones can induce production of lateral rootlets. Soil fumigation is the same as for root-knot nematodes, but rootstocks resistant to root-knot nematodes do not appear to be resistant to the root-lesion nematode.

Daggar Nematodes (Xiphinema sp.). These relatively large nematodes are migratory types that do not penetrate the roots but feed instead on young rootlet tips, which often become swollen, curved, or distorted.

Xiphinema index. This is the vector of the fanleaf, yellow mosaic, and veinbanding virus disease complex of grapes. If an old virus-infested vineyard is pulled out, new plantings can be infected by nematodes even after several years delay in planting. *X. index* is found in coastal areas and in the San Joaquin Valley as far south as Delano.

Citrus Nematodes (Tylenchulus semipenetrans). These are sedentary insects found in southern California and the San Joaquin Valley. Young larvae feed on cortical layers of young roots, and as females develop through successive larval stages, the head penetrates the root but the body remains outside. The body appears as a swollen structure on the root surface.

Other nematodes include various insects that are found in vineyard soil, although little information exists as to their habits and effects on the vine. They include spiral, ring, and pin nematodes.

Control of Nematodes

Soil fumigants with 1,3-D (trade names are DD, Vidden D, or Telone), a mixture of 1,3-dichloropropene and 1,2-dichloropropane, have been the most widely accepted material used before vineyards are planted. Rates up to 80 gallons per acre have been used (Raski et al. 1973; 1976). Unfortunately, the results are usually unsuccessful since the surviving nematodes often attack the vines before the end of the first growing season. Thus within a few years much of the advantage of the treatment is lost.

Recent experiments have shown that long-lasting control of dagger nematode (transmits fanleaf virus) and fanleaf virus can be obtained using high dosages of 1,3-D [up to 250 gall per acre, (2,337 liters per hectare)] in a split application (200 gall per acre at 30 in. depth on 2-ft spacing plus 50 gall per acre at 8–10 in. depth on 1-ft spacing (1,870 liters per hectare at 76.2 cm depth on 61.0 cm spacing plus 467 liters per hectare at 20.3–25.4 cm depth on 30.5 cm spacing). Deep placement of 1,3-D has controlled root knot infestation in the lighter interior valley soils for four years or more. Similar results were obtained with deep placement of methyl bromide at high rates.

Sometimes nutritional deficiencies show up in vines growing in treated soil. This may be due to low populations of mycorrhizal fungi present in the treated soil (Raski et al., 1976). Mycorrhizae are fungi that are associated with the roots of some higher plants including grapes. It is believed these fungi are able to make nitrogen and other nutrients more available to the roots.

Western Grape Rootworm (*Adoxus obscurus*)

This insect is of minor importance. In May adult beetles (about ⅕ in. (5.08 mm) long) emerge from the soil and feed for about 2 weeks on leaves in the lower part of the vine, making slitlike holes about ½ in. (12.7 mm) wide and ¼–½ in. (6.35–12.7 mm) long. In severe infestations, the insects feed on shoots and cut shallow grooves in the berries. In the grub or larval stage they feed on roots of the vine, and can eat small rootlets and gouge holes through the bark and outer wood of roots the size of a pencil or larger.

INSECTS THAT ATTACK RAISINS

Dried Fruit Beetle (*Carpophilus hemipterus*)

These insects attack raisins moist enough for fermentation to start and any other material that can be fermented. Adult beetles are dark brown with light brown or amber spots, are about $\frac{1}{8}$ in. (3.17 mm) long, and have short wing covers with light brown spots. The antennae are knobbed at the tips, and the beetles are strong fliers. The larvae feed on the flesh of the raisin which, together with excreta and cast skins, spoils the dried fruit. They are controlled by fumigation.

Indian-Meal Moth (*Plodia interpunctella*)

This insect attacks stored raisins and other types of dried fruit and nuts. The moth is about $\frac{3}{8}$ in. (9.53 mm) long, with a wingspread of about $\frac{5}{8}$ in. (15.9 mm). At rest the wings are folded around the body. The outer two-thirds of the front wings are dark coppery-brown, and the inner third is cream-colored. Infested raisins can contain living worms, excrement, cast skin, webbing, dead worms, and silk cocoons. Control is by fumigation and cleaning in the various processing steps in the packinghouse.

Raisin Moth (*Cadra figulilella*)

This insect can feed on ripening grapes in the vineyard but does most damage to raisins in storage. This moth is similar to the Indian-meal moth except that the fore wings of the former are gray, the wing markings are obscure, and the hind wings are whitish. Young larvae hatched from eggs laid on the raisins feed chiefly on the ridge crests of the raisins, but can also penetrate the flesh. The worms do not consume raisins, but move about leaving behind excreta and webbing. Control is by fumigation and sanitary cultural practices.

Saw-Toothed Grain Beetle (*Oryzaephilus surinamensis*)

This pest feeds on all dried stored food. It is a slender, flattish brown beetle about $\frac{1}{10}$ in. (2.54 mm) long. Sharp projections stand out on each side of the thorax, resembling saw teeth. The beetle attacks all parts of the raisin. There is no webbing, the excreta are yellowish, and pellets are smaller and more elongated than those of the moth larvae.

INSECTS THAT ATTACK WINERIES

Lead Cable Borer (*Scobicia declivis*)

This insect can bore holes in lead sheathing of telephone cables, can damage wooden wine casks, and is a strong flier. The adult is a black cylindrical beetle about ¼ in. (6.35 mm) long, with a tan or reddish spot on each side of the body near the middle. The insect breeds almost entirely in oak trees, especially the California live oak and black oak. Eggs are laid in healthy wood after it has been cut off from the tree for 2–6 months.

In wineries the adult female beetle bores into wine casks as though intending to lay eggs there. When a hole is bored completely through the cask, wine leaks out. No oak wood that has been felled for 2–6 months should be stored near the winery. Broken limbs from oaks should be burned by May each year before adults can emerge.

Pomace Fly (*Drosophila*)

Control of *Drosophila* consists of placing barriers to the insect at all possible stages of processing and storage. Screens of 24 × 24 mesh can be used, or important rooms can be pressurized by drawing air in through a filter and forcing it out through openings to the room at 7–10 miles (11.27–16.09 km) per hr. Insecticides may also be useful when applied to screens or as aerosols. General sanitation should be good.

Chapter 17
OTHER VINEYARD PESTS

In addition to fungi, bacteria, viruses, and insects there are other larger vineyard pests. Some of these including birds, pocket gophers, rabbits, and deer can inflict severe damage on vines and/or fruit. Detailed information including sources of supply of equipment can be found in the California *Vertebrate Pest Control Handbook* (Clark, 1975).

BIRDS

Birds may eat the ripe fruit before it is picked; early-ripening fruit and isolated vineyards are especially vulnerable. The commonest devices used for controlling birds such as starlings, linnets, jays, crows, magpies, robins, and so on, are automatic *acetylene exploders*. These devices explode with a sound usually three times greater than that of a 12-gauge shotgun blast. One gun will usually protect about 40 acres (16.19 hectares). The units should be operating before bird damage begins, and should be moved around every 3 to 4 days as the birds become more quickly habituated to sound constantly coming from the same point. Intervals between explosions should also be varied. These guns are only partially effective, and the degree of effectiveness varies with the bird variety.

Biosonics, the study of animal communicator systems, is a new and promising area for investigation. Birds have various communication signals that include assembly, distress, and alarm calls. Phonograph records of distress calls are amplified and replayed into the vineyard to frighten the birds away. Currently, however, biosonics is not commercially feasible for large commercial vineyards, although it can sometimes reduce crop losses when used in conjunction with other methods.

Cage traps can be used for starlings, blackbirds, grackles, and cowbirds (Zajanc and Cummings, 1965). Birds enter slots in the center section of a V-shaped cage top, but try to escape through the outer wall of the cage instead of the top. The best baits are foods the birds feed on in the vicinity. Trapping is often used around feedlots, although it is usually ineffective in vineyards since not enough birds are caught to reduce damage significantly.

Use of mesh nets over the vines is the only sure way to keep birds from getting to the fruit, but involves a high initial expense and maintenance problems. With small plantings, ripening grape clusters can be enclosed in paper bags.

Toxic bait should be used only when damage is extensive enough to warrant more drastic steps than the use of other methods, or when noisemakers, scare devices, repellents, traps, or exclusion do not provide adequate control. Prebaiting with nontoxic bait should be done before toxic bait is used. If birds readily consume the nontoxic bait, such as various types of seeds, most of the toxic bait will be eaten over a 24-hr period (Clark, 1975). Toxic bait should not be applied to the vines. A good method is to construct troughs in the vineyard to hold the bait (Fig. 17-1). Strychnine, often used as a toxicant, may be applied only under supervision of the state agricultural commissioner.

DEER

The best way to control deer is to fence them outside the vineyard area; a woven-wire fence will usually do the job (Longhurst et al., 1962). An electric fence may be useful in the summertime.

Taste repellents can also be used, but they are generally impractical because the repellent must be renewed each time it is washed off by rain. Improved Z.I.P. (a commercial ZAC Rhodoplex product with methocel added), Arasan 75, or Arasan 42-5 may be used in winter by brush or spray application (Anonymous, 1968). In the growing season, spray applications of Z.I.P. Arasan 75, or Arasan 42-5 can be employed. It may be necessary to apply repellents frequently to cover new growth.

POCKET GOPHERS

These rodents can girdle vines near the soil level. They dig furrows in the soil about 2–3 in. (5.08–7.62 cm) in diameter and 4–12 in. (10.2–30.5 cm) below the surface. These furrows can cause substantial irrigation water loss and create other watering problems (Cummings, 1971). They must be killed by trapping, or the use of bait prepared with strychnine alkaloid, or other bait (Marsh and Cummings, 1974). A mechanical bait applicator has been developed.

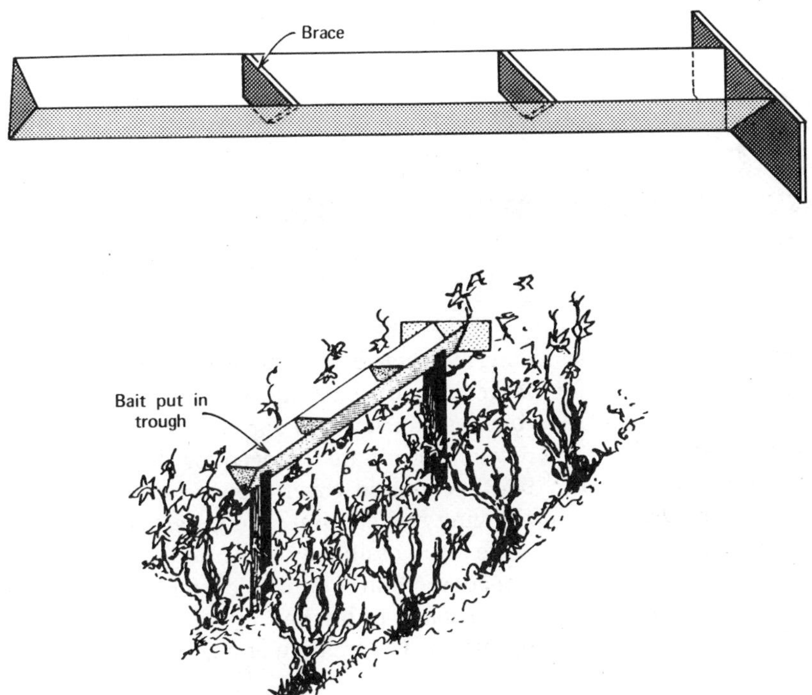

Fig. 17-1. Baiting trough in vineyard, used for holding bait for bird control. (After Clark, 1975.)

RABBITS

Rabbits can kill vines in young vineyards by chewing the bark and leaves. The best protection is to fence off the area with a mesh fence 2½ ft (76.2 cm) high with about 6 in. (15.2 cm) of wire buried in the ground to prevent burrowing (Jacob, 1950).

Chapter
18
WEED CONTROL AND AIR POLLUTION

Control of weeds in young vineyards is necessary because weeds compete with vines for water and nutrients and interfere with harvesting of mature vines. Herbicides can be used for strip weed control. A weedfree strip down the row can eliminate the need for close cultivation in vines (Lange et al., 1974). A mowed cover crop between rows used with a weedfree strip in the vine row can improve water penetration in some soils. Savings in cost are also possible by using herbicides for weed control in vineyards. The reader should obtain a current issue of *Crop Weed Control Recommendations,* published by the University of California for the latest suggestions (Agamalian et al., 1972).

CONTROL OF ANNUAL WEEDS

Several herbicides used to control annual weeds are discussed below.

Aromatic Weed Oils

Young weeds 1–2 in. (2.54–3.81 cm) high are most sensitive to emulsions of these oils. Repeat applications may be required to keep weeds under control since oils usually cause only contact damage. Vine trunks should not be sprayed.

Diuron (Karmex) and Simazine (Princep)

Although these chemicals are structurally unrelated, their herbicidal properties in vineyards are similar. They are applied as wettable powders containing 80 percent active ingredients. These herbicides are applied to the soil surface and must be leached into the soil by irrigation or rainfall to be effective. Excessive irrigation in light soil must be avoided, however, or the chemicals may be leached into the rootzone of young vines and out of the rootzones of weed seedlings (Lange et al., 1974).

Diuron is more effective than simazine on crabgrass and barnyardgrass, while simazine is more effective on wild oats and groundsel. Both chemicals are not toxic to weed seeds until they germinate, when they attack the young seedlings that absorb the herbicides through their roots. Most annual weeds will be destroyed or repressed if the herbicides are applied and leached into the soil before the seeds germinate.

Low rates of these chemicals in soil are effective against germinating weed seedlings from three months to a year or more, depending on the amount used. Both herbicides are decomposed by soil microorganisms, and they break down more rapidly in warm, moist soils. Diuron and simazine are generally ineffective in controlling well-established annual or perennial weeds.

Higher rates of herbicides are required on clay soils and those with high levels of organic matter, usually typical of rainy areas. When diuron and simazine are leached into the soil, most is adsorbed on clay particles and organic matter and this portion is unavailable for absorption by seedling roots. Although diuron generally provides good selective weed control, it is less safe than simazine, especially in light coarse-textured soils low in organic matter.

Diuron and simazine should be applied as a band treatment in the vine row from December 1 to March 1 on bearing vines three or more years old with trunk diameters of 1½ in. (3.81 cm) or more. The rate per acre is based on the area sprayed rather than on the area of the entire vineyard.

Injury to grapevines from excessive applications of simazine or diuron is first indicated by chlorotic leaves, which eventually turn brown on the margins. With extreme injury, leaves eventually die; later, new leaves develop (Lange et al., 1974).

Prior to the first application of the herbicides, soil in the vine row should be tilled into an even ridge sloping slightly upward to the vines. This ridge should not be disturbed by additional seasonal cultivation in the middle of the row.

Napropamide (Devrinol)

This is an extremely safe herbicide in vineyards when applied preemergence and incorporated by rainfall or sprinkler. It controls many of the same weeds as nitralin and is particularly effective on grasses and on some weeds resistant to nitralin and trifluralin, such as sowthistle and yellow nutsedge. Napropamide gives poor control of weeds in the tomato family.

Napropamide can also be applied to soil surfaces in young, established vineyards. Four pounds per acre (4.48 kg per hectare) is required when applied at preemergence. It should be leached into the soil as soon as possible after application. If the proper amount of water is used, good residual control will persist throughout the summer. If there is a delay of 10 days or more before leaching the chemicals into the soil, weed control may be poor.

Paraquat*

This is most effective against young weeds. It is water-soluble and must be applied to weed foliage, which it kills on contact. It is not translocated in grapevines. It may be combined in the spray mixture with a preemergence herbicide to control weeds, and an additional application after preemergence treatment is unnecessary.

Young seedling weeds that germinate in the winter while vines are dormant can be controlled by applying paraquat. A wetting agent (surfactant) should be added to the paraquat at 1 qt/100 gal (1 liter/400 liters) of spray solution. Paraquat should not be applied to green grape shoots and green trunks of young vines. It is tightly bound to the soil and is only slowly available to microorganisms which can degrade the unbound paraquat. In the soil paraquat has almost no phytotoxicity, except in sandy soils when applied at very high rates.

Trifluralin (Treflan) and Nitralin (Planavin)

These herbicides are chemically similar and are insoluble in water. Both are unstable on the soil surface. The rate of surface breakdown is greater with trifluralin. The chemicals must be incorporated into the soil immediately following application. Nitralin is less volatile and can be incorporated by rainfall or sprinkler irrigation. Trifluralin must either be applied by subsurface layering or tilled in within hours after application. Both chemicals can be used with greater safety on sandy soils than either diuron or simazine.

Trifluralin, nitralin, and napropamide inhibit the growing points of

* Paraquat is restricted and requires a permit from the County Agricultural Commissioner.

roots and shoots on contact, and are very effective for control of grass species. Neither roots or shoots of susceptible plants will grow through soil containing these chemicals, and plants die because roots cannot contact soil moisture. However, in heavy soil that cracks, chemical-free shoots can grow through these cracks. When incorporated into the soil, these chemicals can persist from a few months to a year or more. They are degraded by microbial and nonmicrobial means.

Incorporation of ½ lb per acre (0.56 kg per hectare) may be used for short-term control of annual weeds on sandy soils, but 1 lb per acre (1.12 kg per hectare) or more should be used on sandy loam and heavy soils. Since incorporation of the herbicides is difficult in established vineyards, the treatments are of most value in new vineyards immediately prior to planting.

CONTROL OF PERENNIAL WEEDS

Field Bindweed (*Convolvulus arvensis*)

Repeated discing between the vine rows can keep bindweed under control, but it does not control the weeds around the vines. Although the row plow or repeated applications of weed oil can partially control bindweed, some injury to young grapevines may result. Paraquat is much safer than oil if kept off foliage and can be sprayed repeatedly in the same manner as weed oil. To burn back bindweed and to control annual weeds in the row, ½–1 lb of paraquat per acre (0.56–1.12 kg per hectare) with a surfactant is effective (Lange et al., 1974).

The amine form of 2,4-D can be used to control field bindweed in vineyards. However, care should be taken to keep 2,4-D off the grape foliage, since leaf malformation and injury to developing clusters can result. It is safer to apply 2,4-D to bindweed after the normal grape berry shatter following bloom than at earlier stages. Lifting the trailing shoots of vines that might otherwise be sprayed helps to minimize damage. Spray drift can be reduced by using a hooded boom equipped with low-pressure flooding nozzles that deliver coarse droplets. Bouncing of the boom can result in drift, so tractor speed should be adjusted to the terrain. Speeds should not exceed 2 or 3 miles per hr. The 2,4-D should be applied only on days relatively free of wind, using 1½ lb of 2,4-D in 60–80 gal of water per acre (1.68 kg of 2,4-D in 561–748 liters per hectare). This method is often not feasible because it is difficult to obtain a permit to use 2,4-D from the county agricultural commissioner after March, and for successful bindweed control 2,4-D sprays are usually required during the summer months.

A safe and effective method for bindweed control is now possible with the application of trifluralin by subsurface layering. The herbicide is applied by means of a spray blade or row plow at a soil depth of 4–6 in. (10.2–15.2 cm) (Fig. 18-1). The layer of trifluralin destroys the growing tips of the weed shoots when they contact the chemical.

No phytotoxic effects on bearing vines has been observed from the use of layered trifluralin. Annual weed control has been effective in addition to bindweed control in the treated areas. However, some weed species such as those in the cabbage family (mustard, shepherd's purse, london rocket) are resistant to layered trifluralin.

Johnson Grass (Sorghum halepense) and Bermuda Grass (Cynodon dactylon)

Repeated discing of the middle of the rows can control Johnson grass or Bermuda grass, but hand labor, row plowing, or herbicides are required to control them in the vine row. Dalapon is partially effective and may control both grasses if applied monthly at low rates (4 lb dalapon/100 gal of water containing 8–16 oz of a surfactant or wetting agent per acre). This is equivalent to 4.48 kg dalapon per 935 liters of water containing 420–560 gm surfactant.

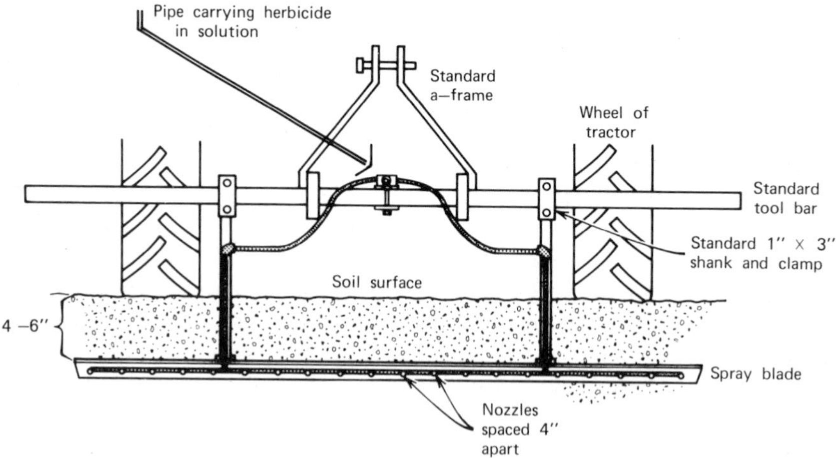

Fig. 18-1. Generalized drawing of back view of a spray blade. In front of the spray nozzles, a shank lifts a layer of soil so nozzles can wet the underlying soil surface. (Redrawn from Lange, 1972.)

Dalapon spray should be kept from contacting grapevines as much as possible, and any spray runoff prevented from soaking into the soil. It should be applied soon after an irrigation to allow as much time as possible between the application and the next irrigation. Dalapon is very soluble and can be leached into the soil, absorbed by grape roots, and translocated to the leaves. Injury is expressed by marginal chlorosis and cupping of the leaves, shoot stunting, and clusters with shot berries. Spot applications on Bermuda grass and Johnson grass rather than spraying the entire vine row can minimize injury.

Incorporation of trifluralin during the dormant season at labeled rates for bindweed also helps control Johnson grass and Bermuda grass. The vine rows should first be plowed to move the soil and rhizomes into the center of the rows, and then the soil should be sprayed with trifluralin. Discing or tilling should follow, as this breaks up the rhizomes and incorporates the herbicide. Although the roots and rhizome pieces in the treated soil do not develop normally and some die, most have a normal appearance after the first year. A second treatment is usually necessary the second year for adequate control. Although neither Johnson grass nor Bermuda grass is eliminated by this treatment, repeated annual use can greatly reduce weed stands with no hazard to vine growth.

Glyphosate (Roundup)

This is a very potent and promising translocated herbicide for the control of perennial weeds in vineyards and other perennial crops (Lange et al., 1975). However, the compound is not yet recommended because there is more plant damage in direct foliar applications or through drift than that caused by other translocated herbicides such as 2,4-D, even though fewer immediate symptoms result.

Glyphosate translocates rapidly in most plants, moves farther into the unsprayed portions of vines and trees than do other translocated herbicides, and slowly kills plant tissues. If foliage is treated rapid wilting results, followed after several days by yellowing and/or dying. Regrowth in woody and some perennial weeds is usually severely stunted.

NONBEARING VINEYARDS

Trifluralin or nitralin incorporated as a preplant treatment 1–2 in. (2.54–5.08 cm) deep can be used to control annual weeds in new plantings. Lower rates should be used for sandy soils than for heavier ones. Grape rootings or cuttings should be planted so that the base is below the treated soil, to insure good root development. Treated soil should not be used to fill in around roots.

These herbicides can also be incorporated around young established vines as they are relatively nontoxic to the vines. High rates can greatly reduce branching of roots of a grape cutting in that part of the soil where the herbicides were incorporated. Nitralin can also be applied to the soil surface and incorporated by sprinkler irrigation or rainfall in young plantings.

APPLICATION OF HERBICIDES

Preemergence herbicides should be applied before rainfall. Before spraying, remove trash from vineyards. It is desirable to prune the vines before spraying. Diuron and simazine are usually applied in the dormant season. Uniform soil coverage in the area to be treated is necessary. Some drift of simazine, diuron, nitralin, or trifluralin onto the vines has not caused injury.

Paraquat and dalapon are usually applied in the growing season, and drift must be minimized. Droplets of paraquat cause necrotic spots on leaves and fruit, and drift of dalapon can cause chlorotic leaves with marginal burn.

2,4-D is usually applied in the growing season and great care must be exercised to minimize drift. Sensitivity of grapes to 2,4-D decreases with age of the shoots; therefore, where 2,4-D permits can be obtained, it is best to delay spraying until 2 weeks after flowering to minimize the effects of accidental drift.

SUBSURFACE LAYERING

For subsurface layering of trifluralin, a spray blade with nozzles spaced 4–6 in. (10.1–15.2 cm) apart at the trailing edge may be used (Fig. 18-1). This allows the herbicide to be trapped under the soil which flows over the blade as it is pulled through the soil. The blade is kept 4–6 in. (10.1–15.2 cm) below the soil surface (Lange et al., 1972).

Winter or early spring is the best time for layering of trifluralin. All trash should be removed, and established annual weeds should be killed to reduce interference with blade operation. Apply trifluralin in 40–80 gal of water per acre (374–748 liter per hectare). Shallow discing or other tillage prevents cracking of soil and will help prevent bindweed penetration through the layer of herbicide. This discing must not penetrate the layer of trifluralin or control will be greatly reduced.

AIR POLLUTION

The rapid introduction of new herbicides and pesticides has led to new problems such as volatilization, drift, soil residues, and misapplication. Other types of industrial emissions and waste can also be pollutants and have an adverse effect on grape growth. Solution of these problems is difficult, but permanent damage to atmosphere and soil must be avoided.

2,4-D Type Injury

In many vineyards of the world 2,4-D type injury can be found on occasional leaves whether or not it has been applied locally, since minute amounts of 2,4-D and related compounds can cause leaf malformation (Weaver, 1970). Grapes are exceedingly sensitive to 2,4-D. One drop containing 0.0001 μg of 2,4-D placed on a young leaf of Tokay grape can later cause formative effects (Kasimatis et al., 1971). Thus 0.8 mg of 2,4-D, when properly placed, could cause a leaf malformation on every vine in the approximately 20,000 acres (8,094 hectares) of Tokay grapes in California.

After spraying with 2,4-D was begun in 1946, it was noted that much drift occurred especially by sprays from airplanes. There was frequent damage to adjoining crops, one of the most sensitive being the grapevine. Drift can occur from airplanes for many miles and also from ground rigs, although these are usually smaller in magnitude. Formative effects from 2,4-D that occur in vineyards early in the season do not ordinarily appear in the following growing season. The effects caused by early sprays of 2,4-D persist, often having formative effects on leaves that develop late in the same season. Vines sprayed with 2,4-D late in the same season may show few or no formative effects because the rate of shoot growth at that time is very slow. A few malformed leaves resulting from small amounts of 2,4-D have not been proven to be detrimental, unless drift occurs at critical times such as flowering.

Smog, Ozone, and Oxidant Stipple

In the 1950s injury to grapes was observed in southern California by Los Angeles type (photochemical) smog. Ozone is the main component of the smog which causes the most damage. A "stipple" of some grape leaves occurs in midsummer and becomes progressively more severe with the season, resulting in a substantial loss of chlorophyll. The primary symptoms consist of small, brown to black, discrete, dotlike lesions, which are easily distinguished from lesions of other grape disorders because the stipples occur only on the upper surface of the leaf. The lesions vary in diameter from about $^{4}/_{1000}$–$^{20}/_{1000}$ inch (0.1–0.5 mm) and are confined to groups of cells bounded by the smallest veins (Richards et al., 1959). Large lesions result from a coalescence of small ones and may measure up to $^{80}/_{1000}$ inch (2.0 mm) in diameter. Aggregates of these minute spotlike lesions produce the typical stippled appearance. Reduced chlorophyll and photosynthesis also indirectly reduces berry weights, sugar content, and cane growth (Thompson et al., 1969). Patches of palisade cells die in mature leaves exposed to toxic concentrations of ozone in the air.

Since the discovery of ozone stipple in southern California the injury has spread to other industrial regions of the state. Widespread oxidant-stippled browning and premature senescence of leaves (ozone injury) has also been observed in grape growing regions near the Great Lakes (Shaulis et al., 1972). Benomyl sprays have been shown to reduce the severity of oxidant stipple. Since ozone entry into the leaf is governed by the degree of stomatal opening, low soil moisture and shading can decrease the degree of stippling (Kender and Shaulis, 1973).

As the population increases, air pollution will probably have a greater impact on grape growers in the future.

Part 5

GRAPE GROWING AND WINEMAKING AT HOME

Chapter 19
GRAPES FOR THE HOME VINEYARD

The home gardener usually desires to grow table grapes. Grapes are produced in most California counties, but they do not ripen well in cold, foggy areas along the coast or in the mountains over 3000 feet (914.4 m). This is particularly true of the *Vitis vinifera.*

Grapevines may be trained on a fence, against the sunny side of a building, or on an arbor or pergola, even after the home garden is filled with trees, berry bushes, and other plants. Generally the *vinifera* type should be grown in the warm valleys, and the hardy eastern varieties are the most satisfactory for plantings near the coast or in the mountains (Kasimatis, 1972).

To protect ripening grapes from birds, which also extends the growing season by several weeks, enclose halfgrown clusters in paper grocery bags. The paper admits enough light for the fruit to ripen normally, and the bag stays in good condition until the berries begin to shrivel or, in later maturing types, until the advent of heavy fall rains.

VARIETIES

Many home gardeners who can plant only a few varieties prefer to plant early ones. Early fruit on the market is usually expensive, but the price comes down with the advent of midseason and late-season grapes. Other gardeners prefer to have early, midseason, and late grapes to ensure a continuous supply of grapes over an extended period of 2 months or more.

Table grapes that grow well in the San Joaquin and Sacramento valleys and adjacent foothills include Perlette, Cardinal, Thompson Seedless, Red Malaga, Niabell, Tokay, Ribier, Muscat of Alexandria, Emperor, Olivette blanche, Delight, Ruby Seedless, Black Monukka, and Early Muscat. Tokay grows better in a cooler climate, and Emperor thrives in a warmer climate. Cardinal, Perlette, Delight, Ruby Seedless, and Black Monukka are good early varieties. Niabell, Golden Muscat, and Pierce are *labrusca* types, suitable for making of juice and jelly.

For moderately cool locations outside the fog belt, Perlette and Delight are recommended. Both are white grapes of good eating quality. In the coastal region within the fog belt and in the mountains, eastern varieties such as Concord, Pierce (California Concord), Iona, Niagara, and Golden Muscat produce excellent fruit.

For coastal locations in southern California, Concord and Pierce are recommended although the latter does better in warm climates. Pierce, Ribier, Golden Muscat, Thompson Seedless, and Red Malaga are suggested for the intermediate climatic districts of the semicoastal and warm inland valleys.

PRUNING AND TRAINING

Grapes must be severely pruned each year. Young vines are usually trained to a single trunk, extending as high as desired for the permanent vine. Two, three, or four short permanent arms are developed at the top of the trunk in the desired directions.

Spurs with 2 or 3 buds are retained on varieties that produce fruitful shoots from buds near the base of the canes. Muscat, Ribier, Red Malaga, Tokay, and most wine grapes are of this type. Thompson Seedless, Olivette blanche, and most eastern varieties such as Concord, Iona, Niagara require long canes instead of spurs to produce good crops.

Cane pruning is an easy method to use for all varieties of homegrown grapes. Retain 1–4 strong canes for the following year's crop. Pick canes that originate close to the old wood left the previous winter, and cut them to 10 or 12 buds in length [about 2 ft. (61.0 cm)]. Then prune other canes to 2 or three buds for renewal spurs. These may be selected near the base of each cane for fruiting.

Although on some varieties renewal spurs will produce shoots but little or no fruit, shoots that arise from renewal spurs will provide fruiting canes for the next year.

If you use a wire trellis, wrap the canes around the wires and tie near the ends. If you have no trellis or fence, pull all the canes upright and tie the ends together like an inverted basket (p. 190). Prune off everything that remains. The vine will then have 1–4 canes about 2 ft long, each having 10 or 12 buds, and 4 or 5 short renewal spurs 2 or 3 buds long. Repeat these steps the following year.

Canes can also be used to grow grapes on top of an arbor (p. 188). Train the trunk up the side of the arbor, and retain up to 4 canes on top of the arbor, and 4 or 5 renewal spurs. If the main trunk branches near the ground, you can train up 2 trunks and treat each branch as a separate but weaker vine. Save up to 2 fruiting canes and 2 or 3 spurs on each branch.

Varieties that bear good crops with spur pruning may be developed into low, self-supporting vines, or be used on a fence or an arbor. Vines requiring long fruiting canes require a trellis or an arbor on which the canes can be tied for support.

THINNING EXCESS CROP

You can improve berry size and rate of ripening by removing some fruit clusters in early summer. Thin the number of clusters to one per shoot after the shatter of berries following bloom, when retained berries are about the size of a match head.

IRRIGATION

Grapes planted in deep soils in moderately warm to hot regions usually require only one to three irrigations each season. On very shallow soils, water may be required as often as every 2 weeks. In the cool coastal areas early summer irrigation is helpful.

PESTS AND DISEASES

Powdery mildew is generally the most serious disease in home vineyards. Mildew can be prevented by dusting with finely divided sulfur. Dust the vines with sulfur when shoots are 6 in. (15.2 cm) long, 12–15 in. (30.5–38.1 cm) long, and then every 2 weeks until the fruit begins to color and soften. Dust the vines lightly from two sides. Delay or reduce the amount of sulfur during hot spells [100°F (38°C) or more]. Vines grown adjacent to lawns or in partial shade require additional dustings.

The grape leafhopper is the most troublesome pest in the hot interior valleys, and can be controlled with malathion during the third sulfur dusting just before bloom.

FERTILIZATION

Grapevines on deep, fertile soils usually do not require fertilizers. However, on sandy soils, vines may respond to nitrogen fertilizers. Chemicals should be applied in the winter months during the dormant season. For bearing vines, about ½ lb (0.227 kg) of ammonium sulfate should be used per vine, applied to the soil surface around the vine but not next to the trunk. A slight pale color of mature leaves at the time of rapid growth in late spring indicates a need for nitrogen fertilizer. Another symptom is less total vine growth.

Other fertilizers may be required in some areas. Deficiency of elements such as potassium and zinc causes distinct leaf and fruit symptoms. The local farm advisor should be consulted for advice.

Chapter 20
WINES FROM GRAPES

The juice from grapes ferments naturally into wine, and no doubt wine was drunk by man before the beginning of recorded history. In the United States during the last decade, winemaking has become a popular hobby.

USES, CLASSIFICATION, AND NOMENCLATURE

Uses of Wine

Wine is a product of nature. If one crushes some grapes in a container the wild yeasts present on the skin will turn the juice into wine of varying quality. The winemaker or enologist uses his scientific ability to help nature produce excellent wines.

Wine has many uses. It is used to complement meals, and in countries where wine is cheap it is served as the normal beverage at mealtime. Wine is also used to improve the flavor of food in cooking. Throughout history it has played an important role in religious ceremonies, and is used to minister to the ill. Wine is also commonly used for the celebration of important occasions, and to welcome guests.

Classification of Wines

Wines can be classified in several ways. One method is based on characteristics that are easily recognized such as color, presence of herbs or flavoring material, amount of carbon dioxide and sugar present, and detection of varietal aromas (Amerine and Singleton, 1965). Geographical origin, or the use to which the wine is to be put are other classification criteria.

In the United States there are five main classes of wines based on use (Anonymous, 1975). They are listed in Table 20-1 along with some of the best known wine types in each class.

Table 20-1. The Five Main Wine Classes and Some of the Important Wine types*

APPETIZER WINES	
Special and Natural Wines	
Sherry (dry or medium)	Vermouth (dry or sweet)
RED DINNER WINES	
Burgundy (dry)	Claret (dry)
Charbono	Cabernet Sauvignon
Gamay	Merlot
Petite Sirah	Ruby Cabernet
Pinot noir	Zinfandel
Rosé (pink; dry to slightly sweet)	
Others: Barbera (dry)	
Grignolino	
Red Chianti (dry)	
Vino Rosso (slightly sweet) (or Mellow Red)	
Concord (sweet)	
WHITE DINNER WINES	
Chablis (dry)	Rhine Wine (dry)
Chardonnay, or Pinot Chardonnay	Emerald Riesling
Folle blanche	Gray Riesling
Pinot blanc	Gewürztraminer
Chenin blanc (dry to sweet)	Johannisberg(er) Riesling
White Pinot	Sylvaner
	Traminer
Sauterne (dry to sweet)	Others:
Fumé blanc	Light Muscat (dry to sweet)
Sauvignon blanc	Catawba (dry to sweet)
Sémillon	Delaware (dry to semisweet)
Haut or Chateau	
SWEET DESSERT WINES	
Port (red, white or tawny)	Sherry (cream or sweet)
Tokay	Muscatel (golden, red, or black)
Others: Angelica	Muscat Frontignan
Marsala	Madeira
SPARKLING WINES	
Champagne (gold or pink)	Sparkling Burgundy (semisweet to sweet
Natural (completely dry)	Cold Duck
Brut (very dry)	Others:
Sec (semidry)	Sparkling Muscat (sweet)
Extra Dry (sweet)	Sparkling Rosé (dry to semisweet)

* Adapted from Anonymous, 1975.

Appetizer wines are best enjoyed before meals. Sherry and vermouth are the main appetizer wines and may be sweet or dry. Sherry made by baking develops a nutty flavor. When made by the flor process, a creamy scum that gives the sherry a special flavor develops in the presence of air. Wines made without flor are called olorosos and are of an entirely different character.

Vermouths are wines that have been flavored with various herbs. Seeds, leaves, flowers, and bark of various plants are used. Vermouth contains 15–20 percent alcohol.

Special and natural wines are made by adding pure, natural flavors to the wine. For example, the retsina wine of Greece is made by adding resin to white wine during fermentation.

Red dinner (table) wines are used to accompany main course dishes, and are usually dry (most or all of the sugar fermented out) with alcohol content ranging from 10.5–14 percent. Some of these wines contain sufficient tannin to produce an astringent (puckery) flavor. Most red table wines belong to the Burgundy or Claret types (Table 20-1). Burgundy wines have a pronounced flavor, body, bouquet, and a deep-red color. Clarets are ruby-red wines and are tart and light or medium-bodied. In California, wines of these two types often bear the name of the grape from which they are made and each has its own distinctive character.

Rosé wines are pink, and contain 10–14 percent alcohol.

White dinner (table) wines vary from dry to sweet. They are usually pale-straw to deep-gold in color, and their alcohol content varies from 10 to 14 percent. The production of white table wine differs from that of red wine in that the former is fermented from the juice only, without the skins. White wines mature quicker than red wines and can be bottled earlier; some attain their optimum quality within a year after bottling.

The most popular white wines are Sauterne, Chablis, and Rhine wine, and most other white wines resemble one of these types. *Sauternes* range from dry to sweet, and in France are traditionally a blend of Sémillon, Sauvignon blanc, and Muscadelle de Bordelais. In the United States many white grape varieties are used, and wines are usually named for the grapes from which they are made.

Chablis has a pale gold color and a delicate flavor. The wine is made from several varieties of white wine grapes, and is often named after the grape from which it is made.

Rhine wine designates a light-bodied, golden or slightly green-golden wine, ranging from tart to semisweet in flavor. Traditionally Johannisberg Riesling has been the chief variety used, but in the United States many other grapes are also used. The wine is usually packaged in tall, tapered bottles (Fig. 20-1). Moselle and Hock also designate Rhine wines; the latter name is used in Australia and England.

Light Muscat (sweet white wine) has a distinct muscat flavor and aroma, and is sometimes named after the grape variety from which it is made. The wine may vary from dry to very sweet.

Delaware, Catawba, Elvira, Dutchess, Steuben, and several other varieties grown in the midwestern and eastern part of the United States can be used to make excellent white wines with distinctive flavors and aromas. Delaware and Elvira are fruity and tart, and Catawba may be dry or semisweet.

Dessert wines are full-bodied, usually sweet wines that are served at the end of the meal. Most of these wines are fortified with brandy, contain around 18 percent alcohol, and, are medium-sweet to sweet and red to pale gold. Popular types include muscatel, port, Tokay, and cream sherry.

Muscatel is a sweet wine made from muscat grapes, with a distinct muscat flavor and aroma. Although Muscat of Alexandria is used to make most of the muscatels, several other varieties, such as Muscat Frontignan (Muscat Canelli), are also used. The color of the wine varies from golden or dark amber to red.

Fig. 20-1. Many wine bottles have traditional shapes, depending on the wine type. Four examples are shown above.

Red Muscatel (Black Muscatel) is a red wine made from Black Muscat (Muscat Hamburg) grapes.

Port wine originated in Portugal and is usually a heavy-bodied red, fruity wine. Most ports are a deep red, but there is also a straw-colored white port and a light-red tawny port. Many grape varieties can be used to make port (Table 20-1).

Tokay is a blend of dessert wines, usually angelica, port, and sherry. The Tokay wine of California should not be confused with the Hungarian Tokaj wines or with the Tokay grape of California. Tokay has a flavor similar to sherry.

Angelica is a white dessert wine produced from Mission, Grenache, and several other varieties. It is straw- or amber-colored and has a fruity taste.

Kosher wines are those certified by a rabbi. Although they may be of any type, the largest production is a sweet red wine made mainly in the eastern United States from the Concord variety.

Sparkling wines are white, pink, or red effervescent wines that are usually sold as champagne, sparkling burgundy, or Cold Duck and contain 10–14 percent alcohol.

Champagne can be made from several grape varieties (Table 20-1), although the traditional ones are Chardonnay, Pinot blanc, or Pinot noir. The driest champagne is termed "naturel"; the very dry "brut"; the semidry "extra-dry," "sec," or "demi-sec"; and the sweet is often labeled "doux" (Anonymous, 1975). Champagne is also made from Emerald Riesling, Burger, Folle blanche, Green Hungarian, French Colombard, Saint Émilion, Sauvignon vert, Delaware, Catawba, Elvira, and other varieties.

Carbonated wines are made from white or red wines (both dry and sweet) by using artificial carbonation to make them more effervescent. They cannot be labeled champagne, sparkling burgundy, or Cold Duck.

Sparkling Muscat is made from light muscat wines. It is also referred to as Muscato Spumante. Sparkling Rosé, made from grapes used for Rosé dinner wines, is pink and is similar in flavor to pink champagne. *Crackling* wines are less effervescent than champagne. In the United States they must be made by a natural second fermentation in a bottle or other container.

Brandy is distilled, fermented grape juice. In the United States only pure distillate of grape wine can be called brandy. The usual types are a straight brandy and one containing sweetening and/or flavoring substances. Two special types are distilled, one from Muscat wines, and a strong-flavored brandy from pomace (crushed grapes after the extraction of juice), called "grappa."

Nomenclature of Wines

Many semigeneric names of wines arose from names of famous Old World viticultural districts; Burgundy, Rhine wine, Sauterne, and Bordeaux are examples. As these wine types became known worldwide, the same names were applied to wines having similar characteristics. For example, early winemakers in United States named many wines after similar ones in their country of origin. Claret was the name given by the English to the light-red wine from Bordeaux, and subsequently, similar wines were known by the same name.

Varietal wines are named for the grapes from which they are made. In California at least 51 percent of a varietal wine must be produced from the grape for which the wine is named. The wine should also have the characteristic taste and aroma of the grape. Some brands of wine are labeled only with a class name (e.g., white dinner wine).

There is also a legal difference among wines. Dinner wines must not contain more than 14 percent alcohol. These are also sometimes referred to as dry wines or light wines. Dessert wines have an alcohol content of 18–21 percent, and sherries contain 17–21 percent.

Some wineries use proprietary names that may not be used by any other winery.

HOW TO MAKE HOME WINE

Besides being an enjoyable pursuit, winemaking can also save a considerable amount of money for the home winemaker. Making wine at home is legal up to 200 gal (757.1 liters) per year for the personal consumption of the family. The winemaker must, however, be the head of the household, and the wine cannot be sold.

Basis for Winemaking

The most important process in winemaking is the fermentation (Webb, 1974), which converts the grape-berry sugars into alcohol and carbon dioxide gas by a series of biochemical reactions. Microscopic one-celled plants called yeast are required for the chemical changes, as the yeasts produce enzymes that are necessary for the reactions (Fig. 20-2). Although many yeasts are collected on the bloom or waxy coating of grape berries and can serve as the fermenting organisms, winemakers use a special yeast for fermentation.

Fig. 20-2. Yeast cells (*Saccharomyces* species) reproduce by budding. These microscopic plants convert sugar into alcohol.

Some procedures used in all wine production (Amerine and Marsh, 1962) are listed below:

1. Crush the fruit to produce juice.
2. Adjust the sugar content so that the juice is from 20 to 24° Brix. California grapes seldom require adjustment, but grapes grown in the midwestern and eastern United States usually require the addition of sugar and/or water. Most other fruits also require some adjustment, which is known as amelioration.
3. Adjust acid content to produce a fermented product with a pleasant taste.
4. Destroy undesirable types of microorganisms in the must by adding an antiseptic agent to it. This will help to prevent production of poor wines.
5. Add yeast so that complete conversion of sugar to alcohol is obtained. If all the sugar is not converted to alcohol (stuck wine) the low-alcohol wine can spoil easily because of bacterial contamination.
6. Remove the pulp, skins, seeds, and other solid matter during or after fermentation by draining and/or pressing.
7. Remove yeasts and other sediments from the wine by siphoning or racking (decantation). This also helps to prevent the development of off-flavors.

Raw Materials

Although the juice of almost any fruit can be fermented into a winelike beverage, the grape is the most suitable. Select clean, fresh fruit that is not overripe; avoid moldy, injured, or rotting fruit. For California grapes around 22° Brix for dry table wines is best. About 12–15 lb (5.44–6.80 kg) of fruit will make 1 gal of wine.

Red, white, or rosé wines may be made; the type depends on the available grape varieties. Red wines are made from black or red varieties. In most varieties the pigment is located in the skins, although some red grapes also have pigment in the flesh of the grape. However, such color is less permanent than pigment from the skins and tends to deposit brown crusts in the bottles during storage. Pink wines can be made by removing skins and seeds at an early stage in the fermentation of red grapes. White wines are usually produced from white grapes; they can also be made from red varieties, if the flesh is not red and skins are removed early.

Wines can also be made from many other fruits. Dried fruits such as raisins, prunes, and figs assume a rather unattractive brown color, but make interesting wines. Honey is a concentrate that produces an acceptable wine and is used in the production of mead.

Materials and Equipment for Home Winemaking

Several essential and/or useful items in home winemaking include small crushers and presses, and other equipment that can be obtained in hardware stores or from mail-order firms.

Fruit Crushers. A roller-crusher is suitable (Fig. 20-3) for crushing the grapes so that the juice can be released during fermentation. Grapes are fed into a hopper at the top, then pass through wooden or aluminum rollers. The rollers of this crusher can be turned by a hand crank. The crushed grapes are then fed into a larger fermenting container, where the first fermentation will take place.

For small quantities of grapes, fruit can be crushed by hand or by pounding with a wooden mallet in a tub. If frozen fruits are to be used crushing is not necessary, because juice readily flows from such fruit when melted. Iron, copper, or galvanized pails, screens, or tubs should not be used.

Pressing. This can be done with a small barrel press (Fig. 20-4), or by using two pieces of wood hinged together at one end (Fig. 20-5) to squeeze the grapes in a cheesecloth bag. Pulp may also be allowed to drain for several hours from a cheesecloth bag suspended above a container. It is best not to squeeze out all the liquid as some bitter taste may develop in the wine.

Grapes that have been fermented on the skins for several days before pressing are much easier to press than freshly crushed grapes. Since red

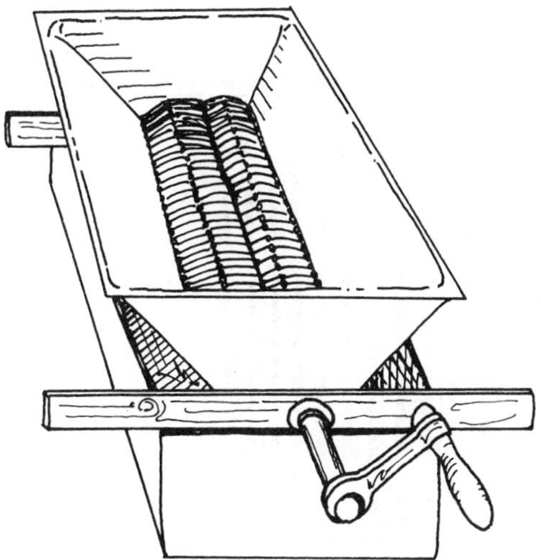

Fig. 20-3. A roller-type crusher in which a crank turns wooden or aluminum rollers. Grapes are dropped into hopper at the top, and stems are caught by screen below the crusher.

grapes are pressed after undergoing fermentation, and white grapes are pressed before fermentation; the former yield about 20 percent more wine per ton of fresh grapes (Amerine and Marsh, 1962).

Containers for Fermentation. Stone, glass, plastic, or wooden containers may be used. Wood is excellent for use with amounts of must greater than 10 gal (37.9 liters) but is not recommended for smaller sizes. Plastic garbage cans or plastic laundry baskets make excellent containers for winemaking and are easy to clean. Plastic containers are also much lighter and easier to handle than crocks or wooden containers. Glass carboys are also useful.

Hydrometer to Measure Sugar. Except for California grapes, most fruits require the addition of sugar and/or acid. A glass or clear plastic cylinder and a hydrometer are necessary to measure sugar content (Fig. 14-1). The higher the sugar content, the higher the bulb will float in the juice or fermenting wine. A reading of 22°–24° Brix will result in wines containing about 12 percent alcohol (Amerine and Marsh, 1962). About 1¼ lb (0.113 kg) of sugar must be added per 10 gal (37.9 liters) of juice and skins for each degree the hydrometer scale registers below 23° Brix.

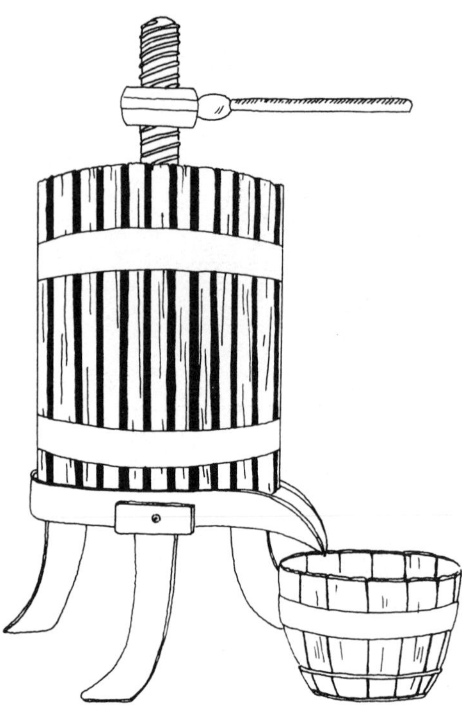

Fig. 20-4. A basket press is suitable for pressing grapes. A ratchet lever system is used to press a plate against the grapes below, and juice comes out through vertical spaces in the barrel.

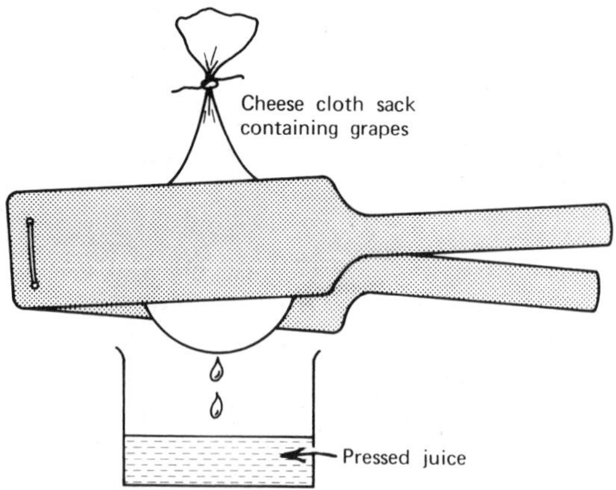

Fig. 20-5. Two pieces of wood, hinged together, can be used to press out juice from grapes enclosed in a cheesecloth bag by squeezing the boards together.

It is best not to add the sugar before the start of fermentation. When fermentation is proceeding actively, add about one-fourth of the required sugar to a small portion of the liquid, let dissolve, and then add it to the fermentation container. The other three one-quarter portions can be added at daily intervals.

Siphon. A siphon tube is needed to transfer wine from one container to another. A plastic or rubber hose about 6 ft. (1.83 m) long with a ½ in. (12.7 mm) inner diameter is satisfactory. The purpose of siphoning from one container to another is to get rid of sediments that have settled out of the wine (Fig. 20-6). The outlet end must be lower than the inlet end.

Containers for Storage. Glass bottles or carboys and plastic containers are recommended for storage containers. Glass bottles should range from 1 to 14 gal (3.79–53.0 liters) and more; 5-gal (15.1 liters) bottles

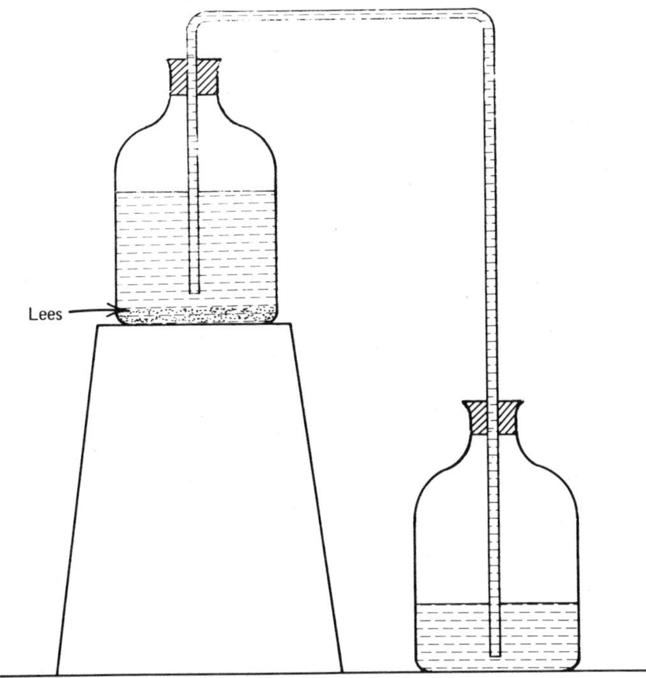

Fig. 20-6. Siphoning can be used to transfer wine from one container to another without disturbing the sediment. Hoses can be made of plastic, rubber, or glass.

are a convenient and available size. White wines are best kept in glass or plastic, while red wines can develop better when stored in wood for 1–3 years in a cool place. However, wood has several disadvantages: evaporation, leakage, possible woody flavored wine, and difficulty of cleaning.

Corks and Bottles. One can buy a small hand-corking machine from a wine supply house. Bottles must be thoroughly cleaned before use, and should be filled so that only a very small air space is present between the cork and the wine. Too much air discolors white wines and causes off-flavors in red wines. After corking, bottles should be stored on their sides at 55°–60°F (13°–16°C).

Acid Testing Kit. Besides being handy for testing acids, this kit can also be used as a sulfur dioxide tester. These are available at most supply houses. A description of testing for acid is given on page 225.

Making Red Wines

Grapes may be bought directly from growers or at markets. First crush grapes in a range of 20–24° Brix, then remove stems and leaves. The crushed grapes may then be placed in the fermentation container, which should not be more than three-fourths full.

Sulfur dioxide (SO_2). Next add sulfur dioxide to the must to kill the wild yeasts present, since they cannot be depended on to cause vigorous fermentation. (Sulfur dioxide is a gas that originates from the breakdown of sodium or potassium metabisulfite, which can be obtained from wine supply houses.) The chemicals can also be obtained in tablet form as "Campden" tablets (Eakin and Ace, 1975). One tablet per gallon (0.26 tablet per liter) yields about 65 ppm (parts per million) of sulphur dioxide (SO_2); 50–100 ppm is needed in the fresh must. Crush and dissolve Campden tablets before using, but do not add yeast for about 4 hr.

One can also add 3 oz (5 level tablespoonsful) of one of the metabisulfite salts in 1 qt of warm water (80.4 g per liter). One teaspoon per gallon will give 65 ppm of SO_2.

Yeast. Four hours after SO_2 has been added, add a special wine yeast. This can be purchased dry, as a liquid, or as a yeast colony on an agar slant. Dry yeast can be added directly to the juice or must.

Fermentation and Pressing. A cloth or a thin sheet of plastic can be used to cover the primary fermenter. Place fermenter at 60°–75°F (16–24°C). In about 24 hr small patches of foam begin to collect, and after 48 hr active fermentation should take place. Allow the fermentation to proceed until sugar content has dropped to 2–4° Brix; this can occur in 4–7 days. It is best not to proceed longer because the wine could become too tannic. The skins and pulp (cap) rise to the top of the must and should be punched down twice each day.

Four to seven days after fermentation in the primary state, bail the must from the fermenter into a press allowing juice to run into a plastic tub, wooden barrel, or other container. Using a siphon tube, transfer the juice into large bottles or carboys. When the press is filled with pulp, slowly apply pressure so that the juice runs into the tub. Release pressure, stir the pomace, and press again. Repeat this procedure until most of the juice is extracted from the pomace.

When using a wooden barrel as a primary fermenter, elevate it and install a spigot near the bottom. You can then drain off the new wine, leaving the pulp in the barrel. This system leaves little juice for pressing.

Secondary Fermentation. The new wine will not be ready to drink for at least a year, and it may continue fermenting slowly for several weeks or even months. Fermentation is complete when all residual sugar disappears as measured by special papers or tablets. During the secondary fermentation, a rubber or standard cork drilled to hold a rubber or plastic tube tightly can be used as a bottle stopper. The free end of this tube should be immersed in water. One can use a bottle that will hold the tube snugly. Special devices called water locks or fermentation valves are more suitable than a rubber tube (Fig. 20-7). They are made of inexpensive plastic or glass and release carbon dioxide but exclude air. They are available at wine supply houses.

Racking. After gas has ceased to evolve from the wine, the wine should be racked or decanted. Deposits at the bottom of the carboy are called "lees." Carefully lift the carboy to a higher place such as a chair or table a couple of days before racking, so that if sediment is stirred up it has time to settle (Fig. 20-6). Tilt the bottle by placing a board or other object under one side, insert the siphon hose, then siphon the wine into another clean container without disturbing the lees. Add 25–30 ppm of SO_2 per gallon of wine (Eakin and Ace, 1975), then fill the containers to within an inch (25.4 mm) of the stopper at all times. Racking should be

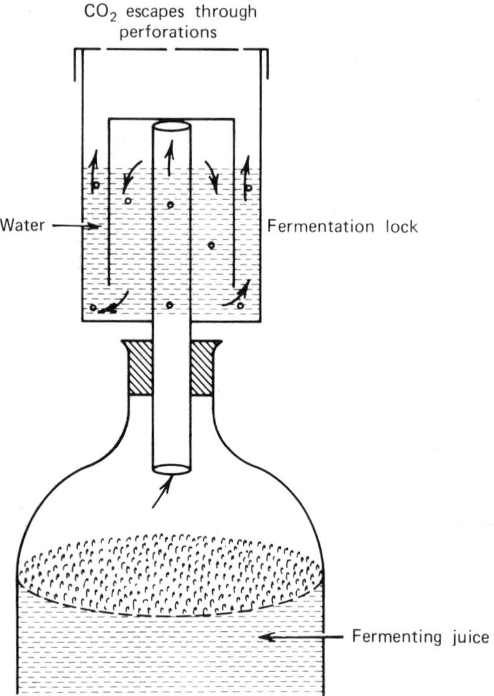

Fig. 20-7. Fermenting juice with a fermentation lock allows CO_2 gas to escape, but prevents oxygen from entering the bottle. Arrows show direction of movement of CO_2.

done two or three times, depending on wine clarity and amount of sediment.

Wine Stabilization. In the spring following the grape harvest store the wine at 16–24°F (−9° to −4°C) for two weeks before bottling. The cold wine will then drop any additional solids and clarify for bottling. It can also be held for several months during the summer and then stabilized in a refrigerator, and bottled in August or September. A temperature of 22–24°F (−6° to −4°C) is recommended (Eakin and Ace, 1975).

If the wine is not clear, a clarifying agent such as bentonite or "Sparkolloid" may be used.

Bottling and Aging. The wine continues to age and improve in the bottle, if it is not stored too long. Most red wines are stored for several years so that the wine may reach its peak quality.

Making Rosé Wines

These wines are made by the same procedure as red wines except that pressing should be done within 12–24 hr after the must begins to ferment. Visual inspection can indicate when the desired amount of red color is attained.

Making White Wines

Since juice does not flow as freely from crushed white grapes as from red grapes, the home winemaker can expect only about 100 gal (378.5 liters) of juice per ton of grapes (Amerine and Marsh, 1962). However, a more complete removal of juice can be obtained if about 0.1 g of pectic enzyme is added per gallon (3.79 liters) of fresh crushed grapes and the must is allowed to stand for 2–4 hr before pressing.

Another commercial technique is to add SO_2 to the must and let the free-run juice drain for about 24 hr. Holes or screens in the bottom of a container can be used. The free-run juice makes the best wine, but the skins and seeds can also be pressed and the resulting juice made into a wine of lesser quality.

SO_2 and pure yeast culture are added to the juice by the technique described under making red wines.

Fermentation. White musts should be fermented at 50°–60°F (10–16°C). Glass carboys make good containers and should be filled about three-fourths full. Use a fermentation trap. Fermentation usually proceeds from 10 to 20 days. The end of fermentation can be ascertained by testing sugar percent with a hydrometer or with diabetic test papers obtainable from drugstores. Other sugar-testing devices are Clintest and Dextrocheck. The latter is a pill that is dropped in the wine and causes a color change which can be matched with a chart. It can be obtained from wine supply houses. The process of fermentation can also be observed by the escape of gas from the fermentation trap and the settling of yeast to the bottom of the container.

Racking. When active fermentation has ceased, fill the container to within about an inch (25.4 mm) of the cork with a wine similar to the one in the container, then install the fermentation lock. When the wine clarifies, transfer the new wine to another container.

Immediately after racking there should be 25–30 ppm of SO_2 in the wine. SO_2 is essential for white wines to prevent oxidation and to keep them from browning. Put the bisulfite into the empty container before adding the wine. The siphon tube should be about half way down in the wine, then gradually lowered as the wine flows out to cause minimum disturbance of the lees at the bottom of the container. The liquid flow can be started by suction. The discharge tube should be in the bottom of the container, as this helps mix SO_2 and wine and reduces aeration. One may have to rack several times before the wine is clear. In all racking operations, care should be taken not to disturb the lees which are very light. After 2 or 3 weeks one can rack again and discard the lees.

Fining agents such as bentonite will aid in precipitating suspended matter from the wine. A good way to test the clarity of a wine is to place a candle behind the carboy while looking through the wine.

New white wines stored at 45°–55°F (7°–13°C) clarify best, and are ready for consumption in 6–12 months.

Wine can be stabilized by subjecting it to cold conditions. Stabilization is the process of storing wine at low temperatures to precipitate tartrate—the crystals that form in the bottom of the wine bottle and detract from the appearance of the wine.

Bottling and Aging. Wine bottles usually have various shapes that correspond to the generic name of the grape (Fig. 20-1). Sweet wines should not start fermenting actively again because the pressure produced can break the bottles. Potassium sorbate or "sorbistat" can inhibit fermentation, and should be used at 200–250 ppm. Use clean bottles that have been immersed in hot water for about 30 min. An additional protection from microorganisms is to add SO_2 to water, shake it in the bottle, then drain well.

Bottles should be filled to one-half inch (12.7 mm) from the cork or cap using a siphon hose. After the bottles have stood upright for about a week to allow the corks to dry out and seat, place the bottles on their sides to age (Eakin and Ace, 1975). Capped bottles can be stored upright because no air can pass through the caps and oxidize the wine.

Many white wines are ready for bottling 6–12 months after fermentation. A cool cellar is a good place to store wine, and storage racks are easy to build. (Fig. 20-8).

Addition of Sugar. Wine grapes in the eastern United States may have insufficient sugar to produce wines with an alcoholic content of 11–13 percent. In these cases sugar must be added. Also, the acidity of the juice may be too high, and water must be added to lower it. With Concord grapes testing from 13–14° Brix it is common to add about 3 gal of water to each 10 gal of juice (12 liters water–40 liters of juice). One should dilute the acidic juices with water first and then determine

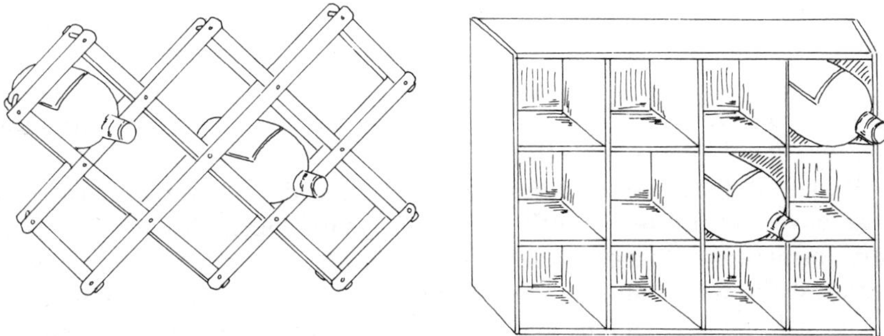

Fig. 20-8. Various types of storage racks can be constructed, such as the cardboard wine case and collapsible rack shown here.

the amount of sugar required to bring the reading to about 23° Brix (amelioration). Subsequent operations are the same as those used for red, rosé, or white wines.

Concentrate Wines

Concentrated juices may also be used to make an acceptable wine. These are available from wine supply firms. Some wineries also sell concentrates of about 70 percent in 5-gal (18.9 liters) containers, enough to produce about 20 gal (75.7 liters) of wine. The container should be placed in hot water 1–2 hr so that the contents will pour easily into a fermentation container. Dilute the concentrate by adding about 15 gal of water for each 5 gal of concentrate (57 liters of water per 19 liters of concentrate) while stirring (Amerine and Marsh, 1962). If the reading is above 23° Brix, add water and stir; if it is below 21° Brix, add more concentrate. Follow directions for making white wines (p. 313).

If the wine is flat after fermentation one can add 1½ oz of citric acid for each 10 gal (43 g acid per 38 liters) of wine. Metabisulfite at about ½ oz/10 gal (15 g per 39.8 liters) should also be added early in the storage period.

Sparkling Wines

A choice white table wine, usually a few months old, is used to make champagne. Champagne yeast and sugar are added to the wine to induce a second fermentation. Then the wine is placed in securely closed bottles to withstand developing pressure caused by fermentation. The pressure is due to the carbon dioxide gas formed that causes the wine bubbles. Sparkling wines require heavy bottles and thick, wired-on corks. The amount of pressure is determined by the amount of sugar in the blend (called the cuvée). Workers in champagne cellars sometimes wear wire face masks and gloves for protection in case the pressure should cause a bottle to break.

Bottles of fermenting champagne are placed in horizontal tiers or large boxes from a few months to several years. During this time secondary fermentation occurs. After its completion, the wine ages in the bottles until the flavor and bouquet are at peak quality. Traditionally, the yeast sediment is removed before the mature champagne can be shipped.

Individual Bottles. When individual bottles are used they are placed upside down on racks. Each day the bottles are lifted and slightly turned so that the yeast and other sediment moves into the neck, then the neck of the bottle is immersed in a refrigerated brine solution to freeze the wine and sediment in the neck. The bottles are then opened so the pressure can force out the frozen wine and sediment (disgorging). Clear champagne and the dosage (sweet syrup and aged wine) are added to refill the bottle. The dosage operation refers to the addition of a syrup to adjust the sweetness of the wine. The bottles are then corked, the corks wired on, and the bottle labeled. The wine usually receives a final aging before it is shipped.

Another technique for making champagne is the transfer system. In this operation the contents are disgorged from the bottles under pressure into a tank. The yeast sediment is then removed by filtering the champagne under pressure into a clean bottle, and the dosage is added to the tanks or to the bottles before filling.

Charmat (Bulk) Process. The Charmat process is a faster and cheaper way to make champagne. The procedure is the same as described above, except that the secondary fermentation takes place in large glass-lined or stainless steel tanks instead of in the original bottles. After fermentation, the wine is removed under pressure through a filter directly into bottles but the sediment remains behind, eliminating the necessity for daily handling and disgorging (Anonymous, 1975).

Dessert Wines

For sweet dessert wines use grapes at their highest degree Brix, but without raisining. After a dry red or white wine of around 12 percent alcohol has been produced, add 1⅗ pints of 90 proof brandy (45 percent alcohol) per gallon of wine (3 liters brandy to 3.75 liters wine). Stir thoroughly, and add sufficient sugar for the desirable sweetness (Amerine and Marsh, 1962). If sediment forms, allow it to settle and then rack and bottle.

Another and inexpensive technique is to begin with the method used for fermenting dry table wines. When the must reaches a 5° Brix and the fermentation is proceeding actively, add sufficient grape concentrate or sugar to raise the sugar percent to 10° Brix. When the degree Brix falls to 5°, add more sugar or concentrate. By making three or four such additions, an alcohol content of 16 or 17 percent can be obtained. Add about one ounce (28.4 g) of potassium metabisulfite per 10 gal (37.9 liters) of wine when the fermentation ceases. Rack the wine several times to get rid of yeasts and other microorganisms.

Sherry

In California, many winemakers use Mission, Palomino, Thompson Seedless, and Tokay grapes to make sherry. After fermentation of the juice to the desired sugar content, brandy is added to stop fermentation. The new wine is called shermat.

The wine is then baked in lined or stainless steel or concrete tanks, or in redwood containers at 100–140°F (38°–60°C) for 2–4 months. Baking may be performed in a heated room, in tanks heated by coils, or by the heat of the sun. Later the sherry is gradually cooled to cellar temperature and is then aged like other wines. The heating, the oxidation caused by contact of the warm wine with air, and the aging in wood barrels combine to result in the pleasant nutty flavor characteristic of California sherry.

Some California wineries produce a "flor" sherry, using the Spanish method by which a film-yeast growth called flor is allowed to form on the surface of the wine in partially filled containers. The more recently developed "submerged flor" process is used by some wineries. This process also gives the wine a distinctive flavor.

Making Fruit Wines

Wines can be made from many kinds of fruit including apples, pears, currants, blackberries, elderberries, raspberries, strawberries, peaches, apricots, plums, cherries, prunes, figs, grapefruits, oranges, bananas, and so on. Nearly all fruit except grapes requires amelioration. Sugar and/or acid must be added.

Select clean, fresh fruit that is not overripe, moldy, or otherwise spoiled; 12–15 lb (5.44–6.80 kg) of most fruits will yield about 1 gal (3.79 liters) of juice. Since many fruit varieties have a stronger flavor than most varieties of grapes, the juice may be diluted with water. As little as 3–5 lb (1.36–2.27 kg) of fruit for each gallon (3.79 liters) of water can be used (Eakin and Ace, 1975).

Must. Place the crushed fruit in the fermenter. Elderberries and blueberries should not be crushed. Since seeds or pits may give a slightly bitter almond taste to the wine, with some fruits such as peaches, plums, or cherries, half to all the pits should be removed. Most of the seed taste from small-seeded fruits can be eliminated by allowing only a 4- 5-five day fermentation period prior to pressing. Fresh apple juice may be used and processed in the same manner as grape juice. Apples make excellent still and sparkling wines.

If no water is to be added, take a hydrometer reading of the juice. If water is added, take a reading following its addition after 5 min of stirring. Degree Brix will be low since the juices of most fruits vary from 5 to 16 percent in sugar content. When water is added, the sugar content of the juice will be low and sugar must then be added.

From 75 to 100 ppm of SO_2 must be present in the must. If dilution is desirable, add water from the hot water faucet after the SO_2 is added. Much of the yeast and bacteria are killed in the water heater. If the fruit is boiled, SO_2 is not required as boiling will kill the undersirable yeast and bacteria. Boiling also releases more of the pectin. Determine the amount of sugar required and add one-half of it to the crushed fruit. Save the rest and add it when the pressed juice is placed in glass carboys for final fermenting (Eakin and Ace, 1975). This procedure may help to prevent a stuck fermentation (wine that has stopped fermenting before all sugar is converted into alcohol).

The requirement of large amounts of sugar, and the fact that the juice may have been diluted with water, reduces the chances for the vigorous growth of yeast. Therefore, addition of yeast nutrient is another preventive measure against the occurrence of a stuck wine.

Many fruits will be low in acid, and one should add tartaric or citric acid powder or acid blends. Low-acid wines will not clear properly, may not ferment completely, will lack a tangy flavor and taste more like flavored water. Some of the high acid fruits will require dilution with one part water to one part juice. A proper acid level for fruit wines is about 0.65 percent.

A pectic enzyme should be added when a fruit wine is produced, especially with pulpy fruits like strawberries, peaches, and plums. The enzyme added to the fresh, crushed fruit will break down the pectin and may result in the greatest possible yield of juice. Another technique is to add the pectic enzyme to the carboy when the juice is pressed from the primary fermenter.

An all-purpose wine yeast may be added 4 hr after the SO_2 is added to the must. If hot water is used for dilution, wait until the temperature drops to 75°F (24°C) before adding the yeast.

Fermentation should be performed at 65°–75°F (18°–24°C). The pulpy mass (cap) should be pushed down into the juice each day, and the must should also be stirred. A sheet of plastic or cloth should be placed over the fermenter to keep out fruit flies. Four to seven days after fermentation starts, press out the juice, and add the remainder of the dissolved sugar. Subsequent steps in winemaking can be done in the same manner as for grape wines.

If fruit wines are fermented to dryness, they are apt to be harsh, bitter, and/or lack desirable flavor. Sugar can make these wines very palatable. It can be added during fermentation, and some may remain as residual sugar that has not been converted to alcohol, or it can be added just before consumption of the wine. It is difficult to stop fermentation in a wine with some remaining sugar. One method is to use pasteurization to stop fermentation, but this technique is slow and tedious. The best technique is to ferment to a dry state and bottle when ready. When drinking, add an amount of sugar syrup (2 cups of sugar dissolved in 1 cup of hot water) that will raise the sweetness to the appropriate taste level. At this time the sugar will dissolve completely. Another method is to add sugar before bottling along with 200–250 ppm of potassium sorbate to inhibit fermentation (Eakin and Ace, 1975).

Dried Fruit Wines

Unsulfured dried fruits such as raisins, figs, and dates produce wines resembling baked sherries. Coarsely grind the dried fruit and add 1 pt of water per pound (0.473 liters per 0.454 kg) of ground material (Amerine and Marsh, 1962). Bring to a simmering temperature and allow to cool. Strain and press out the juice through cheesecloth two or three times, and put the juices into a fermentation container. Adjust the juice to about 23° Brix by adding water or sugar, depending on the degree Brix.

Raisins contain almost sufficient acid to produce a good wine, but citric acid must be added to figs and dates before fermentation; about ⅔ oz/gal (19.97 g/4 liters) is normally required. Add yeast and ferment under a fermentation trap. Siphon the new wine away from the sediment after fermentation is completed. Then follow the instructions given for making fruit wines.

GLOSSARY

ABSCISSION. Dropping of leaves or fruits as a result of breakdown of a layer of thin walled cells.

ABSORPTION. The taking in of nutrients or other chemicals by roots or through stomata or cuticle of the foliage.

ACID SOIL. A soil having a pH less than 7.0.

ACTIVE INGREDIENT. The chemical agent in a formulation that produces the desired effect.

ADSORPTION. The concentration of molecules or ions on the surfaces of colloidal particles or on solid material.

ADJUVANTS. Materials combined with spray materials to act as wetting or spreading agents, stickers, penetrants, etc., to aid in the action of the active ingredient.

ADVENTITIOUS ROOTS. Roots formed in unusual places such as nodes of stems or cuttings.

ADVENTITIOUS BUD. A bud developing in an unusual location. Shoots arising from such buds are termed adventitious shoots.

AFTER-RIPENING. A period of chilling required by seeds and buds before germination will take place.

ALKALI SOIL. A soil with sufficient exchangeable sodium to interfere with the growth of the vine.

ALKALINE SOIL. A soil with a "basic" reaction, a pH greater than 7.0.

ANTHER. The pollen-bearing part of a stamen.

ANTHESIS. The time of full bloom in a flower, just after the calyptra has fallen.

ANTHOCYANIN. A plant pigment that gives leaves and fruit a red, blue, or purple color.

APEX (APICES). The tips of shoots or of lobes of the leaf.

APICAL DOMINANCE. The inhibition of lateral bud growth by the apical meristem. The control is mediated by the production of hormones.

APICAL MERISTEM. Tissue at the apex of a root or stem where active cell division occurs.

ARID. A dry climate with low rainfall and a high rate of evaporation.

ASEXUAL PROPAGATION. Propagation by plant parts other than seeds, such as budding, grafting, layering, and cuttings.

ASTRINGENCY. A puckery taste sensation caused mainly by tannin in the wine.

AUXINS. Plant regulators that can stimulate cell division and enlargement in plants.

AVAILABLE WATER. The part of the soil moisture, mainly the capillary fraction, which the plant can readily absorb.

BACTERIA. Microscopic one-celled plants containing no chlorophyll.

BAND APPLICATION. Application of a chemical to a continuous restricted area, as along a crop row instead of over the entire field area.

BAR. A unit of pressure equal to one megadyne per square centimeter (about 750.076 mm of mercury).

BARK. The tough external covering of a woody stem or root outside the cambium.

BASAL BUD. A small bud that lies at the base of a cane or spur as part of a whorl, which usually does not grow unless the distal buds fail to grow.

BIOLOGICAL CONTROL. Control of pests by disease-producing organisms, predators, and parasites.

BILATERAL CORDON. Vine training in which the trunk is divided into two branches extended horizontally on a supporting wire. Often referred to as cordon training.

BLADE. The expanded portion of a leaf.

BLEEDING. The exudation of sap from cut canes; usually occurs toward end of dormant season.

BLOOM. The delicate waxy or powdery substance on the surface of berries.

BOUQUET. The perfumelike scent, produced by esters and ethers, that is noted after uncorking and pouring wine.

BROADCAST APPLICATION. An application of spray or dust over an entire area, rather than only on rows, beds, or middles.

BUD. An undeveloped shoot usually protected by scales; tissues are mainly meristematic.

BUD SCALES (see prophylls). Protective scale-like leaves of a bud during winter, covered with hairs and impregnated with suberin.

BUD SPORT (mutation). A branch, flower, or fruit that arises from a bud and differs genetically from the rest of the plant. It is usually caused by a spontaneous mutation of cell genes.

CALCAREOUS SOIL. A soil containing $CaCO_3$ or $MgCO_3$ that effervesces when treated with dilute HCl; usually has an alkaline reaction.

CALLUS. Parenchyma tissue which grows over a wound or graft and protects it against drying or other injury.

CALYX. The external part of a flower consisting of sepals.

CALYPTRA. The fused petals of the grape that fall off the flower at anthesis.

CAMBIUM. A very thin layer of undifferentiated meristematic tissue between the bark and wood.

CAP STEM (pedicel). The stem of the individual flowers or fruit.

CAPILLARITY. The elevation or depression of a liquid in contact with a solid (as in a capillary tube). It is dependent on the relative attraction of the molecules of the liquid for each other and for those of the solid.

CAPILLARY WATER. Water retained in the interstices of soil or other porous material after drainage.

CARBOHYDRATE. A compound produced by plants containing chemically bonded energy composed of carbon, oxygen, and hydrogen. Hydrogen and oxygen often have a ratio of 2:1.

CARBON-TO-NITROGEN RATIO. The relative proportion of carbon to nitrogen in the soil, organic matter, or plant tissue.

CATCH WIRE. A wire which serves as an attachment for developing grape shoots.

CATION. A positively charged ion.

CERTIFIED PLANTING STOCK. Grapevine propagation material certified free of known virus diseases by the California Department of Agriculture, under regulations of the Grapevine Certification and Registration program.

CHARACTER. Proper color, taste, and bouquet associated with a certain type of wine.

CHIMERA. A mixture of tissues of different genetic constitution in the same plant.

CHISEL. An implement having one or more soil-penetrating points, with sufficient weight to force the points into the soil to loosen the subsoil.

CHLOROPLAST. Structure in a leaf called a plastid that contains chlorophyll.

CHLOROPHYLL. Green pigment of plants that absorbs light energy and makes it effective in photosynthesis.

CHLOROSIS. Yellowing or blanching of green portions of a plant, particularly the leaves, which can result from nutrient deficiencies, disease, or other factors.

CLARIFY. Clearing up cloudy wine; also referred to as "fining."

CLAYPAN. A dense soil horizon under the upper part of the soil that is hard when dry and plastic when wet.

CLONE. A group of vines of a uniform type propagated vegetatively from an original mother vine.

COMPATIBLE CHEMICALS. Materials that can be mixed together without adversely changing their effects on plants or pests.

COMPATIBILITY. Ability of the scion and stock to unite in grafting and form a strong union. Also refers to male and female nuclei that have capability to unite and form a fertilized egg which will grow to maturity.

COMPLETE FLOWER. A flower having sepals, petals, stamens, and a pistil or pistils.

CONTACT HERBICIDE. A chemical that kills the portion of the weed or plant with which it comes in contact.

COROLLA. The petals of a flower.

CREAM OF TARTAR (POTASSIUM ACID TARTRATE). A white crystalline deposit which settles out from wines as temperature is lowered.

CROWN SUCKERING. The removal of unwanted shoot growth from the vine head and trunk while the shoots are still young; often referred to as suckering.

CUTTING. A severed portion of cane used for propagation.

DEGREES BALLING. See DEGREES BRIX.

DEGREES BRIX. A measure of the total soluble solid content of grapes, approximately the percentage of grape sugars in the juice.

DENTATE. Toothed; usually with the teeth directed outward as in a leaf.

DIOECIOUS. Having the male (staminate) and female (pistillate) organs of the flower on separate plants.

DORMANT. Plants, buds, or seeds which are not actively growing.

DRYING RATIO. The pounds of fresh grapes required to make one pound of raisins.

EMASCULATION. Removal of male parts of the flower, usually by tweezers. Used by breeders to control pollination.

ENATIONS. Very small leaflike outgrowths from the lower surfaces of leaves, usually along the larger veins.

ENOLOGY. The science and study of winemaking.

ENTIRE. Leaves without any indentations or division.

ENZYME. Organic catalysts composed of proteins that control most chemical reactions occurring in living cells.

EPINASTY. Downward bending of leaves caused by some hormone sprays, water stress, etc.

EROSION. The wearing away of land surface, usually by running water or wind.

ETHYLENE. A gas having a chemical structure of $CH_2{=}CH_2$, produced by most tissues in very small amounts. It is often called a fruit-ripening hormone.

EYE. A compound bud of a grape.

FASCIATION. Flattening of the stem as a result of multiple buds growing in the same place.

FEEDER ROOTS. Fine roots and root branches with an unusually large absorbing area. Important in uptake of minerals and water by their root hairs.

FERMENTATION. The change of sugars into alcohol and carbon dioxide in the presence of yeast.

FERMENTATION LOCK. A device which allows carbon dioxide to escape from fermenting wine but keeps out air from the atmosphere.

FERTILITY. A soil quality in which essential plant nutrients are provided in proper amounts and balance for optimum plant growth.

FERTILIZATION. The fusion of one gamete with another to form a zygote in sexual reproduction. Also refers to application of fertilizers.

FIELD CAPACITY. The amount of water retained in a soil against the force of gravity, usually measured 24 to 36 hours after flooding, and about equal to the moisture content at one-third the atmospheric pressure. It is also referred to as water-holding capacity.

FILAMENT. The stalk supporting a stamen.

FLOWER BUDS. Buds containing undeveloped flowers; also referred to as fruit buds.

FOLIAGE WIRE. See CATCH WIRE.

FOLIAR FEEDING. Fertilization of a plant through the surface of the leaves; usually accomplished by spraying.

FORMATIVE EFFECTS. Cupping, stunting, and abnormal venation such as that caused by 2,4-dichlorophenoxyacetic acid (2,4-D).

FOXINESS. The peculiar smell and taste of *labrusca* grapes.

FRUIT. A mature ovary (berry) or cluster of mature ovaries.

FRUIT-SET. The stage of cluster development following the drop of impotent berries after bloom. The small, retained berries are said to be "set."

FUNGICIDE. A chemical used to control the infection and spread of fungi on crops.

FUNGUS. A lower plant form that lacks chlorophyll and subsists on other plants, plant remains, or animals.

GAMETE. A male or female sex cell which fuses with gamete of the opposite sex to form a zygote.

GERMINATION. The sprouting of a seed or other reproductive body.

GENUS. A group of plants comprising a number of closely related species.

GIBBERELLIN(s). A plant growth regulator that promotes cell elongation and growth of shoots. Subscripts such as GA_1, GA_2 indicate specific analogues.

GIRDLING. The removal of a ring of bark from a shoot, cane, or trunk.

GLABROUS. Smooth; not rough, pubescent, or hairy.

GLUCOSE. A simple sugar: one of the monosaccharides.

GRAFTING. The joining of a bud or other part of one vine to a portion of another so that their tissues unite.

GREEN-MANURE CROP. A crop grown and plowed under to improve the soil, especially by adding organic matter.

GUARD CELLS. Specialized cells on the epidermis of leaves that control opening and closing of the stomates.

GUTTATION. Loss of liquid from intact plants.

HARDPAN. A hardened or cemented soil horizon below the soil surface.

HEAD. The pressure due to the difference in elevation between two points in a body or column of liquid, as a measure of the pressure or force causing water to flow. Sometimes referred to as pressure head. Also, that portion of a trunk where arms or cordons originate.

HEADLAND. Unplowed land at the ends of the rows or near a fence; often used for turning space for various equipment.

HEARTWOOD. Wood in the center of the trunk or branch that has changed in character and no longer serves for conduction.

HERBICIDE. Any chemical used to kill plants; some are selective and kill only certain plants.

HERMAPHRODITE. A flower with both male (stamen) and female (pistil) parts.

HIRSUTE. Pubescent, having rather coarse or stiff hairs.

HORMONE. Plant hormones are regulators produced by plants that, in low concentrations, regulate physiological plant processes. They usually move from a site of production to a site of action.

HOST. Any plant or animal attacked by a parasite or predator.

HYBRID. A cross-breed of two species. Hybridizing refers to the practice of crossing species.

HYDATHODE. A structure that releases liquid during guttation; usually located on leaves.

HYDROMETER. A floating device used for measuring the density or specific gravity of liquids.

HYDROPONICS. The culture of plants in a water solution of nutrient elements.

HYPHA. Fungal thread or filament.

IMPERFECT FLOWERS. Flowers in which either functionable stamens or pistils are present, but not both.

INCOMPATIBILITY. Inability of sex cells to unite to form a fertilized egg that will grow to maturity. Failure of the scion and stock to unite and form a union that will continue to grow. Also refers to agricultural chemicals that cannot be mixed or used together because of undesirable reactions.

INCOMPLETE FLOWER. A flower that does not have all four floral organs.

INDEXING. Determination of the presence of disease, such as a virus, in a vine by removing buds or other parts and grafting them into a readily susceptible variety that exhibits symptoms of the virus.

INDIGENOUS. A vine or plant native to the region.

INDOLEACETIC ACID (IAA). An important, naturally occurring auxin-type plant regulator which causes many growth responses.

INFLORESCENCE. The flowering cluster of the grapevine.

INFILTRATION RATE. The rate of downward water movement from the soil surface into the soil.

INTERNODE. That portion of a shoot or cane between two adjacent nodes.

IRRIGATION. The artificial application of water to soil to supply moisture essential for plant growth.

LAND LEVELING. Removing soil from a high place and filling in a low place to level the land.

LATERAL. A branch of the main axis of the cluster; also a shoot arising from the main shoot.

LATENT BUD. A dormant bud, usually hidden, which is over a year old and may remain dormant indefinitely.

LEACH. The process of removal of soluble material by the passage of water through the soil. This is a primary step in reclamation of a saline soil.

LEAF BUD. A bud which develops into a stem, bearing leaves but no flower clusters.

LEAF SCAR. The scar left on the stem after a leaf falls.

LEES. The sediment at the bottom of a wine cask, consisting mainly of tartrate.

LENTICEL. A porelike, slightly raised spot on pedicels and grape berries.

LIGHT SATURATION. The light intensity at which an increase in light does not result in an increase in the rate of photosynthesis.

MATURITY. Stage of fruit development when it has reached the maximum quality for its intended purpose.

MERISTEMATIC. A region of the plant where rapid cell division takes place.

METABOLISM. The sum of processes by which food is manufactured and broken down for energy gain.

MILLERANDAGE. The condition shown by clusters that are too loose because of poor fruit set and too many shot berries.

MONOECIOUS. Having male and female flowers on the same plant.

MULCH. Materials placed on the soil surface or mixed in the soil, such as straw, to promote moisture retention, temperature control, provide cleaner fruit (as in strawberries), weed control and, if the mulch is plant material, to supply nutrients as the mulch decomposes.

MUTATION. Genetic change in a mother plant or stock that may influence the character of the offspring, buds, or cuttings removed from the plant.

MYCELIUM. A group or mass of fungus filaments.

MYCORRHIZA. Symbiotic association of the mycelium or hyphae of certain fungi with the roots of the plant.

MUST. Crushed berries and juice.

NEMATODE. A very small parasitic worm that lives in or on grape roots.

NOBLE ROT. A fungus growth that develops on the skins of ripe grapes in certain locations. The fungus penetrates the skin of the grape without breaking it, so the grape can wither without spoiling. Such grapes have a very high sugar content.

NODE. The area on a stem from which shoots develop.

NONSALINE-ALKALI SOIL. A soil that contains sufficient exchangeable sodium to interfere with the growth of the vine, but which does not contain appreciable quantities of soluble salts.

OSMOSIS. The passage of materials through a semipermeable membrane from a higher to a lower concentration.

OVARY. The enlarged basal portion of the pistil containing ovules or seeds.

OVERCROPPING. The production of more crop than the vine can bring to maturity at normal harvest time. It can result from light pruning, water stress, insects, or disease.

PALISADE CELLS. Elongated chlorophyll-containing cells just below the upper epidermis of leaves.

PARASITE. An organism that lives on or in the body of another organism and obtains food from it.

PARENCHYMA. A tissue composed of living, thin-walled cells with intercellular spaces that often fit rather loosely together. It makes up the soft parts of plants (e.g., the berries).

PARTHENOCARPY. Development of fruit without seed production, as in the Black Corinth grape.

PEDICEL. The stalk of one flower or berry in a cluster.

PEDUNCLE. The cluster stem. Usually refers to stem, from the point of attachment to the shoot to the first lateral branch on the cluster.

PERFECT FLOWER. A flower having both stamens and pistil.

PERMANENT WILTING PERCENT. The moisture content of a soil at which a plant wilts and fails to recover when in a relative humidity atmosphere of 100 percent. The tension is usually about 15 atmospheres.

PETIOLAR SINUS. A cleft in the leaf margin at the attachment of the petiole.

PETIOLE. The leaf stalk attaching the leaf blade to the shoot.

PH. Refers to degree of acidity or alkalinity as a scale of numbers from 1 (very acid) to 14 (very alkaline). pH 7.0 is neutral, representing the reciprocal of the hydrogen ion concentration and expressed in gram atoms per liter of a solution.

PHLOEM. Region of tissue in the plant composed of sieve tubes and parenchyma which translocate food materials elaborated by the leaves.

PHOTOSYNTHESIS. The process by which carbon dioxide and water are incorporated into carbohydrates. Radiant energy is the source for this process.

PHYLLOXERA. A small, yellowish aphidlike insect that attacks roots and/or leaves, but does most damage to the root system.

PHYTOTOXIC. Causing injury or death of plants.

PITH. The tissue in the central part of shoots or stems; usually made up of soft parenchyma cells.

PISTIL. The female part of the flower, consisting of a stigma, a style, and an ovary.

PISTILLATE FLOWER. A flower having a pistil but lacking stamens.

PLANT PROPAGATION. Controlled reproduction of vines and other plants by man, to perpetuate selected individuals or groups of plants of special value to him.

PLANT REGULATOR. Organic compounds other than nutrients which, in small amounts, promote, inhibit, or otherwise modify any physiological plant process. The five main groups are auxins, gibberellins, cytokinins, inhibitors, and ethylene.

PLAQUE. A flattened, visible group of mycelium strands, often arranged in a fan shape.

POLLEN GRAINS. The microspores and developing male gametophytes found in the anthers.

POLLINATION. The transfer of pollen from the anther to the stigma.

POMACE. Crushed grapes after the juice has been extracted.

POROSITY. Refers to pore space. It is the ratio of the volume of air- and water-filled space to the total volume of soil plus water and air.

PORT. A very sweet, full-bodied, rich dessert wine.

PPM. Parts per million. The concentration of a material expressed as the number of units per million units. It is the same as milligram per liter.

PROPHYLLS. Scale-like leaves produced before foliage leaves borne on the first and second node of a bud. Also referred to as bud scales.

PRIMORDIA. Regions in plants, such as the tips of stems and roots, where plant structures are initiated.

PUBESCENT. Covered with hairs.

QUARANTINE. A legal action preventing the sale or shipment of reproductive parts, such as seeds, cuttings, and rootings, to prevent the spread and infestation of a disease such as a virus, or of an insect pest.

QUIESCENCE. Period of nonvisible growth controlled by external factors. Growth will proceed under favorable environmental conditions.

RACKING. The pumping out or siphoning of wine from the sediment (lees).

RESPIRATION. The oxidation of food materials by plants, and the release of energy during which oxygen is absorbed and carbon dioxide is released.

REST. Period of nonvisible growth controlled by internal factors. Growth will not occur even under favorable environmental conditions.

RINGING. See GIRDLING.

RISERS. Pipes used in sprinkler irrigation that support the sprinkler heads above the level of the foliage.

ROOTSTOCK. Stock material to which other varieties of grapes are grafted to produce a commercially acceptable vine. Some rootstock varieties are less susceptible to injury by root parasites, such as phylloxera and nematodes, than are the roots of the scion variety.

ROSETTE. Leaves with a bunched appearance caused by development of short internodes.

RUGOSE. Wrinkled or uneven.

SALINE SOIL. A nonalkali soil containing sufficient soluble salts to injure the crop.

SAPWOOD. Outer, younger, more porous region of the vine beneath the bark where active growth occurs.

SCION. The plant part that is grafted or budded onto the stock.

SEC. Dry.

SEED. A fertilized and ripened ovule containing an embryo which can develop into an individual by germination.

SELECTIVE HERBICIDE. A chemical that can kill some weeds but is only slightly or non-toxic to the crop plant.

SELF-POLLINATED. Pollen placed on the stigma of the same plant or a plant identical to it.

SEPAL. One of the modified leaves of the calyx; the outermost floral organ.

SERRATIONS. Teethlike indentations at the margin of a leaf.

SEXUAL REPRODUCTION. Reproduction by the fusion of gametes.

SHOOT. Current season's stem growth that bears leaves and buds.

SHOT BERRY. Very small berries that fail to develop to normal size; usually seedless.

SHOULDERED CLUSTER. The basal laterals are larger than the other laterals.

SIPHON. A hose for transferring liquids from a higher elevation to a lower one.

SOD CULTURE. Type of management in which a permanent perennial ground cover is kept at all times, and is usually mowed periodically during the growing season.

SOIL STRUCTURE. The aggregate arrangement of individual soil particles.

SOIL TEXTURE. The relative proportion of the percentage of sand, silt, or clay particles in a soil.

SPARKLING WINES. Wines that effervesce when opened.

SPERM. The male gamete.

SPRINKLER IRRIGATION. Irrigation by means of surface pipes and sprinkler heads.

STAMEN. The pollen-producing organ of a flower, consisting of the anther and a filament.

STAMINATE FLOWER. A flower with stamens but no pistil.

STENOSPERMOCARPY. The phenomenon in which fertilization occurs and seeds are produced but soon abort. Mature berries contain only rudimentary seeds, as in Thompson Seedless and Black Monukka.

STIGMA. The upper surface of the pistil, where the pollen grain is received and germinates.

STOCK. The stem or root onto which the scion is grafted.

STOMATA (STOMATE). A pore in the epidermis of a leaf or young stem surrounded by two guard cells.

STRATIFICATION. The subjecting of seeds to an after-ripening period to terminate the rest period. The seeds are usually exposed to low temperatures under moist but well-aerated conditions.

STYLE. The portion of the pistil between the stigma and the ovary.

SUCKER. A shoot arising from the lower part of the trunk, or from the part of the stem below ground.

SUNSCALD. Injury to outer tissue due to excess sun heat, as on berries.

SURFACTANT. A chemical which modifies the surface tension of spray droplets, causing them to spread out on a leaf and form a thin film.

SYSTEMIC. An insecticide absorbed by plants and translocated throughout, which kills certain insects.

SYSTEMIC HERBICIDE. A compound which is translocated throughout a plant and can readily injure or kill the plant.

TANNIN. A component of wine that gives it an astringent taste.

TEMPERATURE INVERSION. A meteorological phenomenon whereby the air temperature becomes warmer, with increasing altitude, instead of becoming cooler as is normal.

TENDRIL. A slender structure on a shoot that can coil around an object and help support the shoot.

TENSIOMETER. An instrument used to measure water tension in the soil.

TOLERANCE. Amount of toxic residue allowable by law in or on edible substances.

TOMENTUM. Composed of densely matted, tiny epidermal hairs; pubescence.

TRANSLOCATION. Movement of water, nutrients, chemicals, or elaborated food materials within a plant.

TRANSPIRATION. Water loss by evaporation from the leaf surface and through the stomata.

TRUNK. The main stem or body of a vine between the roots and the place where the trunk divides to form branches.

TURGIDITY. Pressure caused by fluids in the cell that press against the cell and distend it.

VARIETAL WINE. A wine named after a grape variety. At least 51 percent of the wine must be from the named variety, and the typical varietal flavor must be present.

VARIETY. A group of closely related plants of common origin that have characteristics not sufficiently different to form separate species.

VEGETATIVE PROPAGATION. Plant reproduction, using vegetative organs such as cuttings.

VÉRAISON. The stage of development when berries begin to soften and/or color.

VIGOROUS VINES. Vines with shoots that grow rapidly and produce much growth.

VINTNER. A winemaker.

VITICULTURE. The science or study of vine production and natural drying of raisins.

WATER BERRIES. A grape disorder characterized by watery berries that fail to ripen properly.

WATER-HOLDING CAPACITY. See FIELD CAPACITY.

WATERLOGGED. Soil with poor drainage that lacks sufficient oxygen for proper root functioning.

WATER SPROUTS. Rapidly growing shoots arising from latent buds on branches or trunks.

WETTABLE POWDER. Solid formulation that forms a suspension when added to water.

WILD VINE. A vine that grows in the wild.

WING. A well-developed basal cluster lateral that projects and is separated from the main body of the cluster.

XYLEM. The woody portion of conducting tissue whose function is the conduction of water and minerals.

ZYGOTE. A fertilized egg; cell arising from the fusion of two gametes.

BIBLIOGRAPHY

AGAMALION, H., et al. 1972. *Weed Control Recommendations.* California Agricultural Experimental Station Extension Service.

ALJIBURY, F. K. 1973. The potential use of drip irrigation to improve the water penetration in San Joaquin Valley vineyards and orchards. *The Blue Anchor.* **50:**12–13.

ALJIBURY, F. K., and CHRISTENSEN, L. P. 1972. Water penetration of vineyard soils as modified by cultural practices. *American Journal of Enology and Viticulture.* **23:**35–38.

ALJIBURY, F. K., et al. 1973. *Heat suppression in vineyards using over-vine sequential sprinkling. California Agricultural Extension* MA-48, March 1973.

ALJIBURY, F. K., BREWER, R., CHRISTENSEN, P., and KASIMATIS, A. N. 1975. Grape response to cooling with sprinklers. *American Journal of Enology and Viticulture.* **26:**214–217.

ALLEY, C. J. 1939. Grape virus diseases tackled. *Western Fruit Grower.* April.

———. 1964. Grapevine propagation 1: A comparison of cleft and notch grafting; and, bark grafting at high and low levels. *American Journal of Enology and Viticulture.* **15:**214–217.

———. 1974. Benchgrafting as done in Germany and France. *Wines and Vines.* **55:**34–36.

———. 1975. Research Note: Grapevine propagation VII. The wedge graft—a modified notch graft. *American Journal of Enology and Viticulture.* **26:**105–108.

———. 1975. Research Note: Grapevine propagation VIII. The side whip graft, an alternative method to the split graft for use on stocks 2–4 cm in diameter. *American Journal of Enology and Viticulture.* **26:**109–111.

———. 1975. Improved field budding of grapevines using a modified cut and plastic tape. *California Agriculture.* **29:**6–8.

AMERINE, M. A., and JOSLYN, M. A. 1970. *Table Wines: The Technology of Their Production.* University of California Press, Berkeley, California.

AMERINE, M. A., and MARSH, G. L. 1962. *Wine Making at Home.* Wine Publications, San Francisco.

AMERINE, M. A., and SINGLETON, V. L. 1965. *Wine.* University of California Press, Berkeley, California. (An introduction for Americans.)

AMERINE, M. A., and WINKLER, A. J. 1963. California wine grapes: Composition and quality of their musts and wines. California Agricultural Experimental Station Bulletin 794.

ANONYMOUS. 1968. *Controlling Deer.* U.S. Department of the Interior, Fish and Wildlife Service, Bureau of Sport Fishing, Wildlife, Lafayette, Indiana, No. 404.

———. 1973. *Growing American Bunch Grapes.* Farmers' Bulletin 2123. U.S. Department of Agriculture.

———. 1973. *Muscadine Grapes, A Fruit for the South.* Farmers' Bulletin 2157. U.S. Department of Agriculture.

———. 1974. Raisins on-the-vine: "Possible and practical." *California Grape Grower*: 30–31. September 1974.

———. 1975. *The Story of Wine and Its Uses.* California Wine Institute, San Francisco. (A nontechnical guide to wine.)

BABO, F. VON, and MACH, E. 1909. *Handbuch der Weinbaues und der Kellerwirtschaft.* 1. Bd., 4 Aufl. Parey, Berlin.

BAILEY, L. H. 1933. The species of grapes peculiar to North America. *Gentes Herbarum.* **3:**149–244. Fasc. IV.

BARNES, M. M. 1970. *Grape Pests in Southern California.* California Agricultural Experimental Station Extension Service Circular 553.

BARNES, M. M., et al. 1973. *Pest and Disease Control Program for Grapes.* California Agricultural Experimental Station Extension Service.

BEUKMAN, E. F., and GOHEEN, A. C. 1970. Grape corky bark. In FRAZIER, N. W., et al., *Virus Diseases of Small Fruits and Grapevines.* University of California Division of Agricultural Sciences, Berkeley, California. 207–209.

BOEHM, E. W., and TULLOCH, H. W. 1967. *Grape Varieties of South Australia.* Department of Agriculture, Adelaide, South Australia.

BOLIN, H. R., PETRUCCI, V., and FULLER, G. 1975. Characteristics of mechanically harvested raisins produced by dehydration and by field drying. *Journal of Food Science* **40:**1036–1038.

BRINK, C. V. D. 1974. Predicting harvest date of the "Concord" grape crop in southwest Michigan. *HortScience* **9:**206–208.

BROOKS, R. M., and OLMO, H. P. 1972. *Register of New Fruit and Nut Varieties.* 2nd ed. University of California Press, Berkeley, California.

BUCKMAN, H. O., and BRADY, N. C. 1960. *The Nature and Properties of Soils.* Macmillan, New York.

BURLINGAME, B. B., et al. 1971. *Frost protection costs for north coast vineyards.* University of California, Agricultural Extension AXT-267.

CARLSON, V. 1972. *How to green graft grapes.* University of California Agricultural Extension AXT-115.

CHAMBEAU, A. DE. 1972. *Creative Winemaking.* WWWWW/Information Services, Inc., Rochester, New York.

CHILDERS, N. F., Ed. 1966. *Nutrition of Fruit Crops.* 2nd ed. Horticultural Publications Rutgers, The State University. Somerset Press, Inc., Somerville, New Jersey.

———. 1973. *Modern Fruit Science.* Somerset Press, Inc., Somerville, New Jersey.

CHRISTENSEN, L. P. 1968. *Zinc Deficiency in Vineyards.* Cooperative Extension of Agriculture and Home Economics, U.S. Department of Agriculture, University of California, Fresno, California.

———. 1975. Long-term responses of 'Thompson Seedless' vines to potassium fertilizer treatment. *American Journal of Enology and Viticulture.* **26:**179–183.

———. 1975. Response of 'Thompson Seedless' grapevines to the timing of preharvest irrigation cut-off. *American Journal of Enology and Viticulture.* **26:**188–194.

CHRISTENSEN, L. P., and FISHER, B. B. 1968. *Winter Covercrops in Vineyards.* Cooperative Extension University of California, U.S. Department of Agriculture, Fresno, California.

———. 1972. *Vineyard Tissue Sampling Guide for Plant Analysis.* Cooperative Extension University of California, Fresno, California.

CHRISTENSEN, L. P., KASIMATIS, A. N., KISSLER, J. J., JENSEN, F. and LUVISI, D. A. 1973. *Mechanical Harvesting of Grapes for the Winery.* Agricultural Extension University of California.

CHRISTENSEN, P., LYNN, C., OLMO, H. P., and STUDER, H. E. 1970. Mechanical harvesting of Black Corinth raisins. *California Agriculture.* **24:**4–6.

CLARK, D. O., 1975. *Vertebrate Pest Control Handbook.* Division of Plant Industry, California Department of Food and Agriculture, Sacramento, California.

CLORE, W. J., WALLACE, M. A., and FAY, R. D. 1974. Bud survival of grape varieties at sub-zero temperatures in Washington. *American Journal of Enology and Viticulture.* **25:**24–29.

COOK, J. A. 1960. *Vineyard Fertilizers Cover Crops.* California Experimental Station Extension Service.

———. 1966. Grape nutrition. In *Nutrition of Fruit Crops Temperate, Sub-tropical, Tropical.* N. F. Childers, Ed. Somerset Press, Inc., Somerville, New Jersey. pp. 777–812.

COOK, J. A., and KASIMATIS, A. N. 1959. *Predicting Nitrogen Response in Vineyards: the Petiole Test.* Cooperative Extension University of California, U.S. Department of Agriculture.

COOMBE, B. G., and HALE, C. R. 1973. The hormone content of ripening grape berries and the effects of growth substance treatments. *Plant Physiology.* **51:**629–634.

CUMMINGS, M. W. 1971. *Controlling Pocket Gophers and Moles.* Agricultural Extension University of California, OSA #n30.

DIAS, H. F. 1970. Grapevine yellow mosaic. In Frazier, N. W., et al. *Virus Diseases of Small Fruits and Grapevines.* University of California Division of Agricultural Sciences, Berkeley, California. pp. 228–230.

EAKIN, J. H., JR, and ACE, D. L. 1975. *Winemaking as a Hobby.* Pennsylvania State University College of Agriculture University Park, Pennsylvania.

EINSET, J., KIMBALL, K., and WATSON, J. 1973. *Grape varieties for New York State.* Plant Science. Pomology 7. Information Bulletin 61. Ext. Publ. New York State College of Agriculture and Life Sciences, Cornell University, Ithaca, New York.

EINSET, J. and ROBINSON, W. B. 1973. Geneva's expanded grape varietal program. *New York's Food and Life Sciences Quarterly.* **6:**11–13.

English, W. H., O'Reilly, H. J., McNelly, L. B., DeVay, J. E., and Rizzi, A. D. 1963. *Cytosporina Dieback of Apricot in California.* California Experimental Station Extension Service Leaflet 165.

Esau, K. 1948. Phloem structure in the grapevine, and its seasonal changes. *Hilgardia.* **18:**217–296.

Flocker, W. J. 1973. *Plant Science.* Kendall/Hunt Publishing Company, Dubuque, Iowa.

Frazier, N. W., Fulton, J. P., Thresh, J. M., Converse, R. H., Varney, E. H., and Hewitt, W. B. (ed. committee). 1970. *Virus Diseases of Small Fruits and Grapevines (a handbook).* University of California Division of Agricultural Science, Berkeley, California.

Gardner, M. W., and Hewitt, W. B. 1974. *Pierce's Disease of the Grapevine: The Anaheim Disease and the California Vine Disease.* University of California Press, Berkeley, California.

Goheen, A. C. 1970. Grape Leafroll. In Frazier, N. W. et al. *Virus Diseases of Small Fruits and Grapevines.* University of California Division of Agricultural Sciences, Berkeley, California. pp. 209–212.

Goheen, A. C., and Luhn, C. F. 1973. Heat inactivation of viruses in grapevines. Series IV Vol. 9 Fasc. 3, pp. 287–289. Rivista di Patologia Vegetale.

Goheen, A. C., Nyland, G., and Lowe, S. K. 1973. Association of a rickettsialike organism with Pierce's disease of grapevines and alfalfa dwarf and heat therapy of the disease in grapevines. *Phytopathology.* **63:**341–345.

Gooding, G. V., Jr., and Teliz, D. 1970. Grapevine yellow vein. In Frazier, N. W., et al., *Virus Diseases of Small Fruits and Grapevines.* University of California Division of Agricultural Sciences, Berkeley, California. pp. 238–241.

Graniti, A., and Martelli, G. P. 1970. Enations. In Frazier, N. W. et al., *Virus Diseases of Small Fruits and Grapevines.* University of California Division of Agricultural Sciences Berkeley, California, pp. 241–243.

Hale, C. R., and Weaver, R. J. 1962. The effect of developmental stage on direction and translocation of photosynthate in *Vitis vinifera. Hilgardia.* **33:**89–131.

Harmon, F. N. 1954. A modified procedure for green-wood grafting of vinifera grapes. *Proceedings of the American Society for Horticultural Science* **64:**255–258.

Harmon, F. N., and Weinberger, J. H. 1962. *The Chip-bud Method of Propagating Vinifera Grape Varieties on Rootstocks.* U.S. Department of Agriculture Leaflet 513.

Hartmann, H. T., and Kester, D. E. 1975. *Plant Propagation, Principles and Practices.* 3rd ed. Prentice Hall, Englewood Cliffs, New Jersey.

Hewitt, W. B. 1970. Pierce's disease of *Vitis* species. In Frazier, N. W., et al. *Virus Diseases of Small Fruits and Grapevines.* University of California Division of Agricultural Sciences, Berkeley, California, pp. 196–200.

———. 1970. Spindle shoot of Colombard grapevine. In Frazier, N. W., et al. *Virus Diseases of Small Fruits and Grapevines.* University of California Division of Agricultural Sciences; Berkeley, California. p. 246.

———. 1974. *Rots and Bunch Rots of Grapes.* 1973. California Agricultural Experimental Station Bulletin 868.

HOPKINS, D. L., MOLLENHAUER, H. H., and MORTENSEN, J. A. 1974. Tolerance to Pierce's disease and the associated Rickettsia-like bacterium in muscadine grape. Journal of the American Society for Horticultural Science. **99:**436–439.

ISRAELSEN, O. W., and HANSEN, V. E. 1962. *Irrigation Principles and Practices.* 3rd ed. John Wiley and Sons, Inc., New York.

JACOB, H. E. 1936. *Propagation of Grapevines.* California Agricultural Extension Service Circular 101.

———. 1944. *Factors Influencing the Yield, Composition, and Quality of Raisins.* University of California College of Agriculture, Agricultural Experimental Station, Berkeley, California.

———. 1944. *Vineyard Planting Stocks.* Circular 360. University of California, College of Agriculture, Agricultural Experimental Station, Berkeley, California.

———. (Revised by Winkler, A. J.). 1950. *Grape Growing in California.* Circular 116. California Agricultural Extension Service, College of Agriculture, University of California, Berkeley, California.

JANICK, J., SHERRY, R. W., WOODS, F. W., and RUTTAN, V. W. 1974. *Plant Science and Introduction to World Crops.* W. H. Freeman and Co. San Francisco.

JENSEN, F. 1971. High level grafting of grapevines. *American Journal of Enology and Viticulture.* **22:**35–39.

JENSEN, F., BAILEY, M., and LYNN, C. 1970. *Grafting Grapevines.* Cooperative Extension of Agriculture and Home Economics, University of California, Berkeley, California.

JENSEN, F. L., FLAHERTY, D. L., and LYNN, C. D. 1973a. *Grape leaf folder in the southern San Joaquin valley.* Cooperative Extension, California Agricultural Extension Service. AXT-396. Berkeley, California.

———. 1973b. *Grape Leafhopper in the Southern San Joaquin Valley.* Cooperative Extension of California Agricultural Extension Service. AXT-392. Berkeley, California.

———. 1973c. *Omnivorous Leaf Roller in the Southern San Joaquin Valley.* Cooperative Extension of California Agricultural Extension Service. AXT-395. Berkeley, California.

JENSEN, F., KISSLER, J., LUVISI, D., PEACOCK, B., HALSEY, D., and LEAVITT, G. 1973. Effect of ethephon on table grapes. *The Blue Anchor.* **50:**16–17.

JENSEN, F., KISSLER, J. J., PEACOCK, W. L., and LEAVITT, G. M. 1975. Effect of ethephon on color and fruit characteristics of "Tokay" and "Emperor" table grapes. *American Journal of Enology and Viticulture.* **26:**79–81.

JENSEN, F., LUVISI, D., and LEAVITT, G. 1975. The effects of prebloom shoot treatments on yield and fruit characteristics of "Cardinal" and "Ribier" table grapes. Talk, 26th Annual Meeting of the American Society of Enologists June 27, San Francisco.

JENSEN, F., SWANSON, F., PEACOCK, W., and LEAVITT, G. 1975. The effect of width of cane and trunk girdles on berry weight and soluble solids in table "Thompson Seedless" vineyards. *American Journal of Enology and Viticulture.* **26:**90–91.

KASIMATIS, A. N. 1967. Grapes and berries Part I—Grapes In Irrigation of Agricultural Lands. R. M. Hagan No. 11, series in Agronomy. American Society of Agronomists, Madison, Wisconsin.

———. 1971. *Vineyard Irrigation.* Agricultural Extension, University of California.

———. 1972. *The Home Vineyard.* Agricultural Extension University of California. OSA #18.

KASIMATIS, A. N., CHRISTENSEN, L. P., LUVISI, D. A., and KISSLER, J. J. 1972. *Wine Grape Varieties in the San Joaquin Valley.* Agricultural Extension University of California. AXT-n26.

KASIMATIS, A. N., and KISSLER, J. J. 1974. Response of grapevines to shoot break-out following injury by spring frost. *American Journal of Enology and Viticulture.* **25:**17–20.

KASIMATIS, A. N., and LIDER, L. 1972. *Grape Rootstock Varieties.* Agricultural Extension University of California. AXT-47. 32 pp.

KASIMATIS, A. N., and LYNN, C. 1967. *How to Produce Quality Raisins.* University of California Agricultural Extension Service. AXT-235.

KASIMATIS, A. N., WEAVER, R. J., and POOL, R. M. 1968. Effects of 2,4-D and 2,4-DB on the vegetative development of "Tokay" grapevines. *American Journal of Enology and Viticulture.* **19:**194–204.

KASIMATIS, A. N., WEAVER, R. J., POOL, R. M., and HALSEY, D. D. 1971. Response of "Perlette" grape berries to gibberellic acid applied during bloom or at fruit set. American Journal of Enology and Viticulture **22:**19–23.

KENDER, W. J., and SHAULIS, N. J. 1972. Air pollution injury in New York vineyards. *New York State Horticultural Proceedings* pp. 70–73.

KIMBALL, K., and SHAULIS, N. 1958. Pruning effects on the growth, yield, and maturity of Concord grapes. *Proceedings of the American Society of Horticultural Science* **71:**167–176.

KISSLER, J. J., OUGH, C. S., and ALLEY, C. J. 1973. *Evaluations of Wine Grape Varieties for Lodi.* California Agricultural Experimental Station Bulletin. 865.

KLIEWER, W. M., and SCHULTZ, H. B. 1973. Effect of sprinkler cooling of grapevines on fruit growth and composition. *American Journal of Enology and Viticulture.* **24:**17–26.

KLIEWER, W. M., and COOK, J. A. 1974. Argenine levels in grape canes and fruits as indications of nitrogen status of vineyards. *American Journal of Enology and Viticulture.* **25:**111–118.

KNOTT, J. E. 1962. *Handbook for Vegetable Growers.* rev. ed. John Wiley and Sons, Inc. New York.

KRIEDEMANN, P. E. 1968. Photosynthesis in vine leaves as a function of light intensity, temperature and leaf age. *Vitis.* **7:**213–220.

KRIEDEMANN, P. E., KLIEWER, W. M., and HARRIS, J. M. 1970. Leaf age and photosynthesis in *Vitis vinifera* L. *Vitis.* **9:**97–104.

KRIEDEMANN, P. E., and SMART, R. E. 1971. Effects of irradiance, temperature, and leaf water potential on photosynthesis of vine leaves. *Photosynthetica.* **5:**6–15.

LANGE, A. H. 1972. *Johnson grass Control Study in Vineyards by Layering Techniques.* Agricultural Extension of the University of California. MA-39. Berkeley, California.

LANGE, A. H., FISCHER, B., and PEREZ, J. 1974. *Foliar and Soil Applied Herbicides for Bindweed Control.* Agricultural Extension of the University of California. MA-65.

LANGE, A. H., LIDER, L. A., and LEONARD, O. A. 1974. *Chemical Weed Control in Vineyards.* California Agricultural Experimental Station Extension Service. Leaflet 216.

LANGE, A. H., FISCHER, B. B., ELMORE, C. L., KEMPEN, H. M., and SCHLESSELMAN, J. 1975. Roundup—the end of perennial weeds in tree and vine crops? *California Agriculture* **29:**6–7.

LEHOCZKY, J. 1968. Spread of *Agrobacterium tumefaciens* in the vessels of the grapevine, after natural infection. *Phytopathologische Zeitschrift* **63:**239–246.

LIDER, L. A. 1958. *Grape Rootstocks for the Coastal Valleys of California.* California Agricultural Experimental Station Extension Service. Leaflet 101.

———. 1958. Phylloxera-resistant grape rootstocks for the coastal valleys of California. *Hilgardia.* **27:**287–318.

———. 1959. *Nematode Resistant Rootstocks for California Vineyards.* Division of Agricultural Science, University of California. Leaflet 114.

———. 1960. Vineyard trials in California with nematode-resistant grape rootstocks. *Hilgardia.* **30:**123–152.

———. 1963. *Field Budding and the Care of the Budded Grapevine.* California Agricultural Experimental Station Extension Service. Leaflet 153.

LIDER, L. A., and SHAULIS, N. 1974. Resistant rootstocks for New York vineyards. *New York's Food and Life Sciences Bulletin* No. **45:**1–3. August.

LIDER, L. A., GOHEEN, A. C., and FERRARI, N. L. 1975. A comparison between healthy and leafroll-affected grapevine planting stocks. *American Journal of Enology and Viticulture.* **26:**144–147.

LIDER, L. A., KASIMATIS, A. N., and KLIEWER, W. M. 1975. Effect of pruning severity on the growth and fruit production of 'Thompson Seedless' grapevines. *American Journal of Enology and Viticulture.* **26:**175–178.

LONGHURST, W. M., JONES, M. B., PARKS, R. R., NEUBAUER, L. W., and CUMMINGS, M. W. 1962. *Fences for Controlling Deer Damage.* California Agricultural Experimental Station Extension Service. Circular 514.

MARSH, R. E., and CUMMINGS, M. W. 1974. (Rerun). *Pocket Gopher Control with Mechanical Bait Applicator.* Cooperative Extension University of California. AXT-261.

MAY, P. 1965. Reducing inflorescence formation by shading individual Sultana buds. *Australian Journal of Biological Science.* **18:**463–473.

MAY, P., and KERRIDGE, G. H. 1967. Harvest pruning of Sultana vines. *Vitis.* **6:**390–393.

MCGREW, J. R. 1972. *Control of Grape Diseases and Insects in the Eastern United States.* Farmers' Bulletin 1893, U.S. Department of Agriculture.

MEYER, J. L., and MARSH, A. W. 1972. *A Permanent Sprinkler System for Deciduous Orchards and Vineyards.* Agriculture Extension of the University of California. AXT-n70.

———. 1972. *A Permanent Sprinkler System for Deciduous Orchards and Vineyards.* University of California Agriculture Extension. AXT-n70.

MOLLER, A. N., and KASIMATIS, A. N. 1975. Newly recognized dying arm disease of grapevines. *California Agriculture.* **29:**10–11.

NEJA, R. A., and ALJIBURY, F. K. 1974. Two experts analyze irrigation systems. *Wines and Vines.* **8:**21–23.

NEJA, R. A., AYERS, R. S., and KASIMATIS, A. N. 1974. *Systematic Appraisal of Irrigated*

Coastal Soil for Grapes. Part III. How to Appraise and Manage Chemical Limitations . . . Soil and Water. Agricultural Extension University of California.

Neja, R. A., and Wildman, W. E. 1973. *Systematic Appraisal of Irrigated Coastal Soils for Grapes. Part I. How to Appraise Soil Physical Factors. Part II. Alter Soil Physical Characteristics Before Planting.* University of California Agriculture Extension Service. U.S. Department of Agriculture.

Nelson, K. E., and Baker, G. A. 1963. Studies on the sulfur dioxide fumigation of table grapes. *American Journal of Enology and Viticulture.* **14:**13–22.

Olmo, H. P. 1939. Breeding New Grape Varieties. *Wine Review:* 1–4.

———. 1951. Introduction improvement and certification of healthy grape varieties. *Wines and Vines.* July.

Olmo, H. P., Studer, H. E., Kasimatis, A. N., Baranek, P. P., Christensen, L. P., Kissler, J. J., Luvisi, D. A., and Lynn, C. D. 1968. *Training and Trellising Grapevines for Mechanical Harvest.* University of California Agriculture Extension Service. AXT-274.

O'Reilly, H. J. 1963. *Armillaria Root Rot of Deciduous Fruits, Nuts, and Grapevines.* California Agriculture Experimental Station Extension Service. Circular 525.

Ough, C. S., Alley, C. J., Luvisi, D. A., Christensen, L. P., Baranek, P., and Jensen, F. L. 1973. *Evaluations of Wine Grape Varieties for Madera, Fresno, Tulare, and Kern Counties.* California Agricultural Experimental Station Bulletin. 863.

Petrucci, V., and Canata, N. 1974. Use of oleic acid derivatives to accelerate drying of Thompson Seedless grapes. *Journal of the American Oil Chemists' Society.* **51:**77–80.

Petrucci, V. E., and Siegfried, R. 1975. The extraneous matter content of mechanically harvested wine grapes. Talk, 26th Annual Meeting of the American Society of Enologists, San Francisco, June 26–29.

Pratt, C. 1959. Radiation damage in shoot apices of Concord grape. *American Journal of Botany.* **46:**102–109.

———. 1971. Reproductive anatomy in cultivated grapes—a review. *American Journal of Enology and Viticulture.* **22:**92–109.

———. 1974. Vegetative anatomy of cultivated grapes—a review. *American Journal of Enology and Viticulture.* **25:**131–150.

Raski, D. J., Hart, W. H., and Kasimatis, A. N. 1973. *Nematodes and Their Control in Vineyards.* California Agriculture Experimental Station Circular. 533 revised.

Raski, D. J., Jones, N. O., Kissler, J. J., and Luvisi, D. A. 1976. Soil fumigation: One way to cleanse nematode-infested vineyard lands. *California Agriculture.* **30:**4–7.

Refatti, E. 1970. Asteroid mosaic of grapevine. In Frazier, N. W., et al. *Virus Diseases of Small Fruits and Grapevines.* University of California, Division of Agricultural Sciences, Berkeley, California. pp. 212–214.

Richards, B. L., Middleton, J. T., and Hewitt, W. B. 1959. Ozone stipple of grape leaf. *California Agriculture.* December 1959:4, 11.

Ryall, A. L., and Harvey, J. M. 1959. *The Cold Storage of Vinifera Table Grapes.* U.S. Department of Agriculture, Handbook 159. Washington, D.C.

Ryall, A. L., and Pentzer, W. T. 1974. *Handling, Transportation, and Storage of Fruits and Vegetables. Vol. 2. Fruits and Tree Nuts.* The Avi Publishing Company, Westport, Connecticut. 545 pp.

SCHULTZ, H. B. 1961. Microclimates on spring frost nights in Napa Valley vineyards. *American Journal of Enology and Viticulture.* **12:**81–87.

SCHULTZ, H. B., and LIDER, J. V. 1968. *Frost Protection with Overhead Sprinklers.* California Agriculture Experimental Station Extension Service. Leaflet 201.

SCHULTZ, H. B., WINKLER, A. J., and WEAVER, R. J. 1962. *Preventing Spring Frost Damage in Vineyards.* California Agriculture Experimental Station Extension Service. Leaflet 139.

SHAULIS, N. 1971. Vine hardiness a part of the problem of hardiness to cold in N.Y. vineyards. Proceedings, Annual Meeting, New York State Horticultural Society. **116:**158–167.

SHAULIS, N., DETHIER, B. E. 1964. Minimizing the hazard of cold in New York vineyards. New York State College of Agriculture Cornell Extension Bulletin 1127.

———. 1970. New York site selection for wine grapes. *New York State Horticultural Society Proceedings.* **115:**288–293.

SHAULIS, N., EINSET, J., and PACK, A. B. 1968. *Growing Cold-Tender Grape Varieties in New York.* New York State Agricultural Experimental Station Bulletin 821. Cornell University, Geneva, New York.

SHAULIS, N. J., JORDAN, T. D., and TOMKINS, J. P. 1973. *Cultural Practices for New York Vineyards.* New York State College of Agricultural and Life Sciences, Cornell University. Extension Bulletin 805.

SHAULIS, N. J., and PRATT, C. 1965. Grapes Their Growth and Development. Farm Research Reprint 401. Cornell University, New York State Agricultural Experimental Station, Geneva, New York.

SHAULIS, N. J., KENDER, W. J., PRATT, C., and SINCLAIR, W. A. 1972. Evidence for injury by ozone in New York vineyards. *HortScience* **7:**570–572.

SINGLETON, V. L. 1972. Effects on red wine quality of removing juice before fermentation to simulate variation in berry size. *American Journal of Enology and Viticulture.* **23:**106–113.

SMART, R. E. 1974. Aspects of water relations of the grapevine (*Vitis vinifera*). *American Journal of Enology and Viticulture.* **25:**84–91.

SMITH, C. R., SHAULIS, N., and COOK, J. A. 1964. Nutrient deficiencies in small fruits and grapes. *In Hunger Signs in Crops, a Symposium.* 3rd ed. H. B. Sprague, Ed. David McKay Company, Inc., New York. pp. 327–345.

STAFFORD, E. M., and DOUTT, R. L. 1974. *Insect Grape Pests of Northern California.* California Agricultural Experimental Station Extension Service Circular. 566. 75 pp.

STUDER, H. E., and OLMO, H. P. 1971. The severed cane technique and its application to mechanical harvesting of raisin grapes. *Transactions American Society of Agricultural Engineers.* **14:**38–43.

———. 1973. Effect of cane severence on quality of machine harvested raisins. *California Agriculture.* **27**(8): 3–5.

———. 1973. Vine-drying of Thompson Seedless grapes. *Transactions American Society of Agricultural Engineers.* **16:**944–948, 952.

———. 1974. Parameters affecting the quality of machine harvested raisins. *Transactions American Society of Agricultural Engineers.* **17:**783–786, 792.

SWANSON, F., CHRISTENSEN, P., and JENSEN, F. 1974. *Preventing Vineyard Frost Damage.* Cooperative Extension Agricultural Home Economics, United States Department of Agriculture, University of California.

TAYLOR, R. H., and WOODHAM, R. C. 1972. Grapevine yellow speckle—a newly recognized graft-transmissible disease of *Vitis. Australian Journal of Agricultural Research.* **23:**447–452.

THOMPSON, C. R., HENSEL, E., and KATS, G. 1969. Effects of photochemical air pollutants on Zinfandel grapes. *HortScience* **4:**222–224.

TISDALE, S. L., and NELSON, W. L. 1966. *Soil Fertility and Fertilizers.* The Macmillan Company, Collier Macmillan Limited, London.

VAADIA, Y., and KASIMATIS, A. N. 1961. Vineyard irrigation trials. *American Journal of Enology and Viticulture* **12:**88–98.

VAUGHN, R. H., and MRAK, E. M. 1954. Protective Farm Storage of Raisins. Cooperative Extension Agriculture Home Economics, College of Agriculture University of California.

VEIHMEYER, F. J., and HENDRICKSON, A. H. 1960. Essentials of Irrigation and Cultivation of Orchards. California Agriculture Experimental Station Extension Service Circular 486.

VUITTENEZ, A. 1970. Fanleaf of grapevine. *In* Frazier, N. W., et al. *Virus Diseases of Small Fruits and Grapevines,* University of California Agricultural Sciences, Berkeley, California, pp. 217–228.

WAY, R. D., and SHERBURNE, J. A. 1971. Bird damage to fruit crops. *Plants and Gardens.* **27:**79–83.

WEAVER, R. J. 1955. Relation of time of girdling to ripening of fruit of Red Malaga and Ribier grapes. *Proceedings of the American Society for Horticultural Science* **65:**183–186.

———. 1970. Some effects on grapevine of exogenous plant regulators and herbicides. *In* Frazier, N. W., *Virus Diseases of Small Fruits and Grapevines,* University of California Division of Agricultural Sciences, Berkeley, California, pp. 247–254.

———. 1972. *Plant Growth Substances in Agriculture.* W. H. Freeman and Company, San Francisco. 594 pp.

WEAVER, R. J., and MONTGOMERY, R. 1974. Effect of ethephon on coloration and maturation of wine grapes. *American Journal of Enology and Viticulture.* **25:**39–41.

WEAVER, R. J., and NELSON, K. E. 1959. Improving Grape Quality by Thinning Girdling Plant Regulators. California Agricultural Experiment Station Extension Service. Leaflet 120.

WEAVER, R. J., and POOL, R. M. 1965. Relation of seededness and ringing to gibberellin-like activity in berries of *Vitis vinifera. Plant Physiology.* **40:**770–776.

———. 1971. Effect of (2-chloroethyl) phosphonic acid (ethephon) on maturation of *Vitis vinifera L. Journal of the American Society of Horticultural Science.* **96:**725–727.

———. 1973. Effect of time of thinning on berry size of girdled, gibberellin-treated "Thompson Seedless" grapes. *Vitis.* **12:**97–99.

WEBB, A. D. 1974. The chemistry of home winemaking. In *Chemistry of Winemaking.* A.

D. Webb, Ed. "Advances in Chemistry" Series **137**:278–305. American Chemical Society, Washington, D.C.

Weinberger, J. H., and Loomis, N. H. 1962. A Rapid Method for Propagating Grapevines on Rootstocks. Agricultural Research Service, U.S. Department of Agriculture. ARS-W-2. Fresno, California.

———. 1974. "Fiesta" grape. HortScience **9**:603.

White, E. D., and Petrucci, V. 1974. *Consumer Evaluation of Mechanically Harvested Sun Dried Raisins.* Marketing Transportation Situation. February. Economics Research Service, United States Department of Agriculture. pp. 28–33.

Wildman, W. E. 1974. Diagnosing Soil Physical Problems. Cooperative Extension, University of California, Berkeley, California.

Wildman, W. E., Meyer, J. L., and Neja, R. A. 1974. *Managing and Modifying Problem Soils.* Cooperative Extension University of California, Berkeley, California.

Winkler, A. J. 1959. *Pruning Grapevines.* California Agricultural Experimental Station Extension Service. Circular 477.

———. 1959. *Spacing and Training Grapevines.* California Agricultural Experimental Station Extension Service. Leaflet 111.

Winkler, A. J., and Kasimatis, A. N. 1964. *Supports for Grapevines.* California Agricultural Experimental Station Extension Service. Leaflet 119.

Winkler, A. J., Lamouria, L. H., and Abernathy, G. H. 1957. Mechanical grape harvest—problems and progress. American Journal of Enology and Viticulture. **8**:182–187.

Winkler, A. J., Cook, J. A., Kliewer, W. M., and Lider, L. A. 1974. *General Viticulture.* University of California Press, Berkeley, California.

Zajanc, A., and Cummings, M. W. 1965. *A Cage Trap for Starlings.* University of California Agricultural Extension Service. OSA #129.

APPENDIX

Conversion Factors*

Multiply	By	To Get
Acres	43,560.	Square feet
Acres	160.	Square rods
Acres	4,840.	Square yards
Acres	0.4047	Hectares
Bushels	2,150.42	Cubic inches
Bushels	4.	Pecks
Bushels	64.	Pints
Bushels	32.	Quarts
Centimeters	0.3937	Inches
Centimeters	0.01	Meters
Centimeters	10.	Millimeters
Cubic centimeters	0.03382	Ounces (liquid)
Cubic feet	1,728.	Cubic inches
Cubic feet	0.03704	Cubic yards
Cubic feet	7.4805	Gallons
Cubic feet	59.84	Pints (liquid)
Cubic feet	29.92	Quarts (liquid)
Cubic inches	0.000465	Bushels
Cubic inches	16.39	Cubic centimeters
Cubic inches	0.004329	Gallons
Cubic inches	0.5541	Ounces (liquid)
Cubic inches	0.02976	Pints (dry)
Cubic inches	0.0346	Pints (liquid)

Conversion Factors*

Multiply	By	To Get
Cubic inches	0.01488	Quarts (dry)
Cubic inches	0.0173	Quarts (liquid)
Cubic meters	1,000,000.	Cubic centimeters
Cubic meters	35.31	Cubic feet
Cubic meters	61,023.	Cubic inches
Cubic meters	1.308	Cubic yards
Cubic meters	264.2	Gallons
Cubic meters	2,113.	Pints (liquid)
Cubic meters	1,057.	Quarts (liquid)
Cubic yards	27.	Cubic feet
Cubic yards	46,656.	Cubic inches
Cubic yards	0.7646	Cubic meters
Cubic yards	202.	Gallons
Cubic yards	1,616.	Pints (liquid)
Cubic yards	807.9	Quarts (liquid)
Feet	30.48	Centimeters
Feet	12.	Inches
Feet	0.3048	Meters
Feet	0.060606	Rods
Feet	1/3 or 0.33333	Yards
Feet per minute	0.01667	Feet per second
Feet per minute	0.01136	Miles per hour
Gallons	3,785.	Centimeters
Gallons	0.1337	Cubic feet
Gallons	231.	Cubic inches
Gallons	128.	Ounces (liquid)
Gallons	8.	Pints (liquid)
Gallons	4.	Quarts (liquid)
Gallons of water	8.3453	Pounds of water
Grains	0.0648	Grams

Conversion Factors*

Multiply	By	To Get
Grams	15.43	Grains
Grams	0.001	Kilograms
Grams	1,000.	Milligrams
Grams	0.0353	Ounces
Grams per liter	1,000.	Parts per million
Hectares	2.471	Acres
Inches	2.54	Centimeters
Inches	0.08333	Feet
Inches	0.02778	Yards
Kilograms	1,000.	Grams
Kilograms	2.205	Pounds
Kilometers	3,281.	Feet
Kilometers	1,000.	Meters
Kilometers	0.6214	Miles
Kilometers	1,094.	Yards
Liters	1,000.	Cubic centimeters
Liters	0.0353	Cubic feet
Liters	61.02	Cubic inches
Liters	0.001	Cubic meters
Liters	0.2642	Gallons
Liters	2.113	Pints (liquid)
Liters	1.057	Quarts (liquid)
Meters	100.	Centimeters
Meters	3.281	Feet
Meters	39.37	Inches
Meters	0.001	Kilometers
Meters	1,000.	Millimeters
Meters	1.094	Yards
Miles	5,280.	Feet
Miles	63,360.	Inches

Conversion Factors*

Multiply	By	To Get
Miles	320.	Rods
Miles	1,760.	Yards
Miles per hour	88.	Feet per minute
Miles per hour	1.467	Feet per second
Miles per minute	88.	Feet per second
Miles per minute	60.	Miles per hour
Ounces (dry)	437.5	Grains
Ounces (dry)	28.3495	Grams
Ounces (dry)	0.0625	Pounds
Ounces (liquid)	1.805	Cubic inches
Ounces (liquid)	0.0078125	Gallons
Ounces (liquid)	29.573	Milliliters (cubic centimeters)
Ounces (liquid)	0.0625	Pints (liquid)
Ounces (liquid)	0.03125	Quarts (liquid)
Parts per million	0.0584	Grains per U.S. gallon
Parts per million	0.001	Grams per liter
Parts per million	8.345	Pounds per million gallons
Pecks	0.25	Bushels
Pecks	537.605	Cubic inches
Pecks	16.	Pints (dry)
Pecks	8.	Quarts (dry)
Pints (dry)	0.015625	Bushels
Pints (dry)	33.6003	Cubic inches
Pints (dry)	0.0625	Pecks
Pints (dry)	0.5	Quarts (dry)
Pints (liquid)	28.875	Cubic inches
Pints (liquid)	0.125	Gallons
Pints (liquid)	0.4732	Liters

Conversion Factors*

Multiply	By	To Get
Pints (liquid)	16.	Ounces (liquid)
Pints (liquid)	0.5	Quarts (liquid)
Pounds	7,000.	Grains
Pounds	453.5924	Grams
Pounds	16.	Ounces
Pounds	0.0005	Tons
Pounds of water	0.01602	Cubic feet
Pounds of water	27.68	Cubic inches
Pounds of water	0.1198	Gallons
Quarts (dry)	0.03125	Bushels
Quarts (dry)	67.20	Cubic inches
Quarts (dry)	0.125	Pecks
Quarts (dry)	2.	Pints (dry)
Quarts (liquid)	57.75	Cubic inches
Quarts (liquid)	0.25	Gallons
Quarts (liquid)	0.9463	Liters
Quarts (liquid)	32.	Ounces (liquid)
Quarts (liquid)	2.	Pints (liquid)
Rods	16.5	Feet
Rods	198.	Inches
Rods	5.5	Yards
Square feet	144.	Square inches
Square feet	0.11111	Square yards
Square inches	0.00694	Square feet
Square miles	640.	Acres
Square miles	27,878,400.	Square feet
Square miles	102,400.	Square rods
Square miles	3,097,600.	Square yards
Square rods	0.00625	Acres

Conversion Factors*

Multiply	By	To Get
Square rods	272.25	Square feet
Square rods	30.25	Square yards
Square yards	0.0002066	Acres
Square yards	9.	Square feet
Square yards	1,296.	Square inches
Temperature (°C.) + 17.78	1.8	Temperature, °F.
Temperature (°F.) − 32	5/9 or 0.5555	Temperature; °C.
Ton	907.1849	Kilograms
Ton	32,000.	Ounces
Ton	2,000.	Pounds
Yards	3.	Feet
Yards	36.	Inches
Yards	0.9144	Meters
Yards	0.000568	Miles
Yards	0.01818	Rods

* After Knott, 1962.

INDEX